# Shaping the Urban Landscape
## Aspects of the Canadian City-Building Process

**Edited by**
**Gilbert A. Stelter and Alan F.J. Artibise**

CARLETON UNIVERSITY PRESS
OTTAWA - CANADA
1982

Shaping the urban
landscape

HT
169

48,865

© Carleton University Press Inc.,
Ottawa, Canada.

32

ISBN 0-88629-002-3 (paperback)

Printed and bound in Canada.

Distributed by:

Oxford University Press Canada
70 Wynford Drive
DON MILLS, Ontario, Canada, M3C 1J9.
(416) 441-2941

**Canadian Cataloguing in Publication Data**

(Carleton library: no. 125)

ISBN 0-88629-002-3

1. City planning — Canada — History. I. Stelter,
Gilbert A.,          — II. Artibise, Alan F.J.,          —
III. Series: The Carleton library: no. 125.

GF125.S53          307.7'6'0971          C82-090115-6
1982

# CONTENTS

# Preface

This collection of essays includes material prepared expressly for this volume and recently published articles. Our goal is to provide readers ready access to the best work currently available on a major approach to the study of Canada's urban past; an approach which emphasizes the process of city-building. The major concern in the study of the city-building process is environmental development: the specific decisions by institutions, groups, or individuals which influenced urban form and structure, and with broader social, economic, and technological trends which shaped the nature of these decisions.

The essays have been organized into sections in order to clarify several aspects of this process. Section I is a general overview. Section II concentrates on a series of factors in urban growth, including the role of business organizations and groups and the municipal corporation. Section III deals with the formation of the social and physical urban environments by examining land speculation and development, planning, architecture, and the construction of circulation networks and buildings. (For a fuller analysis of the place of planning in shaping the urban entity, readers are directed to a companion volume: A.F.J. Artibise and G.A. Stelter, eds., *The Usable Urban Past: Politics and Planning in the Modern Canadian City* [1979].)

The publication of this volume involved the contributions of several people. We wish to thank the President and the Dean of the College of Arts of the University of Guelph for generous financial support. We are also grateful to Gerald Bloomfield, Chairman of the University of Guelph's Geography Department, for advice with many of the maps; to Gail Lange for assistance with translations; to Donna Pollard, Gloria Orr and June Belton for their help in typing the manuscript.

We dedicate this volume to the memory of the late H.J. Dyos of the University of Leicester, one of the prime forces in the development of urban history at the international level. His publications have made a major impact on the writing of urban history, but perhaps what made him stand out most was his role in developing ties among those interested in the urban past. Many Canadian scholars came to know him personally, especially because of his trips to Canada in 1974 and 1977. To a large extent, the future of urban history depends on the foundations he laid and the signposts he built.

*Gilbert A. Stelter*, Guelph
*Alan F.J. Artibise*, Victoria

# The City-Building Process in Canada*

Gilbert A. Stelter

Cities and towns are multidimensional phenomena and are not easily definable or explainable. Therefore we should not be suprised when those of us who try to explain them seem to fall into the practice of the proverbial blind men, who, when examining an elephant, each seized upon a different feature and proclaimed it as the only true characteristic. In spite of the difficulties inherent in studying the urban past, there are growing signs that some definite and useful paths have been charted through the complexity. In fact, the study of urban history in Canada and elsewhere has made tremendous strides in the past decade. One measure of this is the sheer volume of research and publication activity in the field. By 1980 there were more than three hundred individuals in Canada engaged in research on some aspect of the historical development of particular cities or on the broader aspects of the urban past.[1] The most recent bibliography lists over seven thousand publications most of which were produced in the last ten years.[2] The *Urban History Review*, which began as a newsletter in 1972, has become a significant journal in the field.

Another yardstick of the development of urban history is the extent to which there is a tendency to take an interdisciplinary approach seriously.

*I with to thank James Simmons, Keith Cassidy, John Taylor and Alan Artibise for helpful comments on an earlier draft of this paper.

1. Gilbert A. Stelter, "Current Research in Canadian Urban History," *Urban History Review* 9 (June 1980), 110-28.
2. Alan F.J. Artibise and Gilbert A. Stelter, *Canada's Urban Past, A Bibliography to 1980 and Guide to Canadian Urban Studies* (Vancouver, 1981).

Some of us regard urban history as a field of knowledge in which many disciplines converge; the term "urban historian" thus could be applied to anyone, regardless of disciplinary affiliation, who works on the urban past.[3] In terms of disciplinary influences, Canadian urban history has developed slightly differently from that in the United States. American urban history is often closely associated with social history and the related social sciences, while in Canada, as in Britain, the influence of the more physically oriented disciplines – especially geography, architecture and planning – are more apparent.[4] Perhaps the relatively small size of the Canadian academic community encourages, even necessitates cooperation among various disciplines studying the same general topic. As an example, the executive of the Urban History Committee of the Canadian Historical Association is made up of historians, geographers, sociologists, an architectural historian, a planner and a city archivist.

A further measure of recent Canadian urban history is the extent to which there is a sense of direction in terms of conceptualization and methodology. In this regard it is possible to distinguish between those urban historians who make the city the unit of study and those who use the city merely as a convenient setting for the study of other issues such as capitalism and labour relations. Drawing on this distinction and on the suggestions of scholars such as Eric Lampard and Theodore Hershberg, Canadian urban historical writing can be placed into one of three categories.[5] The first is *urban as entity*, in which the aim is to explain the formation of the urban environment in terms of both people and place. A variety of independent variables – political economy, population movements, technology, economic growth – are assessed for their impact on the final product, the city as a dependent variable. A second is *urban as process*, in which the urban environment itself becomes the independent variable, affecting the people and events contained within the boundaries of a particular community. Just how the city influenced social organization and behaviour has not been established in any definitive way in Canada, but the work of Peter Goheen and Michael Katz is instructive in this regard. A third approach is *urban as setting* in which the city or town is an incidental location for the study of other exlanatory categories such as class and social mobility.[6]

3. The best argument in this respect is H.J. Dyos, *Urbanity and Suburbanity: An Inaugural Lecture* (Leicester, 1973).
4. For a summary of Canadian urban literature in this regard, see Stelter, "A Sense of Time and Place: The Historians' Approach to Canada's Urban Past," in Stelter and Artibise, eds., *The Canadian City: Essays in Urban History* (Toronto, 1977), 420-41.
5. The basis of this approach is outlined in Eric Lampard, "The Dimensions of Urban History: A Footnote to the 'Urban Crisis'," *Pacific Historical Review*, 39 (August 1970), 268; Theodore Hershberg, "The New Urban History: Toward an Interdisciplinary History of the City," *Journal of Urban History*, 5 (1978), 3-40.
6. For an extended version of this outline see the Introduction to Artibise and Stelter, *Canada's Urban Past*, xiii-xxxii.

The essays in this volume usually represent *urban as entity*. Although most authors do no make it explicit, they regard a city as a dependent variable, whose form and structure are determined by large-scale economic, political and social forces as well as by thousands of individual and corporate decisions. A key feature is a sense of *place*, a recognition of the particular qualities of a city or town. This includes a sensitivity to geographic site and to situation – the local, regional and national context – but also an appreciation of the complex mix of people who make up the population and the way in which the people are organized in that particular place.[7]

While progress in the study of the Canadian urban past is evident on several fronts, it must also be conceded that there has been little forward movement toward any larger scale explanation of Canadian urban development. Do Canadian cities have anything in common? Are commonalities less significant than differences based on regional diversity? We don't have answers to these questions as yet and it could be argued that we need a good deal more research on detailed topics before we attempt to generalize. The problem is that we don't really have a good way of even "posing" the questions. The current approach of isolated case studies does not appear to promise much in this regard beyond a series of disconnected local histories or at best, regional studies. It seems, however, that the urban historian's task, as well as special opportunity, is to walk that difficult ground between the search for general patterns on the one hand, and the concern for the unique and particular on the other. Based on these dual concerns of pattern and complexity, I wish to suggest a rough categorization of urban environments based on what I regard as the dominant characteristics of particular historical eras and to show how the essays in this volume can be more meaningful when read from this point of view.

A description of stages in the evolution of cities is not a new idea, of course, for several frameworks of this kind have been proposed for American urban development. Most are heavily indebted to the scheme devised by Patrick Geddes and Lewis Mumford who emphasized technological eras.[8] More recently, geographers John Borchert and Allan Pred, and historians Sam Bass Warner, David Goldfield, Blaine Brownell

---

7. The idea of the city-building process as used here has been heavily influenced by the major statement on the approach: Roy Lubove, "The Urbanization Process; An Approach to Historical Research," in Alexander Callow, ed., *American Urban History* (New York, 1973), 659-62. For other discussions of factors involved in the creation of the "built environment," see David Harvey, "The Urban Process Under Capitalism: A Framework for Analysis," *International Journal of Urban and Regional Research*, 2 (March 1978), 101-32; John Friedmann and Robert Wulff, *The Urban Transition: Comparative Studies of Newly Industrializing Societies* (London, 1976), 6-15; Simon Kuznets, "Modern Economic Growth: Findings and Reflections," in Kuznets, *Population, Capital and Growth: Selected Essays* (New York, 1973), 165-66.

8. Lewis Mumford, Glossary, in his *The Culture of Cities* (New York, 1966), 495-96.

and Theodore Hershberg have proposed classifications of urban environments.[9] In the Canadian literature, the notion of successive environments was the central theme in the work of N.S.B. Gras and variations of it have been built into the work of a variety of scholars.[10] There is no question that categorization can become a dangerous operation, especially if a Procrustean framework is imposed which violates the complexity of reality.[11] On the other hand, the main value of classifying urban environments is the possibility of integrating and seeing the connections among a wide range of phenomena such as political power, economic growth, population movements, technological change and trends in the actual construction process of city building.[12]

To suggest that Canadian cities have certain common characteristics in each of several eras does not discount the significance of differences based on scale, regional location and the ethnic composition of the population. Rather, it emphasizes the fact that what is common focuses on the nature of the national political and economic systems. This becomes clearer when Canadian cities are compared to those in the United States and other countries. The relevant questions in this regard involve the nature of decision-making in regard to city building: Who had the power and on whose behalf was it used? How was the power structure of society related to the economic orientation of urban places? To what extent were the large-scale population movements, such as the various stages of European immigration, related directly to the political sphere?

9. John Borchert, "American Metropolitan Evolution," *Geographical Review* 57 (1967), 301-32; Alan Pred, *City-Systems in Advanced Economics* (London, 1977); Sam Bass Warner, " 'If All the World Were Philadelphia': A Scaffolding for Urban History, 1774-1930," *American Historical Review* 74 (1968), 24-43; Warner, *The Urban Wilderness: A History of the American City* (New York, 1972); David Goldfield and Blaine Brownell, *Urban America: From Downtown to No Town* (Boston, 1979); Hershberg, "The New Urban History."

10. Norman S.B. Gras, *Introduction to Economic History* (New York, 1922). Donald C. Masters, *The Rise of Toronto, 1850-1890* (Toronto, 1947) is a direct application of the Gras model. The concept of categorizing environments is explicitly built into Peter Goheen, *Victorian Toronto, 1850-1900: Pattern and Process of Growth* (Chicago, 1970); and Michael Katz, *The People of Hamilton, Canada West: Family and Class in a Mid-Nineteenth Century City* (Cambridge, Mass., 1975).

11. Dean C. Tipps, "Modernization Theory and the Comparative Study of Societies," *Comparative Studies on Society and History*, 15 (1973), 199-226; Raymond Grew, "More on Modernization," *Journal of Social History*, 14 (Winter 1980), 179-87.

12. John Mercer, "On Continentalism, Distinctiveness, and Comparative Urban Geography: Canadian and American Cities," *Canadian Geographer* 23 (1979), 119-39.

13. For a general discussion of this approach, see Anthony D. King, "Exporting 'Planning': The Colonial and Neo-Colonial Experience," *Urbanism Past and Present*, 5 (1977-78), 12-22. Quebec's early history has been interpreted from this stance in Denis Moniere, *Le développment des idéologies au Québec: des origines à nos jours* (Montreal, 1977), especially chapters 2 and 5 on the British colonial period.

With these questions in mind, I have tentatively outlined a rough sort of periodization which I think is relevant to the stages of Canadian urban development. I have labelled the earliest period the "mercantile" phase; this lasted until the early nineteenth century and was characterized by imperial control over urban location, function and growth. Functionally, urban places tended to be administrative or military centres while their economic orientation was that of the entrepôt – collection agencies for colonial staples and distribution centres of manufactured goods from the mother country. A second phase is less definable for it served as a sort of transition between the mercantile and the industrial era which began in earnest in Canada during the 1870s. I have categorized this as the "commercial" era, for it was marked by the increasing control of commercial interests over urban development as the former colonies won a degree of autonomy from imperial domination. Economically this period saw an increase in interregional trade and the growth of small-scale manufacturing related to the commercial role of urban places. The third phase was represented by the "new industrialism" which created a national network of communications and transportation and tended to centralize metropolitan power in the major central Canadian cities.

Perhaps a word of caution about these labels and dates would be in order. All urban places did not systematically pass through these three phases in some deterministic fashion. For example, the modern single-enterprise towns share some qualities with the entrepôts of the earliest years, yet in other respects they are the products of an industrial age which determines their function and form. In a similar vein, the towns of the west essentially functioned as commercial towns although they grew up in a generation dominated by industry and its products. A detailed typology of places – an assessment, for example, of which places were truly "commerical" cities in the mid-nineteenth century or definitely "industrial" cities at the turn of the century – is beyond the scope of this essay even if such a typology were possible at this stage of our knowledge about Canadian cities. The point to be made here is simply that certain periods of our past were characterized by a particular political economic milieu and that cities and towns, regardless of scale, function, and regional location, were shaped to a great extent by the milieu.

## URBAN DEVELOPMENT IN THE MERCANTILE ERA

The mercantile towns were tiny outposts of imperial and mercantile expansion, and this central fact determined their functions and form. It is possible to view this imperial-colonial relationship from the perspective of the recently developed "dependency" literature, for the Canadian experience represented something between two extremes. The usual type described is that in which the imperial power exploited a colony primarily as a market for its commodities, as a source of raw materials and as a sphere for capital investment. This kind of European imperialism in what is today known as the Third World tended to stifle the emergence of an indigenous capitalist class.[13] The other type was that of Britain's American colonies which served as colonizing regions for its surplus population and which became an extension of its system. The towns in these colonies developed a high degree of economic and local municipal autonomy long before the Revolution, with a merchant class as a significant element of the power structure. Canadian development combined some features from each type. While the colonies were regarded as extensions of the empire and a source for the imperial country's population, direct control of the colonies, and hence of urban development, lasted much longer than in the American colonies. The motivation for this continued control was partly a response to the results of American rebellion which imperial officials assumed was the product of too much autonomy. It was also the product of the pressure of a colonial elite which benefited from the mercantile system and was reluctant to see the phasing out of direct political rule. As in colonial Latin America, elite membership often seems to have been determined by access to power rather than by the ownership of the means of production.

Towns were key elements in the process as they became vanguards of French and British imperial expansion from the seventeenth to the early nineteenth centuries. Most of these towns were "planted" in the sense that they were consciously conceived to precede and stimulate more general settlement. In this regard, the military and administrative functions were important, more so than in most American towns. While the French colonial towns of Quebec and Montreal were originally founded by private companies for fur-trading and missionary purposes, Louis XIV's government took over direct control after the establishment of royal control of the colony in the 1660s. Quebec became the centralized headquarters for a far-flung colony which eventually extended to the Gulf of Mexico. The military function of towns was best exemplified by the eighteenth-century military fortress, Louisbourg, designed as the Atlantic bulwark in the bitter French-English imperial rivalry for control of the northern portion of the continent. The first of the British towns, Halifax, was founded for strategic purposes to counteract the threat of Louisbourg. The military function of towns became particularly evident after the American Revolution with the increased fortification of Halifax and Quebec and with Governor Simcoe's system of founding semi-military towns on the western frontier, Upper Canada.

In economic terms, the towns were usually entrepôts, collecting staples from their region for shipment to the metropolitan centre for final processing and, in turn, distributing the manufactured goods from the metropolis. A characteristic of these mercantile towns was their lack of significant connections with other towns in the colonies, for the primary connection and area of interest was the overseas metropolis. Quebec and Montreal were early exceptions in that they played complementary roles in the French mercantile system, Quebec providing the administrative infrastructure and the link with Europe (the French connection), Montreal the link with the interior (the Indian connection). After the Conquest, and particularly in the aftermath of the American Revolution, Quebec's and Montreal's mercantile functions as entrepôts grew, especially with the development of the new export staples of lumber and wheat. As the essay by Lafrance and Ruddel on Quebec City in this volume indicates, the Conquest produced a new ruling elite of anglophone merchants who established a new set of metropolitan connections for these cities. In the Atlantic colonies, the towns were also dependent on their connection with the overseas metropolis. The economy of St. John's, Newfoundland, was entirely based on fishing; that of the towns in Nova Scotia and New Brunswick was more diversified in the sense that several staples were exported. While there has been little study of the nature and extent of the connections of these mercantile towns, it appears that each staple – fur, fish, lumber, wheat – generated its own urban network and created towns to function specifically as an entrepôt for a particular staple. In addition to their function as entrepôts, what these places in various regions shared was the imperial political economy and the presence of a similar class of imperial officials.

Another common feature of towns during this period was their role as ports, because the overriding technological characteristic of the period was a dependence on water transport, powered by wind and sail. Overland transport remained undeveloped with few roads (except in Upper Canada) through the densely forested interiors. The state of technology thus tended to set the limits on the capacity to settle and exploit the colonies, channelling settlement along the St. Lawrence–Great Lakes corridor or to the coastal outposts of the Atlantic colonies. Because they preceded general settlement, the towns usually represented a high proportion of the total population. Although they continued to grow, their proportion dropped during a process of ruralization which lasted until the 1820s. During this second phase, newer secondary centres emerged and developed a good deal of autonomy because the primitive transportation system left them isolated and because the original major towns did not yet have the necessary facilities to dominate all aspects of the life of the region. In New France, three towns represented one-quarter of the population, while villages, as half-way points between town and country, hardly existed, presumably because of the weakness of a largely self-sufficient rural economy not directly integrated into the export economy of the towns. In the Atlantic colonies, the initial phase also represented a high proportion of the total population, especially during the 1780s and 1790s with the building of the instant loyalist towns. Only in Upper Canada, on the western frontier, was the situation reversed, with many small towns constituting only a minuscule percentage of the total (3.6 per cent by 1821) and dispersed widely throughout an essentially agricultural economy.

Like the function of the mercantile town, its form was in large measure determined by imperial needs and designs, whether French or British.[14] During the seventeenth century, the French state became more directly involved in the building or redesigning of towns, as in Colbert's founding of Rochefort as an arsenal and naval base, or in the rebuilding of La Rochelle, the main port for Canada, as a fortified site along the lines recommended by the great military planner, Sebastian Vauban.[15] The desire for order and control led to a series of regulations concerning the future alignment of streets and houses in Quebec and Montreal after the royal takeover of the colony in the 1660s.[16] In spite of official attempts to impose regularity, the look of the towns remained essentially that of the medieval French provincial town, with houses added to the next with no concern for unity of scale or form. The result was a lively, diversified townscape of steep-roofed houses topped by a variety of dormers and other projections. But the town which most closely reflected official ideas of town building was Louisbourg, planned and built in the early eighteenth century according to the principles of Vauban, with elaborate fortifications surrounding a compact grid.

The British-built towns, beginning with Halifax and Charlottetown, were characterized from their origin by the regularity and symmetry of their Georgian plans, with their central squares for parades and their grid-iron layout drawn up by imperial officials. As in Britain at the time, where new town planning in places like Bath, Edinburgh, Aberdeen, Dublin and Limerick had caught everyone's imagination, individual streets and houses were subordinated to a general scheme. Not surprisingly, the government's hastily surveyed loyalist towns like Shelbourne, St. John, Kingston and Niagara were based on the regular grid. In what might be termed a post-loyalist phase, the planning of early Toronto by imperial officials, and the nature of land assignment within the town by Simcoe, represented the desire to re-establish the British class system on the frontier. The largest and most favourably located lots were granted to a privileged few, as was land immediately to the north of the town site. As well, officials reserved considerable portions of the outlying area for public purposes to ensure their continued control over the direction

14. I have outlined this in some detail in "The Political Economy of the City-Building Process: The Case of Early Canadian Urban Development," in Anthony Sutcliffe and Derek Fraser, eds., *The Pursuit of Urban History: Essays in Honour of H.J. Dyos* (London, 1982).

15. Josef Konvitz, *Cities and the Sea: Port City Planning in Early Modern Europe* (Baltimore, 1978), 112-22.

16. Marc Lafrance, "Evolution physique et politiques urbaines: Québec sous le régime française," *Urban History Review*, 3-75 (February, 1976), 3-22; Peter Moogk, *Building a House in New France* (Toronto, 1977), 13-18.

of development.[17] In contrast to the French-built towns, the built environment of the British towns reflected the Georgian values that emphasized symmetry and solidity. The classical lines of the official public buildings were reflected in the brick and stone dwellings of the elite and even in the more modest wooden homes of the rest of population.

While American town councils led by successful businessmen promoted economic order and growth, most Canadian towns during this mercantile phase were not granted charters of incorporation, on the grounds that local autonomy had precipitated the American Revolution. In New France, commercial regulation and community control were under the jurisdiction of the colony's Sovereign Council; after the Conquest and in the British colonies, provincial assemblies dominated. As shown by the essay on Quebec City, the form of local government imposed was the traditional English system of magistrates, closely tied to a centralized power structure in each colony's capital through patronage. Although this form of government proved totally inadequate to meet the needs of growing communities, petitions from urban places for more autonomy were consistently turned down before the 1830s.[18]

The social landscape of the mercantile town was the product of its dependence on the political and economic connection with the imperial power. Town and country were different worlds in New France. The rigid social stratification in the towns, with an elite headed by the nobility, was in sharp contrast to the relatively egalitarian society of the rural areas. A possible reason for this urban-rural distinction was the fact that cheap land combined with poor markets for agricultural products did not attract merchant investment in land as was the case in France.[19] In most British towns, the elite was headed by political and military leaders while merchants also became important, particulary in the young loyalist towns and in Quebec and Montreal after the Conquest when the old elite was replaced by recent British arrivals. The spatial organization of these towns reflected their social structure. The spatial segregation by class which Lafrance and Ruddel have found for Quebec City appears to have been typical of other major centres. In Halifax, for example, the

---

17. Gilbert A. Stelter, "Urban Planning and Development in Upper Canada," in Woodrow Borah, Jorge Hardoy and Gilbert Stelter, eds., *Urbanization in the Americas: The Background in Comparative Perspective* (Ottawa, 1981), 143-55.

18. D.C. Harvey, "The Struggle for the New England Form of Township Government in Nova Scotia," *Canadian Historical Association Annual Report*, 1933, 15-22; J.H. Aitchison, "The Development of Local Government in Upper Canada, 1783-1850," Ph.D. thesis (University of Toronto, 1953). For the contrast with the American experience, see Jon C. Teaford, *The Municipal Revolution in America* (Chicago, 1975).

19. R. Cole Harris and Leonard Guelke, "Land and Society in Early Canada and South Africa," *Journal of Historical Geography*, 3 (1977), 135-53; Harris, "The Simplification of Europe Overseas," *Annals of the Association of American Geographers*, 67 (December 1977), 469-83; Louise Dechêne, *Habitants et marchands et Montréal au XVIIe siècle* (Paris, 1974).

elite and its activities were concentrated at the centre of town while the lower classes spread into the outskirts.[20] In Montreal by the late eighteenth century, the old city's boundaries marked the business and residential section for the merchant elite, while craftsmen and labourers were pushed to the suburbs because of high property values in the central area.[21] These spatial distinctions may have been less obvious in the towns of the western frontiers, although in Toronto the elite originally had been granted the best sites near the public institutions facing the lakefront.

## URBAN DEVELOPMENT IN THE COMMERCIAL ERA

Several features distinguished towns and cities in the commercial era from those of the preceding period. It does not seem possible at this point to define a "commercial" town as precisely as was done with "mercantile" towns and yet it is clear that commercialism – defined here as an emphasis on local, regional and international trade and small-scale artisanal production – was the context in which urban society operated. For example, leadership tended to be concentrated in the hands of a Canadian entrepreneurial elite dedicated to commercial growth through improved transportation and manufacturing. Regional and interregional commerce increasingly dominated the functions of urban places, although the international trade in staples continued to be important. Cities became considerably larger (in population and space covered) with their central cores dominated by a concentration of commercial functions. Their growth and expansion were no longer controlled by imperial officials; decision-making had become more decentralized, with commercial interests the dominant factor.

The political economy of this commercial era was dominated by two significant sets of events. The first was the passing of the old colonial system, with its rigid domination of colonial government and its economic policies of tariff preferences and shipping monopolies. Despite the wails of the Halifax and Montreal entrepreneurs whose wealth and position depended on the system, they were able to adjust effectively to the new situation by concentrating their efforts more directly on the provincial level. Very few Montreal leaders became directly involved in provincial politics but their influence on governments led to government-built canals and harbour improvements, government guarantees for railways, and subsidies for steamship lines and for industrial complexes.[22]

20. Based on my analysis of the 1792 assessment rolls for Halifax and vicinity, located in the Public Archives of Nova Scotia, Halifax.
21. Dechêne, "La croissance de Montréal au XVIIIe siècle," *Revue d'historie de l'Amerique française*, 27 (September 1973), 163-79.
22. Gerald Tulchinsky, *The River Barons: Montreal Businessmen and the Growth of Industry and Transportation, 1837-53* (Toronto, 1977), chap. 12.

The second set of events was the Confederation of the British North American colonies in 1867, representing a victory of the commercial interests whose activities had been limited by previous political divisions, especially the splitting of Lower and Upper Canada. Certain areas of economic development were put under federal control by the British North America Act, which was to become particularly useful for entrepreneurial leaders in the following era. While Confederation represented new economic opportunities, it also represented a new aggressiveness on the part of the former colonies, characterized by the growing economic maturity and strength of the major cities. The stage was set for the newly created entity becoming an imperial power in its own right, by expanding to and taking over the British northwest.[23]

But the threat of the powerful rival to the south kept Canadians tied to British apron strings in several ways. The maintenance of territorial integrity and the acquisition of the West could only be done with the aid of British capital, British markets and the British military. In fact, the commerical era in Canadian cities is also very much the era of massive British fortifications and burgeoning British garrisons. Several cities became heavily fortified after the War of 1812 on the well-founded fears of further American invasions. The Citadel and other works at Halifax were greatly strengthened during the nineteenth century, for Halifax was to serve as the source of the main offensive thrust in any future war with the United States.[24] Quebec City was regarded as a defensive site, built to withstand a major assault, with ramparts, walls, gates and a citadel which together occupied more than one-quarter of the land in the city.[25] Kingston became the great military centre of the west with a complex of fortifications built across the harbour at Fort Henry.[26] When the British troops were withdrawn from the garrisons at Kingston and Quebec City in 1871, Halifax was not abandoned by the British but was maintained as an "imperial station," Britain's North American version of Gibralter or Malta.

Political changes at the federal and provincial level were paralleled by changes in the structure and the leadership of local government. Although the changes in some places were probably more apparent than real, in general there was a move away from oligarchic control by the 1830s and 1840s as cities won acts of incorporation. The new leaders,

23. Donald Creighton, "The Decline and Fall of the Empire of the St. Lawrence," Canadian Historical Association, *Historical Papers* (1969), 14-25.
24. C.P. Stacey, "Halifax as an International Strategic Factor, 1749-1949," Canadian Historical Association, *Annual Report* (1949), 46-56.
25. In addition to Lafrance and Ruddel in this volume, see also E.H. Dahl, H. Espesset, M. Lafrance and T. Ruddel, *La Ville de Québec, 1800-1850: un inventaire cartes et plans* (Ottawa, 1975), 10-27.
26. John W. Spurr, "Garrison and Community, 1815-1870," in Gerald Tulchinsky, ed., *To Preserve and Defend: Essays on Kingston in the Nineteenth Century*, (Montreal, 1976), 103-181.

the entrepreneurial elite, did not differ in terms of wealth or power from their predecessors, but their positions tended to be based on achievement, not birth.[27] In Montreal, entrepreneurial wealth constituted the prime political resource, with nine of the twelve mayors from 1840 to 1873 being important executives, industrialists or merchants. The make-up of the councils was similar.[28] In Hamilton, and probably in most other places, the entrepreneurial class represented an over-lapping elite which governed not only political but economic and cultural life as well.[29] Toronto may not be such a clear-cut case; after all, William Lyon Mackenzie served as the first mayor.

The activities of municipal government reflected the goals of its commercial leadership. Only minimal local services were provided in efforts to keep down taxes. In some cases, as in the building of Montreal's water and sewer systems, action seems to have been a response to the very real fear of disease which also threatened the lives and property of the leadership. In other cases, such as Hamilton, leaders promoted the building of utilities and schools as modern innovations which would add to the prosperity of the city and, hence, to their own fortunes. But the greatest area of interest was directed to anything that would directly stimulate growth, especially transportation facilities. This emphasis was not confined to the largest places, although they took the lead. Small frontier towns like Guelph, as Leo Johnson shows, had entrepreneurially dominated councils which granted generous bonuses to road and railroad companies in order to foster commercial relations with their hinterlands.[30]

In the sphere of economic activity, the outstanding characteristics of the commercial era were a move away from an exclusive reliance on staples exports as the basis of a community's existence to a new concern for regional and interregional commerce and small-scale artisanal production for a local or regional market. Several related aspects of this shift were particularly important for community development. First, commerical expansion did not lead to the incredible growth rates experienced by American cities in the 1820-1860 period. Urban growth rates were higher than that of the total population, however, with the urban proportion (based on those in places of 1,000 or more) rising from about

27 .David Sutherland, "The Merchants of Halifax, 1815-1850: A Commercial Class in Pursuit of Metropolitan Status," Ph.D. thesis (University of Toronto, 1975); Sutherland, "Halifax Merchants and the Pursuit of Development, 1783-1850," *Canadian Historical Review* 59 (1978), 1-17.

28. Guy Bourassa, "The Political Elite of Montreal: From Aristocracy to Democracy," in L.D. Feldman and M.D. Goldrick, eds., *Politics and Government of Urban Canada* (Toronto, 1969), 124-34.

29. Katz, *The People of Hamilton*, 177.

30. See also, Leo Johnson, *History of Guelph, 1827-1929* (Guelph, 1977); E.J. Noble, "Entrepreneurship and Nineteenth Century Urban Growth: A Case Study of Orillia, Ontario, 1867-1898," *Urban History Review*, 9 (June, 1980), 64-89.

8 per cent in 1821 to 18.3 per cent in 1871.[31] The number of places over 5,000 grew from five to nineteen in the same period, but the largest cities remained relatively small by North American standards, Montreal reaching 107,000 in 1871 while Toronto achieved about 56,000 by the same date. The major cities of the Atlantic region, increasingly left on the periphery of development after the middle of the century, achieved only moderate growth to about 30,000 in the same period. Even modest growth, however, was to provide a new consumer market for each city's trade goods and manufactured products and also represented the creation of an urban workforce which made possible the beginnings of industrialization.

A second aspect of commercial expansion in relation to cities was the extent to which these relatively small cities began to assume metropolitan functions. By dominating other immediate regions and smaller places, they established primitive urban systems centred on themselves. Montreal replaced Quebec City as the major city of British North America by the 1830s and became a sort of regional metropolis for Upper Canadian development.[32] In a similar fashion Toronto replaced Kingston as the major city of Upper Canada during the 1830s. Montreal and Toronto both benefited from a general westward movement of population, which put them in a central position in relation to a sizable new market.[33] The size and prosperity of the agricultural hinterland which developed around each of them was an important factor in their early success. But if the ingredients for urban growth were present, the question of which city would ultimately benefit often depended on the skill and initiative of the urban-based entrepreneurs who helped shape the metropolis-hinterland system of relationships. It is possible, for example, to distinguish three levels of this entrepreneurial group in Montreal: an upper level, national in orientation, English-speaking, with interests in banking and transportation; a middle level of both English- and French-speaking, involved in regional commerce and artisanal manufacturing; and a lower group of shopkeepers, mostly French-speaking.[34]

A third aspect of the economic reorientation of this period was the use of new technologies in transportation, notably the application of steam to shipping, to railroads and to the means of production. Steamboats opened up the inland waterways to heavy traffic before 1850; after 1850, the railway was the key, with the mileage of lines built growing from 66 in 1850 to 6,868 by 1880. Montreal's entrepreneurs contributed

31. Statistics cited here and elsewhere calculated from the various published census returns of Canada.
32. J.M.S. Careless, "Metropolis and Region: The Interplay Between City and Region in Canadian History Before 1914," *Urban History Review*, 3-78 (February 1979), 99-118.
33. William Smith, *The History of the Post Office in British North America, 1639-1870* (New York, 1973 reprint of 1921 edition).
34. Paul-André Linteau, "Quelques réflexions autour de la bourgeoisie québecoise, 1850-1914," *Revue d'histoire de l'Amerique française*, 30 (Juin 1976), 55-56.

to their city's hegemony by first working for the improvement of the great river to the west and later getting railway links to the west and to the Atlantic region. Toronto's rise to regional hegemony within Ontario can partly be attributed to the success of its entrepreneurs in securing a complex system of radiating rail lines to other parts of the province from the 1850s on. The location of extensive railway shops in Montreal and Toronto also stimulated the growth of large-scale manufacturing in these cities. In the Maritimes, on the other hand, the emphasis on the sailing vessel in the trans-Atlantic carrying trade continued; in fact, the golden age of sail was not reached until the 1850s. The new technology of steam power and structural iron also affected the nature of manufacturing in basic ways. The source of water power had dictated the location of industry, often to sites outside of urban places, but steam power allowed the centralization of manufacturing activities in cities near the growing unskilled labour market. Factory production of this sort appeared in most cities, and even in small places like Berlin/Waterloo by the 1850s, but steam power was introduced very unevenly, even in larger cities like Toronto where hand technology still greatly surpassed mechanized production in 1871.

A final aspect of economic development that should be discussed is the complicated relationship between staples exports, local and regional commerce and early manufacturing. To a large extent, staples exports set the pace for economic growth by a sort of spread effect which was quite unlike that of the mercantile era when fur and fish exports had not contributed to large-scale urban growth. The difference was in the nature of the new staples, timber and wheat, which required financial institutions, transportation equipment such as railways and ships, storage facilities and tools. The export industries also stimulated the production of consumer goods for the workers in these industries, with wheat probably having the greatest impact, for the agricultural section of Upper Canada/Ontario required the greatest range of goods and services. The production of timber led to an extensive shipbuilding industry in Quebec City and Saint John.

In several important respects Saint John was the first and became the quintessential Canadian commercial city. From its founding as a loyalist town it had been granted a charter which included "the several powers and privileges usually granted to mercantile towns for the encouragement of commerce." Throughout the commercial era it was larger and more prosperous than its regional rival, Halifax. The basis of its commercial strength centred on its timber ships carrying wood to the British market and its coasting vessels carrying a variety of local goods. The dominant group within the commercial community was made up of about thirty "great merchants" who owned the wharfs and the ships in the transatlantic timber trade. This group tended to oppose diversification of the economy and used their political and economic power to limit credit facilities for shipbuilders and other manufacturers. A lesser group of merchants had more regional horizons and actively promoted efforts to produce ships, woolen mills and coal oil refineries. By the 1870s, both groups were in decline because of the contraction of the lumber industry and the failure of the shipbuilders to diversify to metal ships, the failure itself partially the product of lack of control and of entrepreneurial skills.[35]

It would appear that those cities which early began to produce consumer goods were to benefit most from the staples system. In Quebec City, lumber merchants, shipbuilders and importers dominated the entrepreneurial class, but in Montreal entrepreneurial interests were more diversified. For example, a variety of small shops produced goods such as confections, drugs and paints, rope, soap and candles from the 1830s. By the mid-nineteenth century the production of ready-made clothing and shoes led the way to mechanization.[36] Toronto's commercial and industrial growth was less closely related to the expansion of staple exports than was the case in other major centres, for Toronto was only one of several lake ports exporting wheat and potash to Britain or the United States. Rather, Toronto was an importer for a growing agricultural hinterland, and by the 1840s and 1850s a growing centre of craft industry in the clothing and metal-working trades, directly related to the city's expanded commercial activity with its hinterland. The growing commercial function was symbolized by the early 1870s with the establishment of what would become the retail giants of the city and province, Timothy Eaton and Robert Simpson.[37]

The form of cities in the commercial era is not readily definable, but a number of features distinguish them from both their predecessors and successors. The first characteristic was an absence of any central direction in shaping cities. No imperial official or municipal officer planned or regulated development; rather the form was determined by the decisions of thousands of private individuals, or, in some cases, private corporations. As Michael Doucet outlines the process in Hamilton, the subdivision of lots was usually done on a small scale by many speculators, often far in advance of actual need, resulting in fragmented patterns of development and increased cost of services.[38] The entrepreneurial

---

35. T.W. Acheson, "The Great Merchant and Economic Development in Saint John, 1820-1850," *Acadiensis* 8 (Spring 1979), 3-28; Carl Wallace, "The Saint John Boosters and the Railroads in the Mid-Nineteenth Century," *Acadiensis* 6 (1976), 71-91; J.M.S. Careless, "Aspects of Metropolitanism in Atlantic Canada," in Mason Wade, ed., *Regionalism in the Canadian Community, 1867-1967* (Toronto, 1969), 117-29.

36. Tulchinsky, *River Barons*, chap. 12.

37. Goheen, *Victorian Toronto*, chap. 2; F.H. Armstrong, "Metropolitanism and Toronto Re-examined, 1825-1850," in Stelter and Artibise, *The Canadian City*, 37-50; D.C. Masters, *The Rise of Toronto, 1850-1890* (Toronto, 1947), 104-105; John McCallum, *Unequal Beginnings: Agriculture and Economic Development in Quebec and Ontario until 1870* (Toronto, 1980), chaps. 5-7.

38. See also Michael Doucet, "Building the Victorian City: The Process of Land Development in Hamilton, Ontario, 1847-1881," Ph.D. thesis (University of Toronto, 1977).

elite was extremely active in this area; gambling in land held the place
that the stock market later would assume.[39] In Montreal, members of
the elite such as Redpath laid out the prestitious "New Town" on the
slopes of Mount Royal.[40] When viewed at the scale of an entire city, the
process of subdivision resulted in an extremely complex pattern, as is
described by Isobel Ganton's essay on subdivision.Like other cities of
this period, Toronto was not plotted in any rigid progression of surveys
out from the centre, but as her maps clearly show, in piecemeal and
often widely separated portions. Much of the subdivision in mid-
nineteenth century Canadian cities was done on a very small scale; by
the 1870s, however, as Ganton demonstrates, land and building companies (or financial institutions) became more commonplace in the
process.

In spite of the lack of any centralized planning or official controls,
there was a remarkable semblance of order and regularity to cities in
this era. The original man-made lines on the ground were usually perpetuated as streets. As Montreal expanded beyond its early nineteenth-
century suburbs, the new street pattern followed the seventeenth-century
division of agricultural land into long narrow strips laid out perpendicu-
lar to the river.[41] In Toronto, as Ganton points out, suburban expansion
was framed by the north-south lines of the original part lots. Transportation routes into the cities also shaped the direction of development.
Subdivisions were drawn to these corridors and internal transportation
– during this era, the horse-drawn streetcar – followed, creating a kind
of spoke-shaped urban form. As well, the provision of services such as
water and sewers by private or public means determined the direction
and density of development.

Building practices were closely related to land development but little
research has been done on this subject for mid-nineteenth century cities.
Susan Buggey's article in this volume is a notable exception and is an
excellent model of how to approach the subject. The transformation of
Halifax from a city of wood to one of brick and stone accompanied
rapid growth and the demand for building services from three distinct
groups. Governments at several levels required everything from additional military barracks, a courthouse, a hospital and markets. Commercial prosperity led to a consolidated business district of three and
four-storey streetscapes made coherent by common heights, patterns
and architectural details. Residential construction beyond the old central core resulted in new suburbs, including a fashionable suburb to the

39. Katz, *The People of Hamilton*, 191-93.
40. David B. Hanna, "Creation of an Early Victorian Suburb in Montreal," *Urban History
    Review*, 9 (October 1980), 38-64.
41. Jean-Claude Marson, *Montréal en evolution* (Montreal, 1974), chap. 3; Phyllis Lambert,
    "The Architectural Heritage of Montreal: A Sense of Community," *Artscanada*,
    202-203 (Winter 1975-76), 22-27.

south. The demand for new building led to substantial changes in the organization and methods of construction and design. The new public buildings were designed by trained architects whereas previous building was usually done by skilled local builders who used published pattern books. During the 1840s and 1850s, building firms were operated by skilled craftsmen who did part of the construction themselves and contracted other parts to other skilled craftsmen. By the 1860s, a master builder tended to erect an entire building with his own permanent staff of specialists.

A second characteristic of urban form in this period was its size and density. Although populations had increased substantially during the first half of the nineteenth century, spatial expansion continued to be limited by the lack of internal transportation beyond the private carriage and the occasional public omnibus. The introduction of the horse-drawn street railway system in Montreal and Toronto in 1861, and Saint John in 1866, removed this limitation. By achieving speeds of six to eight miles an hour and by maintaining definite schedules and rates, the new system literally allowed cities to double their size, for the possible radius from the centre increased from two to about four miles. The streetcars usually came to areas that had already been surveyed and subdivided, but they preceded intense development.[42]

The third feature of urban form was the complex fashion in which a city's functions and people were sorted out. Notable in this regard was the commercial and industrial domination of the central core. The ability of railways to carry bulky goods between cities had greatly outdistanced the ability to move goods within the cities, and so industry tended to concentrate along rail lines or, in the case of Montreal, along the Lachine Canal. In Toronto, city council actually aided the railway takeover of the waterfront area at public expense.[43] In both major cities the commercial cores migrated slightly north by the 1870s away from the concentration of industry and railroads along the waterfront.

In terms of residential distribution, another feature of cities in this era was the extent to which the population was differentiated on the basis of class and ethnicity. This process seems to have been more pronounced in the older cities. By 1830, Quebec City had developed a working-class suburb associated with the shipbuilding industry, St. Roch. Other distinctions were on the basis of streets, not sections, but merchants began to leave Lower Town for the more salubrious climes of Upper Town after the devastating cholera epidemic of the 1830s. The core of Montreal was abandoned by the elite to commercial purposes by the 1870s as

42. Louis H. Pursley, *Street Railways of Toronto, 1861-1921* (Los Angeles, 1958), 5; J.I. Cooper, *Montreal, A Brief History* (Montreal, 1969), 104-105; Carl Wallace, "Saint John, New Brunswick, 1800-1900," *Urban History Review*, 1-75 (June 1975), 20-21.
43. Francis Mellon, "The Development of the Toronto Waterfront During the Railway Expansion Era, 1850-1912," Ph.D. thesis (University of Toronto, 1974).

they moved to the new upper middle class suburbs of the northwest, while Sainte-Anne's suburb became a working-class district. Montreal's characteristic ethnic division was apparent by mid-century as well, with French Canadians concentrated to the east of St. Lawrence Street. A somewhat similar ethnic division existed in Ottawa from its origins. In the cities of the western frontier, however, class and ethnic distinctions do not seem to have been as clearly drawn on the maps.[44] But a degree of segregation by ethnicity and class nevertheless existed, especially on the part of the Irish Catholic immigrants, who not only concentrated in several shanty-town sections of Toronto and other cities but also consciously created a parallel structure of education, health and welfare, and even of employment in order to resist assimilation into the predominant Protestant milieu of these cities.[45]

## URBAN DEVELOPMENT IN THE INDUSTRIAL ERA

The new industrialism, usually dated from the 1870s in Canada, permeated every aspect of urban life. It became the basis for a national network of communications and transportation which centralized metropolitan power in the major central Canadian cities. Whereas the leading cities of each region had been relatively equal to those in other regions in the two previous eras, a growing disparity between those of central Canada and those of the Maritimes became increasingly apparent in the industrial period. The new cities and towns of the west, on the other hand, were founded and grew up within an established metropolis-hinterland context. The new industrialism also stimulated a new scale of population growth and of physical and spatial expansion. Various functions became separated in a more definite manner than had been the case in the commerical era. In the larger cities, differences between people based on class, ethnicity and religion were accentuated by segregating them into distinct neighbourhoods.

Not all urban places could be designated as "industrial" towns and cities during this era, although many such places emerged in the Maritimes and central Canada, often growing out of a commercial base. Montreal and Toronto, which grew to metropolitan proportions, depended heavily on industry for their success in the urban hierarchy but their functions became highly diversified beyond industry. The cities of the west made strenuous efforts to attract industry, especially to process staples, but they tended to remain commercially oriented.

The political economy of the industrial era was marked by the emergence of industiral capitalism and its counterpart, the industrial work-

---

44. Goheen, *Victorian Toronto*; Ian Davey and Michael Doucet, "The Social Geography of a Commercial City, c. 1853," Appendix I of Katz, *The People of Hamilton.*
45. Murray Nicolson, "The Roman Catholic Church and the Irish in Victorian Toronto, 1850-1890," Ph.D. thesis (University of Guelph, 1981).

ing class. It would appear that the new industrial elite merged with the older commerical elite during the later nineteenth century, polarizing society into those who owned capital and purchased labour and those who sold their labour as a commodity. A second phase of industrialism after the turn of the century modified this dichotomy to the extent that the development of corporate capitalism was accompanied by the creation of bureaucratic forms of organization, the growth of a managerial elite divorced from ownership, and the emergence of trained experts in a variety of areas of life.

The changes in the nature of this industrial leadership have been analysed by several scholars. During the 1870s, the leading industrial firms were small, highly personal, family or partnership concerns with assets of $100,000 to $200,000. By 1910 the dominant form was the joint stock company, owned by seemingly anonymous stockholders and impersonal directorates and managed by career executives. While the proprietors of the family firm often were located in towns and small cities, the corporate industrialist gravitated to the biggest four or five cities and particularly Montreal and Toronto. Some qualitative differences between the business leaders in these two cities reflect more basic characteristics of these two leading industrial centers. Montreal's leaders were more cosmopolitan, drawn either from the English-speaking community within the city or from outside the province. Toronto's elite was largely made up of migrants from small-town southern Ontario, giving a staid and somewhat puritan cast to the city's business and cultural pursuits. Although regional elites had quite distinct origins and tastes, by the early twentieth century their business interests brought them together in board rooms and exclusive social clubs in what amounted to an integrated national industrial system.[46]

In the conception and implementation of national development strategies during the industrial era, the activities of business and the state were closely intertwined, as they had been in the earlier era.[47] One example was the Canada Bank Act of 1871, the product of intense lobbying by Toronto and Montreal financiers, which centralized banking in a few large banks with many branches instead of adopting the American system of many independent banks. The result was the centralization of financial control in Montreal and Toronto, for by 1890 almost three-quarters of all Canadian banking assets were headquartered in these two cities.[48] The way banking policies could affect the pattern of industrialization is outlined in Ronald Rudin's chapter on Montreal banks in this

---

46. Especially T.W. Acheson, "Changing Social Origins of the Canadian Industrial Elite, 1880-1910," in Glenn Porter and Robert Cuff, eds., *Enterprise and National Development* (Toronto, 1973), 51-79.

47. Hugh, G.J. Aitken, "Defensive Expansionism: The State and Economic Growth in Canada," in Aitken, ed., *The State and Economic Growth* (New York, 1959); Herschel Hardin, *A Nation Unaware, The Canadian Economic Culture* (Vancouver, 1973).

48. George Nader, *Cities of Canada*, vol. I (Toronto, 1975), 215.

volume. Banks provided an essential stimulus to industrial development through short-term credit for working capital, freeing an enterpreneur's own resources for longer term or fixed investments. Rudin argues that the major Montreal banks, directed by anglophones, refused to open branches in small Quebec francophone towns, thereby inhibiting the industrial development of these places.

Another example of business-state connections was even more significant for centralization of economic power in the cities of central Canada. This was the federal government's adoption of a National Policy in 1879, designed in part to protect domestic manufacturing. With strong support from Boards of Trade and manufacturers' associations, the act probably symbolized the declining influence of the old commercial elite over government policies. In political terms, the act was an attempt to make a nation out of a group of isolated provinces and territories. In addition to the protective tariff this policy included the building of transcontinental railways to tie the regions together and the acquisition and settling of the prairie and coastal west. In the practical terms of city-building, the National Policy strengthened the metropolitan hegemony of Montreal and Toronto by accentuating the earlier development of industry in a manufacturing corridor from Montreal to Hamilton. There was also a short-lived flowering of manufacturing as a result of the policy in the Maritimes. The policy insured an east-west trade orientation, centred on the two main cities by precluding the development of separate north-south relations with the United States on the part of the Maritimes and the west. In spite of what seemed to be defensive expansionism by central Canadian metropolitan interests, however, business and government officials actually welcomed American branch plants which jumped the tariff barrier, for they believed that this added investment would promote rapid growth. The forces shaping Canadian cities thus became increasingly continental, resulting in a new set of core-periphery relationships and a new form of dependency. What the west and Atlantic provinces became in relation to central Canada, all of Canada tended to become in relation to the American northeastern/midwestern core – a marginal area particularly vulnerable to the negative effects of an era of corporate consolidation and rationalization.[49]

The changing nature of the urban system can also be demonstrated in the case of resources development on the northern edge of settlement and in the creation of a multitude of instant, single-enterprise towns which became a feature of the Canadian scene in the industrial era. In one respect these regions and towns became the hinterland of the major

49. James Simmons, "The Evolution of the Canadian Urban System," in Alan Artibise and Gilbert Stelter, *The Usable Urban Past: Planning and Politics in the Modern Canadian City* (Toronto, 1979), 9-33; W.T. Easterbrook, "The Entrepreneurial Function in Relation to Technological and Economic Change," in B.F. Hoselitz and W.E. Moore, eds., *Industrialization and Society* (Paris, 1963), 57-73.

cities in the southern portion of most provinces. For example, Toronto's growth was spurred by the development of "New Ontario" from the 1880s after the transcontinental railway opened up potentially valuable nickel deposits in the Sudbury region. As one speaker declared in 1912, "Men are now delving 500 feet or 1,000 feet below the ground that business in Toronto may prosper."[50] In another respect, however, these new mineral deposits and the towns they spawned were outposts of American economic power. Prime Minister Macdonald welcomed and encouraged the American company which developed the Sudbury ore body and built several company towns such as Copper Cliff.[51] The same pattern of American involvement was present in the expansion of the pulp and paper industry after American publishers succeeded in having the tariff removed on pulpwood. The mill and town of Espanola in northern Ontario, for example, was taken over by an American firm, as part of a process whereby Canada supplied an increasing proportion of United States newsprint requirements.[52]

While government at the federal level helped shape some of the larger contours of industrialization, the specific location of a particular industry and the extent of its expansion often depended on local promotional activity channelled through municipal governments. This promotional drive – boosterism – was particularly intense in the new cities of the west because there the final hierarchical system of cities had not yet been established. As Alan Artibise demonstrates in an article in this volume, the municipal governments of each major western city were dominated by a business elite partly because of a restricted franchise which effectively limited opposition and because businessmen were able to convince many others that their interests and those of the general community coincided. They were therefore able to spend large sums of public money promoting railways, attracting industry, and building huge public works in their pursuit of growth, often on the basis of deficit financing, without spending much on community services such as sanitation, health, or welfare.[53]

In central and eastern Canada, manufacturing was regarded as the key to a community's success in much the same way that railways had

50. Quoted in Gilbert Stelter, "Community Development in Toronto's Commercial Empire: The Industrial Towns of the Nickel Belt, 1883-1931," *Laurentian University Review*, 6 (June 1974), 5.

51. Gilbert Stelter, "The Origins of a Company Town: Sudbury in the Nineteenth Century," *Lauentian University Review*, 3 (February 1971), 13-18.

52. Eileen Goltz, "Espanola: The History of a Pulp and Paper Town," *Laurentian University Review*, 6 (June 1974), 75-104.

53. The best studies of boosterism in the west have been initiated by Alan Artibise, in his *Winnipeg: A Social History of Urban Growth, 1874-1914* (Montreal, 1975), and more recently in Artibise, ed., *Town and City: Aspects of Western Canadian Urban Development* (Regina, 1981), which contains several useful articles in this regard by William Brennan, Barry Potyondi, Paul Voisey, Max Foran and others.

dominated the thinking about growth during the commercial era. Virtually every community tried to attract industrial firms because of the presumed multiplier effect on the local economy. The vehicle used was the "bonus" which could involve direct cash grants and loans, or less direct aid such as exemptions from taxation, free or cheap sites, free water, and other urban services. Councils either acted directly through by-laws or referred the issue to ratification by the local property owners. The conditions for grants usually included agreements to provide a specified number of new jobs. It is still unclear whether communities were actually able to attract substantial outside investment. In some towns like Berlin/Waterloo which carried on a very active bonusing program, most grants were given to existing local businesses to allow them to expand. These local entrepreneurs were not necessarily council members, but through the local Board of Trade they managed to promote the "proper climate" for council and ratepayer decisions.[54] It would appear, based on a number of case studies in this volume (see articles by Johnson, Linteau and Artibise), that the practice of bonusing had the desired effect in instances where communities became heavily involved before regional rivals took similar steps. After the turn of the century, however, economic power and decision-making were increasingly located in the board rooms of the national corporations. The implications for individual city growth were enormous. The result was the loss of local initiative, for corporate decisions were made in terms of the needs of the corporation, not the needs of a particular community. This applied whether the corporation's branch firm looked to a headquarters in Montreal or Toronto, or was a branch plant of an American firm. Thus, for example, the business elite of Winnipeg could effectively use local initiative to win a transcontinental railway and then attract industry and immigrants, but this kind of localized activity was increasingly unlikely to be effective after the significant changes in the nature of continental business organization.

The context of corporate capitalism provided one of the early motivations for a move to muncipal ownership of some utilities in places like Guelph, according to Johnson. The pro-business Guelph *Mercury*, for example, argued in 1902 that "growth of trusts and monopolies, and the extension of the power and control of corporations over public necessaries, has rapidly educated the people to the necessity of public ownership in larger things."[55] The result in Guelph and many other places was municipal ownership of gas and power utilities, street railways, and in the province of Ontario particularly, public power from Niagara Falls. The so-called "gas and water socialists" were not opposed to a society dominated by private ownership; rather, they simply sought to make it

54. Elizabeth Bloomfield, "The City-Building Process in Berlin/Kitchener-Waterloo, 1870-1930," Ph.D. thesis (University of Guelph, 1981).
55. Quoted in Leo Johnson's article in this volume.

work more effectively in their own interests.[56] Support for public ownership came from Boards of Trade and leading businessmen because lower utility rates benefited local business directly and because of the booster factor – lower rates would attract new business to the community.

Public ownership of some services placed new responsibilities on civic government; so too did the expansion of private services during this period of rapid urban growth, but the existing system of government did not appear to be adequate to meet these needs, according to a growing number of observers. Reforms in the late nineteenth century had stressed moral issues and corruption, but reformers after 1900 called for basic changes in the structure and leadership of municipal government. One solution was to divide policy-making and administration by putting complex new technical matters in the hands of experts, thereby reducing the power of the amateur politicians and making government more efficient. The second was to operate municipal government on business principles in terms of policy. A noted businessman-reformer, Morley Wickett, clearly outlined this approach when he proudly pointed out that "throughout Canada the municipality is regarded as a species of joint stock company, only those contributing the capital being allowed to share in the direction of its affairs."[57] The right voters would support the correct business-oriented policies and this might not always include organized labour or recent immigrants. Efficiency and business principles became the basis for changes which included the abolition of the ward system, and the use of Boards of Control and commissions in many cities. The result, as in the United States at the time, was a narrowing of "the actors in the decision-making process and the range of alternatives and debate."[58]

The widespread belief that industrialization was the key to growth was borne out by the growth statistics of the era, for industrialization had followed urbanization and had stimulated it to a greater extent than had been the case with commerce. The urban proportion of the population rose considerably from the 18.3 per cent of 1871 to represent almost half of the population (49.5 per cent by 1921), a figure comparable to that in the United States (51 per cent in 1920), although Canada had

56. James Anderson, "The Municipal Government Reform Movement in Western Canada, 1880-1920," in Artibise and Stelter, *The Usable Urban Past*, 101. For similar interpretations, see John Weaver, "Tomorrow's Metropolis Revisited: A Critical Assessment of Urban Reform in Canada, 1890-1920," in Stelter and Artibise, *The Canadian City*, 393-418, and P.H. Wichern, "Winnipeg's Civic Political History and the Logic of Structural Urban Reform: A Review Essay," *Urban History Review*, 1-78 (June 1978). 111-21.

57. "City Government in Canada," 1902, reprinted in Paul Rutherford, ed., *Saving the Canadian City: The First Phase, 1880-1920* (Toronto, 1974), 292.

58. Sam Hays, "The Changing Political Structure of the City in Industrial America," *Journal of Urban History*, 1 (November 1974), 28.

been far behind in that regard at the end of the commercial era.[59] The most spectacular growth was achieved by the two largest cities, Montreal and Toronto, who outdistanced their nearest regional rivals by four or five times. By 1921 Montreal's population, exclusive of several contiguous cities, was 618,506 and Toronto's was 521,893, while Quebec City's was 95,193 and Hamilton's was 114,151. Perhaps the most dramatic change to the old hierarchical system was the rather sudden emergence of the western cities during this era, for by 1921 Winnipeg was the third city of the country with a population of 179,087 and Vancouver was fourth, with 117,217. They were followed in eighth and tenth places by two other young giants, Calgary and Edmonton. In sharp contrast, the three main cities of the Atlantic region had grown only slowly, reflecting a decline of the old Atlantic trading system and the growth of a continental economy dominated by the central regions. A major component of urban population growth was foreign immigration (over two million came to Canada between 1896 and 1914) and much of this went to the largest cities of the central and western regions. Rural to urban migration also became significant during this era, with the mechanization of farming decreasing the necessary size of the agricultural workforce.

The extent and nature of urban development depended in part on major improvements in the technological capacity of Canadian society during the industrial era. Science and engineering were systematically applied to transportation, communications, building methods and production. National railway networks, based on improved steel rails and heavier equipment, tied the regions together with 38,369 miles of line by 1917, much of this built on the prairies. The use of structural steel led to the high-rise and made possible the enormous concentration of office space in the central core. A new energy source, electricity, virtually replaced water power and proved to be much more flexible than steam. For example, electrification transformed the effectiveness of internal transportation by increasing the possible speed and carrying capacity of the street railway systems. The new technology also made possible the final stages of what really was an industrial revolution, for fully mechanized factory production became the hallmark of the era, with the machine becoming the norm not the exception in manufacturing.

The process of industrialization appears to have accentuated certain centralizing tendencies in the evolution of the urban system. An increasing proportion of manufacturing workers were located in Montreal and Toronto between 1881 and 1911, but Toronto's workforce grew much more quickly than Montreal's. For example, Montreal's industrial workforce was more than twice as large as Toronto's in 1881, but by 1901 they were virtually equal. Toronto's industry was more heavily

59. Leo F. Schnore and Gene Petersen, "Urban and Metropolitan Development of the United States and Canada," *Annals of the American Academy of Political and Social Sciences*, 316 (March 1958), 61.

oriented to machinery such as farm implements for the southern Ontario agricultural region and for the west, whereas Montreal concentrated on products such as textiles, tobacco and leather. At least some of Toronto's increased capacity came at the expense of its own region, with the moving of major firms like the Massey farm implement company from smaller places. In 1871, Toronto's proportion of southern Ontario manufacturing stood at 10 per cent, but by 1901 it was 34 per cent.[60] Only its regional rival, Hamilton, with its new emphasis on heavy industry, had a greater proportion of its population in manufacturing. Another factor in the growth of central Canadian manufacturing and of that in Toronto in particular, was the opening of the west, with its wheat producers providing the base for a vast new market for central Canadian industrial products. Cities like Winnipeg developed significant industry in their own right, but they remained more commercially oriented than cities in the central region.[61] In the Maritimes, entrepreneurs took advantage of the national tariff policies to introduce industry to Halifax, Saint John and a host of small places like Amherst and New Glasgow, with textile mills, refineries and steel mills based on Nova Scotia coal. L.D. McCann maintains that this promising start was not maintained because of the essential core-periphery relationship of central Canada and the Maritimes. Major markets were located in central Canada and so too was financial strength, making the Maritime region's industries vulnerable to take-over by central Canadian interests.[62]

The outstanding physical characteristics of cities during this period were the enormous spatial expansion of the suburbs and the tall office towers of the central core. Suburban growth which had taken place beyond the city's boundaries was incorporated into cities in a wave of annexations beginning in the 1880s. Toronto, for example, increased its area by almost four times in the years from 1880 to 1914. Western cities like Edmonton and Vancouver incorported large new sections into their jurisdictions, often long before they were required for actual development. Montreal, on the other hand, failed to annex some contiguous communities like Westmount and Outremont, with the result that much of Montreal's elite took little political responsibility for the larger community in which it lived.[63]

The electrification of the streetcar system was a major factor in the decentralization process of most cities for it made possible rapid spatial expansion beyond the compact, walking-distance community. The impact of internal transportation was actually the reverse of that of

60. Jacob Spelt, *Urban Development in South-Central Ontario* (Toronto, 1972), chap. 5.
61. Artibise, *Winnipeg*, 305-308.
62. See also W.T. Acheson, "The National Policy and the Industrialization of the Maritimes, 1880-1910," in Stelter and Artibise, *The Canadian City*, 93-124.
63. J.I. Cooper, *Montreal: A Brief History*, 98-102, 127-30; Terry Copp, *The Anatomy of Poverty: The Condition of the Working Class in Montreal, 1897-1929* (Toronto, 1974), 147.

26 GILBERT A. STELTER

intercity transportation. Railways led to a centralization of industrial and commercial functions in cities at strategic locations along the rail lines because bulky goods could not be moved easily within cities. While the radius of towns and cities in the mercantile eras had been about two miles, the horse-drawn railway from the 1860s increased this to about four miles. From the 1890s, with electrification, this was increased to almost ten miles. In some cases such as Montreal and Vancouver, street-car line extensions and inter-urban radial railroads stimulated development and helped create the settlement pattern. Norbert MacDonald's article on Vancouver suggests that the British Columbia Electric Railway Company's decisions to lay tracks well beyond existing settlement were based on the expectations of a rapid rise in real estate values. The opposite situation prevailed in Winnipeg and Toronto where the Mackenzie and Mann owned transit companies refused to build lines to suburban areas because they were not also in the real estate business. Michael Doucet's essay on the Toronto streetcar system describes the company's refusal to build beyond the heavily built-up parts of the city in order to ensure a good profit from heavy ridership, the highest except for New York City, in North America. The company's policies actually forced Toronto to remain a compact city, making possible the later public transportation successes.

The scale of land development during the industrial era increased substantially from earlier periods with developers in several cities assembling large parcels of suburban land. The relationship between development, planning and promotion is brought into clear focus by Paul-André Linteau's work on Maisonneuve, an industrial suburb of Montreal. He shows how the promoters reaped windfall profits on large tracts of land they owned by controlling the municipal council and bringing about the adoption of an industrial policy favouring population growth. Railway companies developed large suburban tracts in most large cities, the most ambitious project being the Canadian Pacific's laying out of a six-thousand-acre site in the southern part of Vancouver. MacDonald's article on Vancouver outlines how the company played a major role in determining the city's street layout and general land-use patterns in the nineteenth century.[64] In his study of a Hamilton suburb during the twentieth century, John Weaver demonstrates how smaller developers produced their own measure of control, employing restrictive convenants to force social or racial conformity in land ownership and stylistic conformity in the construction of buildings.

Development practices were roughly similar in the large cities, but the types of housing built varied considerably by region. According to the statistics in the 1921 census, the single-family detached house, built of wood, was the predominant form in the cities of the Maritimes and the

64. See also Deryck Holdsworth, "House and Home in Vancouver: Images of West Coast Urbanism, 1886-1929," in Stelter and Artibise, *The Canadian City*, 186-211.

west. In Toronto, the brick semi-detached house was the most common form, while in Montreal the terrace or row house, once the preserve of the elite, became the standard type, especially in the enormous new working-class suburbs. Montreal also led in making the apartment house a significant form of housing, with 13.5 per cent of its dwellings of this type.[65] A construction boom in housing and in central city office towers was accompanied by escalating land prices, particularly in the central portions. But wages did not rise as rapidly as land prices or rents. The situation for the working class and the poor was most desperate in Montreal, as described by Herbert Ames in 1897 in his famous *The City Below the Hill*. The degree of homeownership in the newest western cities reached as high as 80 percent but in Montreal the great majority of residents were tenants. The lowest levels of the working class in Montreal were housed by dividing older single-family houses into flats located at the fringe of the commercial and industrial core.

From the late 1890s, urban leaders began to call for solutions to the threat which rapid growth seemed to pose for the future of the nation. Contemporary planning concepts from the United States and Britain were proposed for adaptation to the Canadian scene. From the 1890s to before 1914, the dominating theme was the City Beautiful movement with its grand designs for aesthetic improvement, but most of these plans never left the drawing board. For a short time before 1914 the emphasis switched to the British-inspired Garden City idea with its concerns for the health and housing of workers. One of the results of this interest was the importation of a leading British planner, Thomas Adams, to a federal government position, and through his persuasion, the introduction of provincial planning statutes. But the hopes for directly involving municipal governments in the provision of housing proved largely illusory. The role of government in this area was to be primarily regulatory, through zoning and housing codes. The result, as in the United States, was to put the increasing segregation of land uses on a legal-administrative basis.[66] The hopes for "master" plans in most cities fared no better. The relationship between planners' ideals and the hard, practical realities of actual city building is described in detail by Elizabeth Bloomfield's history of planning in Kitchener-Waterloo. Only in the planning and building of resource towns could planners apply their current ideas without worrying about any existing infrastructure as was the case in most cities. In this regard, the authors of the final article in this volume suggest that successive planning styles of resource towns are an accurate reflection of planning concerns in vogue when particular towns were built.

---

65. *Census of Canada*, 1921, vol. 3, p. 39.
66. For more detail on trends in planning during the early twentieth century, see the essays by Thomas Gunton, P.J. Smith, Walter Van Nus, Shirley Spragge, Oiva Saarinen, and Peter Moore in Artibise and Stelter, *The Usable Urban Past*.

The social landscape of cities in the industrial era was greatly affected by the changing scale of the cities. A kind of giantism prevailed, from the size of new suburbs and the height of the buildings in the central core to the organization of new business enterprises and the building of enormous factories. Another major characteristic was an increased specialization in land use, as industrial, commercial and residential functions became more definitely separated from each other. Retailing, wholesaling and financial functions dominated the central core, with industry radiating from the centre along major transportation routes. A significant social result of this specialization was the two part city, with residence separated from workplace. But perhaps the most characteristic feature of the social landscape of industrial towns and cities was the way in which society increasingly sorted itself out on the basis of ethnicity or class. The prototype of the segregated city was Winnipeg, with its elite located in beautifully landscaped suburbs in the south and the non-English-speaking working class crowded into a notorious ghetto north of the railway tracks. The social divisions in Winnipeg provided the setting for the dramatic general strike of 1919, the most complete general strike in North American history. The conflict was not a sudden outburst of local labour discontent, but represented the most dramatic example of decades of class polarization heightened by depression and war which were present in all major Canadian cities.[67]

Since 1920, Canadian cities have entered the modern phase characterized by the technology of the automobile and the truck, an economic orientation away from industry to service functions, and massive spatial decentralization of population and activities. City-building, however, is a cumulative process, with successive building on the structures produced by previous generations. The pre-1920 city still exists as the urban core of the modern, dispersed metropolitan area. There has been relatively little historical research on urban development for the period since 1920, but there are notable exceptions. Some examples are the volumes being done for the History of Canadian Cities Series which will bring this story of Canadian cities up to the present. As they are consciously designed in a comparative context, they should allow us to see recent general and regional patterns which are not yet clearly understood.[68]

This effort to draw some guidelines for an understanding of the Canadian city-building process not only tells us what we know know, but also points to what we don't know. The dynamic of the building process still

67. Artibise, *Winnipeg*; David Bercuson, *Confrontation at Winnipeg: Labour, Industrial Relations and the General Strike* (Montreal, 1974).
68. The general editor of this series is Alan Artibise; it is co-published by the National Museum of Man and James Lorimer. Volumes published to date include Alan Artibise, *Winnipeg* (1977); Max Foran, *Calgary* (1978); Patricia Roy, *Vancouver* (1980) and many more volumes are projected.

remains vague, as does the general framework of change. There are several aspects of the process which obviously require further work by those studying the urban past. Among the most promising is the question of political and social power at the societal level in general and at the municipal level in particular. Involved here is the question of who in effect ran municipal government in different periods and the extent to which municipal government became relatively autonomous during the commercial era, only to have this autonomy decline by the early twentieth century.[69]

Another aspect of the city-building process requiring further analysis is that of the evolution of an urban "system." The theoretical work of J.M.S. Careless and James Simmons has provided some of the necessary concepts, but we still have not found a way of empirically measuring connections between places in the earlier periods. Related to this is the need to define the essential characteristics of communities by type which will require a model that not only takes economic orientation into consideration but is sensitive to the community's place in the larger system. A third aspect of the building process which could be strengthened is the emphasis on the actual construction of the physical artifact. Susan Buggey's article in this volume is a useful example of what can be done for one city in one time period. Finally, we need to begin to ask the question of what successive environments do to the people and their activities in those environments. This is based on the assumption, in Hershberg's words, that "people living . . . in cities at different stages in their economic development – commercial and industrial, for example – can be hypothesized to display different behavioural and additional characteristics, and their collective patterns should lend themselves to documentation and measurement."[70]

69. This subject has been thoughtfully explored by John Taylor, "Canadian Urban Autonomy: Its Evolution and Decline," paper presented at the Canadian Historical Association, Dalhousie University, May 1981. The paper will be published in a revised and expanded version of Stelter and Artibise, *The Canadian City*, which will appear in late 1982.

70. Hershberg, "The New Urban History," 14-15.

# Ideology and Political Economy in Urban Growth: Guelph, 1827-1927

LEO A. JOHNSON

The development of Guelph as a significant secondary urban centre represents the triumph of a business community over an environment which yielded few natural advantages over its commercial and industrial rivals. Lured initially to the location by the extravagant promises and clever promotions of the Canada Company and its founder, John Galt, Guelph's businessmen found themselves engaged in a century-long struggle to prevent the loss of their business investment and to develop an economic environment which would enable them to prosper and expand.

Critical to their success was the self-conscious development of a theory of political economy which, although it originated with the interests and aspirations of a mercantile community serving an agricul-tural hinterland, was sufficiently sophisticated and flexible to enable Guelph's businessmen to perceive the necessity for and to undertake the creation of an industrial city as a strategy to enhance their mercantile prosperity. In so doing, they found it necessary to reject theories of free trade and to create an ideological basis for large scale government intervention in economic life. Although this ideological development was by no means unique to Guelph, the community's businessmen were amongst the first in Canada to accept it wholeheartedly. By 1914, Guelph was widely proclaimed a leader to be emulated by other centres. This short paper, then, will examine the economic circumstances in which Guelph's businessmen found themselves compelled to develop their theories and strategies, the main arguments used, and the methods and results of their application.

The history of Guelph's political economic development falls generally into four stages which coincide roughly with the following chronological periods: 1827-1847, the period of non-intervention; 1847-1880, the trans-

30

portation era; 1880-1900, industrial subsidization; and 1900-1927, municipal ownership. Of course, there were overlaps of activity (for example, the Guelph Junction Railway was built in 1887, and industrial subsidization continued unabated during the entire municipal ownership period), but the periods presented are those in which the new ideologies and strategies of economic development were at the forefront of public consciousness.

I

It was John Galt's development strategy which created the the first major problem to be resolved by Guelph's business community. Galt's strategy was based upon an unusually perceptive insight into the process of frontier economic development. As he saw it, the usual pattern of land settlement was one in which the pioneer settlers began life by creating a virtually self-sufficient primitive community and, only gradually as this community matured, did it demand, or could afford, the services of an urban centre. Once the maturation process was under way and urban services could be used, village growth occurred, and farm land near the villages rose rapidly in value. Because wild land was generally in a huge surplus in Upper Canada, most land entrepreneurship had consisted in jobbing off wild land at low prices (generally on credit) to poor but ambitious British immigrants (the Talbot Settlement provides a prime example of this method).[1] Galt's strategy was designed to maximize the return to the Canada Company by reversing this process. It was his strategy to acquire the highest price for the company's agricultural land by providing ready-made villages, and thereby not only commanding the "unearned increment" which proximity to village services returned to nearby land owners, but, through a vigorous program of subsidization of village services and largescale promotion, to attract to the Guelph area the "better class" of immigrant who could start life in Canada as farmers rather than as pioneers.[2]

In accordance with this strategy, Galt laid out a large urban centre on the Guelph townsite, built roads, cleared streets and provided employment for a large number of workmen at top wages.[3] In addition, he subsidized the immigration of skilled artisans (carpenters, bakers, blacksmiths, shoemakers, etc.), built a school, donated land for churches, and built houses which he sold, on credit, to those who were employed developing

---

1. Fred C. Hamil, *Lake Erie Baron* (Toronto, 1955).
2. John Galt, *Autobiography* (London, 1833), vol. II. See especially pp. 54-98 for Galt's discussion of his strategy.
3. Robert Thompson, *A Brief Sketch of the Early History of Guelph* (Guelph: Mercury Steam Printing House, 1877), p.2

Figure 1:    The town plot and township of Guelph about 1831.
SOURCE: Canada Company Papers, Public Archives of Ontario.

the townsite. With the town apparently booming, a large number of immigrants flocked to the area to take up land or to set up stores, taverns and workshops in the town.[4] Within two years, however, the directors of the Canada Company, unsettled by Galt's expenditures, fired him and ended the period of large scale capital investment in the Guelph area.[5]

Immediately the capital flows stopped, the village found itself in the grip of a severe depression. Unemployment was endemic, building stopped, businesses closed and, as one observer noted, "the dark days of Guelph succeeded, and brooded o'er the land for three years, during the currency of which, a mad bull might have rushed through the streets of Guelph without the risk of hurting many people; in truth, every person that could, left the apparently doomed locality."[6]

The general depression which followed Galt's departure was relieved for a short period after 1830, when the Canada Company, concerned by lagging sales of farm land, built a large grist mill which cost some £2,000 to erect. It was not until 1832, however, that prosperity returned. The reason for this upsurge was, once again, the large scale expenditure of capital, this time by a group of wealthy English and Anglo-Irish immigrants,[7] who were fleeing the expected rebellion in Great Britain. These immigrants set out to recreate in Guelph the genteel lives they had enjoyed in England and Ireland, and spent freely to do so. As one contemporary observer noted:

> Sovereigns then were plentiful, and champagne bottles strewed the streets; and although the former soon chang ed masters, still that they were not lost was soon apparent in the numerous buildings which sprang up, and the extensive clearings that gladdened the eye. Indeed, the rich stream irrigated the whole locality. . . .[8]

The period from 1832 to 1836, therefore, was marked by a good deal of growth in population and a general optimism prevailed. In the town and township of Guelph (the town was not incorporated until 1851) population grew from 703 in 1830 to 1,854 in 1835, while taxable houses increased from 17 in 1831 to 97 in 1836, and merchant shops increased from 3 to 10 in the same period. In Guelph's market area, the figures were equally impressive. In the townships of Puslinch, Eramosa, Nichol,

---

4. Canada Company Circular, February 1, 1828, copy in the archives of the Guelph Historical Society.

5. Galt, *Autobiography*, II, pp.99ff. Acton Burrows, *Annals of the Town of Guelph* (Guelph, 1877), pp.30ff.

6. From a talk given by James Hodgert, November 30, 1850, reported in the *Guelph Advertiser*, Dec. 5, 1850, microfilm, P.A.O.

7. Acton Burrows, *Annals of Guelph* pp. 47-48

8. James Hodgert, *op. cit.*

Erin and Garafraxa, population increased from about 700 in 1830 to 3177 in 1835.[9] Thomas Rolph reported of Guelph in 1836 that:

> During the past year no less than 16 frame and two brick houses made their appearance in our streets, and there are at present two large taverns in progress, a chapel, and seven or eight frame houses building or contracted for in the town, and buildings of all sorts and descriptions daily rising out of the wood, if I may use the expression, in the country. . . .
>
> Stores, seven or eight in number, hotels, taverns, watchmaker, saddler, chairmaker, and mechanics of every description. The vicinity is greatly celebrated for the quantity and quality of Barley grown—and sleighs well laden with it, are brought during the winter months to the respective breweries for sale.[10]

It was a rosy picture, but one based upon insecure grounds.

The years 1836-1838 dealt a staggering blow to Guelph's economy. First of all, the harvest of 1836 was almost entirely ruined by continuous rain which sprouted wheat in the sheaf and rotted crops in the ground. During the winter of 1836-37 large quantities of flour had to be imported at greatly inflated prices.[11] No sooner had spring arrived than the effects of the 1837-38 worldwide depression began to be felt. With immigration dropping and few new capital projects being undertaken, unemployment once again became a severe problem. The unsettled conditions following the rebellions made matters even worse. As one local observer remarked:

> For months the young men and mechanics of the place were without employment, quite a number joined the incorporated batallions for the permanent defence of the province, and those who stayed at home might be seen whiling away their time or amusing themselves at cricket and other sources of amusement or recreation. The state of things in the whole province assumed such a threatening aspect that many families returned home or removed to the States and many more would have gladly done so had they possessed the funds or could have realized on their landed property.[12]

This situation appears to have continued until the return of international prosperity in 1839-40.

The underlying problem that made Guelph so vulnerable to such economic fluctuations was that the development of an economic base for the town had not kept pace with its urban growth. In 1831, Adam Fergusson had pointed out the dangers involved in the manner in which the town of Guelph had been developed:

9. "Provincial Secretary's Papers," Upper Canada, P.A.C., R.G. 5, B. 26, vols 1-7; and *Journals of the Legislative Assembly of Upper Canada*, 1823-1842, Appendices, P.A.O. microfilm. Figures for 1830 are estimates based upon earlier and later data.

10. Quoted by James Innes, "History of Guelph," Instalment XV, *Guelph Mercury*, March 15, 1866. Innes' history was reprinted in the *Guelph Evening Mercury* in 1963.

11. *Guelph Evening Mercury*, Centennial Edition, July 20, 1927.

12. Innes, "History of Guelph," Instalment XV, March 15, 1866.

When farms become numerous and a mill is erected in a convenient situation, a town soon grows up; but here the town has been hurried forward, in the hopes of settling the land. A vast deal of capital has been expended upon roads, etc., which must have so far benefited labourers, and tended, in some measure, to enable them to purchase lots; but, at present, a very desolate complexion marks Guelph, as a city which may be very thankful to maintain its ground, and escape desertion.[13]

Guelph's severe economic difficulties between 1829 and 1831 and after 1836 demonstrated that Fergusson's concern was well founded. Until Guelph's economic base grew to a sufficient size to support the town no long-term cure for depression was available.

Between 1836 and 1845, two developments took place which put the village on a much more secure footing: first, the agricultural economy continued to mature; and second, Guelph was made the administrative seat for the newly-created District of Wellington.

In spite of the violent economic fluctuations experienced in Canada between 1836 and 1845, immigration continued and Wellington District farmers continued to clear land and to accumulate livestock. For example, improved acreage in Wellington District increased from 15,805 in 1836 to 53,971 in 1845,[14] while in Guelph Township population increased from 1,854 to 3,400 (including the village), and average improved acres per farm increased from 18.3 to 36.1.[15]

The significance of the establishment of district government in Guelph in 1841, as a stimulus to local growth, was not lost upon the local business community. As the editor of the *Guelph Advertiser* remarked in 1853:

The rebellion threw the town back, but the setting off of the Wellington District and the establishment of Guelph as the County Town again imparted a vigor and spirit of progress which has never flagged since that period. . . .[16]

By being named district seat Guelph's economy not only acquired the large salaries and fees associated with its governmental status, but this status helped concentrate in local businesses what agricultural trade there was in the area. The stimulus provided by these changes was felt immediately. In 1843 the population of the town was estimated at 700,[17] and by 1845 it had increased to 1,240.[18]

13. Adam Fergusson, *Tour in Canada. . . in 1831* (London, 1833), pp. 279-80.
14. "Provincial Secretary's Papers" and *Journals of the Legislature Assembly of Upper Canada*.
15. "Provincial Secretary's Papers" and "Appendices" of *J.L.A.U.C.* Calculations are based upon an average of 100 acres per farm, the average size in 1851.
16. *Guelph Advertiser*, June 30, 1853.
17. From a map of Guelph drawn by Donald McDonald, October 1847, Guelph Historical Society Archives.
18. W.H. Smith, *Smith's Canadian Gazetteer* (Toronto, 1846), p. 72.

## II

The development of the northern townships in Guelph's market area did not, however, go unnoticed by the town's commercial rivals. No sooner had Guelph won the battle to be named the district seat, than Galt and Berlin began a rapid improvement of their road system with the intent of undercutting Guelph's newly–won advantage. In 1838 Absalom Shade and James Crooks, the local elected members to the Legislative Assembly and both owners of mills along the Dundas and Waterloo road, managed to persuade the government to vote a substantial sum for that road's improvement. The Dundas and Waterloo road was made a government work and major businessmen along the route were named trustees responsible for its management and the collection of tolls from users.[19] In the same year a group of businessmen from Berlin and Preston incorporated the Waterloo Bridge Company to build a first-class bridge across the Grand River, thus removing the last major barrier to traffic between those villages.[20] By so doing, merchants and millers along the Dundas and Waterloo road began to be able to undercut the prices of Guelph's businessmen and to threaten control of its market area. Unfortunately for Guelph's business community, the shorter direct route from Guelph to Dundas was in such bad condition that most travellers found it necessary to take the much longer route around by Galt. The main problem, however, was not the inconvenience and discomfort of the much longer trip, but the cost:

> The principal part of all necessary inportations had to be brought by the circuitous route past Galt, and the cost of teaming was a most important consideration in the minds of our merchants and mechanics. One dollar per hundred weight was the regular charge for teaming goods from Hamilton and Dundas to Guelph and for heavy goods, such as salt, iron, etc., the increase on the regular cost was very considerable.[21]

Thus when improvements were made on the Dundas and Waterloo road below Galt, all the rival towns – Galt, Guelph and Berlin – benefited. But once the road north of Galt began to be improved, Guelph's relative disadvantage increased.[22]

In 1838, Guelph's businessmen had attempted to create a privately owned toll road company to build a direct road from Guelph to Dundas, but the attempt failed. A survey was made, but the estimated cost, more than £31,000, or £1,285 per mile, scotched the attempt.

In the early 1840s, a second, even more serious threat to Guelph's control of its market area developed, this time from the east. In 1840, the merchants of Bronte incorporated themselves as a harbour company,

19. William IV, cap.79.
20. Victoria, cap. 32.
21. Innes, "History of Guelph," Instalment XIV, March 8, 1866.
22. Acton Burrows, *Annals of Guelph* p. 54.

and undertook to raise £5,000 for harbour improvements.[23] Fortunately for Guelph, scarcity of money prevented substantial improvements from being made on the road from Bronte to Esquesing Township.

In 1846, however, Guelph's business community faced a much more serious threat from the same direction. The Chisholm interests in Oakville undertook to open a major traffic link between Oakville and Owen Sound which was intended to cut directly through Guelph's northern market area. The project, as envisioned by the Oakville interests, sought to combine a privately owned toll road from Oakville to Fergus, with a government subsidized road to run north from that point to Owen Sound. Capitalized at £20,000, the Trafalgar, Esquesing and Erin Road Company[24] posed a formidable threat. As the Guelph *Herald* said in 1847:

> The effect of the adoption of this route would have been to divert the travelling and traffic altogether from Guelph, thereby reducing it to the insignificant inland town of a third or fourth-rate District, and left without the means of ever rising to anything greater.[25]

Faced with this challenge from Oakville, the merchants and millers of Guelph, Hamilton and Dundas banded together to preserve their dominance of the Wellington District trade. On October 4, 1846, they announced the application for a charter for a toll road company from Flamborough West Township to Guelph,[26] and a few weeks later, announced a second road company which was to build a road from Guelph north to Arthur.[27] Both of these charters were granted in July 1847.[28]

As Guelph's newspaper editors saw it, the two projected roads held the potential to guarantee Guelph's commercial future:

> With the [Guelph and Dundas] Road once finished and the Owen's Sound Road in a tolerable condition, we need entertain no fears for the future prosperity of Guelph. The whole travel will be on this road from Hamilton and the adjacent towns, as it is by far the shortest and most expedicious route, and if properly macadamized, it will be one of the best roads in the Province. The Beverly Swamp [on the Dundas and Waterloo road] will be avoided with its break-bone logs and holes. We have been kept behind too long for the want of a little public energy, whilst our more fortunate neighbours of Galt have been

23. 3 Victoria, cap. 33.
24. 9 Victoria, cap 97. See the description of this road in Hazel B. Mathews, *Oakville and the Sixteen* (Toronto, 1953), pp. 193-94.
25. Guelph *Herald*, October 7, 1847.
26. *Guelph and Galt Advertiser*, January 15, 1847.
27. *Ibid.*, February 26, 1847.
28. 10-11 Victoria, cap. 88 and cap. 91. See also the *Galt and Guelph Advertiser*, September 3, 1847.

reaping the advantages of our ineptness. . . All that is required is unanimity of action and sameness of purpose, and we must effect a large change for the better.[29]

In spite of the optimism displayed by the editors concerning the beneficial effects of these roads, the promoters of the Guelph and Dundas Road Company found that potential investors in the company's stock were most reluctant to risk their savings in such an untried project. Although Guelph's businessmen vigorously supported the project and unanimously proclaimed its necessity, after nearly a month of intensive canvassing only about £2,000 worth of shares had been subscribed in Guelph.[30] When the sales campaign was extended to Dundas and Hamilton, it met with a stony refusal by potential investors. As a piece of private enterprise, the Guelph and Dundas Road Company was a dead letter. With the town's economic life threatened by the Trafalgar, Esquesing and Erin project, it was time for drastic action.

The use of municipal government to further the collective economic welfare of the local community in 1847 was by no means a novel idea. In his inaugural address to the Wellington District Council in 1842, the first Warden, Alexander Dingwall Fordyce, had made it clear that that body was to leave pressing social and political issues of the day to other levels of government. Its duty was to concentrate upon economic goals, particularly the provision of good roads:

> I do not flatter myself or you gentlemen by supposing, that one shall be able to satisfy everybody, even if we use our utmost endeavors, the envious and discontented will say, that some other plan would have been better, the miser or sluggard will talk of the dreadful nature of assessments, - but the *honest* Farmer trotting along, on a Plank or Macadamised Road, with 50 or 100 bushels of Grain in his team and an additional sixpence or shilling per bushel in his eye, will pour forth blessings on the district Council.[31]

From Guelph's point of view, it was time to put these sentiments to the test.

The technique by which Wellington and Gore districts took over the Guelph and Dundas Road Company was simplicity itself. After a series of joint meetings the two councils agreed to purchase all of the shares of the company, and to pay for them by issuing debentures.[32] The Wellington District Council, composed of the area's millers, merchants and wealthiest farmers, passed the bylaw with little debate on December 15, 1847.[33]

29. Guelph *Herald*, September 2, 1847.
30. *Galt and Guelph Advertiser*, September 3, 1847.
31. Guelph *Herald*, June 4, 1842.
32. See the *Guelph and Galt Advertiser*, October 15 and 29, 1847.
33. Minutes, Wellington District Council, December 15, 1847.

Immediately that the road contract was let, money, once again, began to pour into Guelph. Several new subdivisions were laid out by speculators,[34] and significant growth in population followed. The road, however, was never a commercial success. Under constant pressure (including, on one occasion, a violent night attack against one of the toll houses) by the merchants and millers of Guelph and Dundas to reduce tolls in order to maximize their competitive position versus Galt, Berlin, Bronte and Oakville, the Road Commissioners kept the tolls uneconomically low.[35] When the commissioners objected to further decreases in 1852, Guelph's businessmen, using town council funds and volunteer labour, built a second bridge across the Speed River so that concession roads could be used to by-pass the tollgates in good weather.[36] Facing such competition, by 1856 losses of the Guelph and Dundas road exceeded £1,600 per annum. The toll roads north of Guelph showed a similar pattern.

Although, because of the intervention of Guelph, the toll roads were failures as businesses, they clearly had, from Guelph's point of view, the desired results. The challenge to Guelph's monopoly of its market area had temporarily been beaten back, while the local agricultural community, with easier access to markets and services, converted to a cash basis at an accelerating rate.

No sooner had the Guelph and Dundas road been completed than the businessmen of Galt once again attempted to seize the initiative. During the summer of 1850 rumours kept circulating that Galt's businessmen had made a deal with the Great Western Railway Company whereby the company would build a branch line to Galt if that community would purchase £25,000 worth of shares in the branch line company. By the spring of 1851, news reached Guelph that the money had been raised and the line would be built. Guelph's business community was thrown into a panic: if Galt acquired a railroad and Guelph did not, Guelph would be ruined. The *Guelph Advertiser*, in advocating the necessity of a railroad from Guelph to Toronto remarked:

> If this railroad, or some other line leading to Toronto or Hamilton, be not processed with, the branch of the Great Western, terminating at Galt will cause the almost total annihilation of Guelph as a place of business: its carrying trade will be ruined, by the cheaper rates at which flour will be carried to the lake, by way of Galt; and the higher price which can then be given at Galt for farm produce, will divert almost the entire current of business from the locality of Guelph. Whilst Galt remains without a railroad, Guelph may get on comparatively well without one also; but as the construction of a railroad to Galt is now

34. See the advertisements in the *Guelph Advertiser*, January 17 and 31, 1850.
35. See, for example, the controversy over tolls in the Guelph and Dundas road outlined in my book, *The History of Guelph, 1827-1927* (Guelph: Guelph Historical Society, 1977), chapter IV.
36. "Minutes of the Town of Guelph," 1852, Appendix.

a matter of certainty, there is no choice for Guelph other than to get a railroad also; or to sink as a place of business, into utter insignificance.[37]

The case in favour of building a railway was not, however, expresssed merely in the negative. There was also a firm belief, with T.C. Keefer, that railways had the capacity to completely transform local society.[38] The owner of the *Advertiser*, in a letter to the editor asserted that without a railway,

> I behold Guelph as a way-side tavern, accommodating transient travellers, without business, energy or enterprise — mills standing, stores vacant, property daily sinking in value, and every one bewailing his position in despair. With a Railway, I can conceive the store-keeper extending his stock and premises, the capitalist his means, the property holder his income, the manufacturer his resources, and the farmer the whole combined. . . .[39]

Unlike the original movement to build the Guelph and Dundas road, there was now no pretence that private means should be used to bring a railway to Guelph. Instead it was proposed that local municipal councils led by Guelph and Guelph township should purchase shares as a means of attracting foreign investors who would build the railroad.[40] Once the local railway committee had secured the overwhelming approval of the local ratepayers for votes of £25,000 from the town and £10,000 from the township, a delegation of Guelph's leading merchants, millers and "men of capital" travelled to Toronto to get that city's backing in the railway project. They were met with open arms.[41] Within weeks the Toronto city council had committed itself to an investment of £100,000 in the enterprise, and two abandoned railway charters were renewed and amalgamated to create the Toronto and Guelph Railway Company.[42] With Toronto interests firmly in control (only three residents of Guelph were elected to the directorate of the railway) that company now turned to British sources for further financing. After more than a year's delay, the charter of the Toronto and Guelph Railway Company was amended to extend the line to Sarnia, and the whole was amalgamated into the Grand Trunk. Construction started in 1853 and the line was opened June 14, 1856.[43]

37. *Guelph Advertiser*, June 26, 1851.
38. For a penetrating analysis of the ideology of railway-building, see the introduction by H.V. Nelles to T.C. Keefer, *Philosophy of Railroads and other Essays* (Toronto, 1972). Keefer's major essay was published in 1850.
39. Letter to the editor by John Smith, *Guelph Advertiser*, June 26, 1851.
40. See the discussion by the "Toronto Railway Committee" reported in the *Guelph Advertiser*, October 16, 1851.
41. *Toronto Patriot*, July 7, 1851, reported in *ibid*, July 10, 1851.
42. These two railway companies, the Toronto and Goderich and the Toronto and Huron, had created a stir in Guelph in 1847 but quickly died when no large scale financing could be found.
43. For a description of the curious and amusing events surrounding the opening ceremonies, see my *History of Guelph*, chapter VII.

The decision by the Toronto-dominated directorate of the Toronto and Guelph Railway Company to extend that line to Sarnia was bitterly resented in Guelph. It was one thing to be the terminus of a railway where farmers and merchants for scores of miles on all sides would have to come to ship their grain and pick up merchandise, and quite another to be merely another way station on a through line. Guelph's businessmen had proclaimed from the first that they intended that the railway should "make their town 'the great centre of attraction and radiation,' to all the adjoining townships,"[44] and now this advantage was being taken from them. In response to Guelph's objections, Mayor Bowes, of Toronto, the president of the railway company, explained Toronto's interest in the matter:

> The city of Toronto wished to continue the road to Sarnia, for when they got there they were in a direct line to obtain the whole business of the north part of the State of Michigan, of Chicago, and of the state of Iowa. Are the people of Toronto prepared to expend £100,000, and then stop at Guelph, and lose the whole of the immense business, and allow the Great Western to obtain it?[45]

In order to placate Guelph's business community who threatened to oppose the amendment of the railway's charter, the town of Guelph was allowed to reduce its stock subscription to £10,000.[46] In 1853 both the town and township sold their shares to the promoters of the Grand Trunk,[47] thereby escaping the loss of their investment when that company collapsed.

The construction of the Galt and Guelph railway between 1853 and 1857 was similar in purpose, rhetoric and events to the Toronto and Guelph. The town of Guelph took £10,000 worth of stock in the line and, when the line went broke in 1855, lent the railway £20,000 (borrowed from the municipal fund) in order to complete construction.[48]

The promotion and construction of the Toronto and Guelph and the Galt and Guelph railways had an immediate and dramatic effect upon Guelph's economic life. With millions of pounds sterling flowing into Canada to build railways, coupled with the rapid inflation caused by the Crimean War, optimism pervaded the Canadian economic climate, and with this optimism, speculation ran riot. In Guelph, as elsewhere, the

44. *Guelph Advertiser*, March 4, 1852.
45. *Ibid.*, January 6, 1853.
46. *Ibid.*
47. Ibid., May 12, and May 19, 1853. See also "Byerly (A.E.) Papers," 1853, P.A.O. for the original of the letter from S. Thompson, Secretary-Treasurer of the Toronto and Guelph Railway Company to the Township of Guelph, May 5, 1853, offering to purchase that municipality's stock.
48. If anything, the construction of the Galt and Guelph Rilway was more scandalous and filled with intrigue and broken promises—although at a smaller scale—than was the Grand Trunk. See chapter VII of my *History of Guelph*, for a description of the "McCracken Affair."

main promoters of the various transportation and land speculation schemes developed a rhetoric which called upon "progress"[49] as the guarantor of the success of their enterprises:

> Year has succeeded year in engendering a spirit of public enterprise; first the Government commenced with a gigantic scale of canals to connect our lakes with the ocean, and almost no sooner were these in operation than the contagion—if I may be allowed that expression—took hold of the minds of the people. Plank, gravel and macadamized roads were constructed to a considerable extent; still the spirit of improvement marched onward until the railroad mania got hold of the minds of both Government and people —until railroads of an unprecedented extent have been and are on the way of being constructed, —and still onward the spirit of improvement. . . .[50]

Editors of the local newspapers delighted in calling attention to every evidence of "progress" in the town:

### Progress in Guelph

> The material progress of the country was never more fully evidenced than at the present time [September 1853], whilst the increase of agricultural produce, the facilities for conveying it to market, the spread of manufacturers, and the increase of population in Town and country, all mark the onward progress of the Province. The building everywhere going on has increased the wages of the mechanic, the railroads are paying 5s. to 6s. 3d. to the labourer, whilst Town Lots are rising in value and money everywhere is plentiful. The reference from time to time in our contemporaries of the erection of a block of buildings here, the opening of new streets there, and the founding of villages in other places, all tend to encourage and elevate our expectations of Canada's future. . . .
>
> All who take into consideration what has been done during the past ten years, and the prospects now opening for the future will admit that in three years or less from this date few inland towns in the Province, without manufacturers, will compare with Guelph for a moment. . . .[51]

And grow Guelph did. Long before the railways were opened, merchants, artisans and labourers flocked to town, both to capitalize on the local construction boom and to be first on the spot once the railways began to operate. From about 2,000 in 1853, the population grew to some 4,500 in 1857.[52]

But if Guelph had managed to establish its commercial dominance over the Wellington County area by the promotion of railways, its situation was equally vulnerable when larger competing centres pushed their

49. For a useful discussion of the "Idea of Progress," see R. V. Sampson, *Progress in the Age of Reason* (Cambridge, Mass., 1956), and L.S. Fallis, "The Idea of Progress in the Province of Canada: 1841-1867" (Ph.D. thesis, University of Michigan, 1966).

50. *Guelph Advertiser*, September 22, 1853.

51. *Ibid.*, September 15, 1853.

52. Robert W. Mackay, *Canada Directory* (Supplement to 1851) (Montreal, 1953), pp. 76-78 and John Lovell, *Canada Directory*, 1857-58 (Montreal, 1858).

railways into the area. The construction of the Toronto, Grey and Bruce and the Wellington, Grey and Bruce lines between 1867 and 1871 would severely undercut the town's commercial base.

Throughout the 1856-1866 period a series of proposals were made to build railways into the area between Guelph and Owen Sound. Although Guelph was cool to the idea, the towns and villages to the north were anxious to acquire railway service, thereby escaping Guelph's dominance. In 1858 and again in 1862, the Wellington County Council placed itself on record as supporting such a project, and in the latter year suggested that it might give an outright bonus of $400,000 if a rail line from Guelph to Owen Sound were completed.[53] In 1864 a group of promoters led by Hon. John McMurrich and Francis Shanly acquired a charter to construct the Wellington, Grey and Bruce Railway from Guelph to Owen Sound. Capitalized at $1,500,000,[54] the project created a local flurry, but the simple fact that it could not pay its way cooled investor interest. Guelph's failure to support the Wellington, Grey and Bruce was due largely to the fact that should such a line be built, the northern farmers and businessmen would no longer have to come to the town to sell their grain and produce or to purchase supplies. The loss in trade would be enormous.

In 1866, however, two Toronto-based railway projects were promoted which carried the potential of cutting off Guelph's grain trade and directing it to Toronto. The first, being pushed by William Gooderham and associates, proposed to extend the Northern Railway to Owen Sound, with a branch (the Grey and Simcoe) to run from Angus to Durham. The second, promoted by John Fowler (an itinerant railway contractor from Port Hope) was to run from Weston or Brampton stations to Arthur or Mount Forest and thence to Owen Sound. With these indications of Toronto interest, merchants and millers in the north Wellington towns began again to agitate for a county bonus. With the threat of a Toronto-based line being built, the Hamilton business community took over the Wellington, Grey and Bruce Railway charter and began to tout it as a solution to Toronto's threatened invasion of its hinterland. The Toronto railway projects, argued the *Hamilton Evening Times* in May 1866, were intended to "monopolize the whole trade of the district,"[55] and should Toronto succeed, the "ambitious city" would be dealt a crippling blow.

From Guelph's point of view, there was an important tactical question to be kept in mind. Although any railway was bound to be injurious, if either the Hamilton or Toronto interests were to build a railway into Guelph's market area, it was desirable that such a line should have its terminus at Guelph. That way, at least the construction expenditures

53. Toronto *Globe*, February 15, 1862.
54. 27-28 Victoria, cap. 93.
55. *Hamilton Evening Times*, May 30, 1866.

and transfer trade could be salvaged. Moreover, too vigorous opposition to the desires of the northern towns' wishes might result in driving a considerable part of that trade to other towns in retaliation. Thus, once the Toronto-based narrow-gauge railway, the Toronto, Grey and Bruce, was off the ground, Guelph's business community threw its weight behind the Wellington, Grey and Bruce. In the end both lines were constructed, although the Wellington, Grey and Bruce, from lack of funds, never managed to reach Owen Sound.[56]

For Guelph's commercial community, the construction of the Toronto, Grey and Bruce and the Wellington, Grey and Bruce, was catastrophic. Although the prosperity of the 1870-1874 period masked the most serious effects of the loss of the northern trade, by 1880 the effects, according to the city's leading businessmen, were clear. In his inaugural address to the city council in 1880 George Sleeman, the mayor, stated:

> I feel it incumbent upon me to call your [ernest] attention to the suffering of the mercantile interests caused in a great measure by the network of railways with which we are now surrounded. . . .[57]

With Guelph no longer the "great centre of attraction and radiation" for the northern farmers, it had become necessary to abandon the old strategy based upon roads and railways, and to look for quite a different kind of customer.

### III

The transformation from a strategy based upon transportation to one based upon industrial development was by no means as difficult as one might imagine. The roots of the necessary ideological elaborations were already laid in the interests and social aspirations of the independent artisan class, although not seized upon as their primary ideology by the merchants until their more narrow and primitive strategies had failed them.

The artisanal assertions concerning the dignity of labour had been, for some decades, popular with small-town newspaper editors who, as "master printers," were members of that class. For example, in 1856 the *Guelph Advertiser* carried this article reprinted from an American journal:

#### The Age of Industry

> Art and industry are the agents of civilization; and guided by religious instruction, are the elevators of humanity over barbarism. The races of non-producers and plundering conquerors. . . are specimens of men heartily and sincerely

---

56. For a more detailed description of the protracted campaign for bonuses see my *History of Guelph.*
57. Guelph Daily Mercury January 19, 1880.

despising labor. They are but a higher type of highwaymen; burglars, pirates and other non-laboring classes, for whose magnanimous contempt of labor, society has very little respect. The old features of the Saracens are passing away before civilization. The Norsemen are humanized into law abiding citizens. The feudal lords have been conquered by the burghers, who were building cities and founding nations, while the barons were hunting and fighting. The aborigines of this country, the most impracticable race which the earth has ever seen, suddenly die out rather than labor. The races which unite intellect and industry shall ultimately possess the earth. . . .

The free industrious classes are the life of a nation. Whatever permanent prosperity it possesses is derived from them. . . .[58]

The frequent praise heaped upon the "industrious classes" was not without point. Not only were the "habits of industry" indispensable to the "development of civilization," but the application of these habits to a particular form of enterprise - manufacturing - represented the "highest stage." In 1847 the *Advertiser* argued as follows:

### Home Manufacturers

Perhaps there is nothing in a secular point of view so conductive to human happiness and prosperity, as industry and economy. . . . To attain true greatness, and develop and increase its resources is the imperative duty of a nation. It is for its prosperity and independence to encourage every kind of useful home manufacture. The nation which neglects that, and bestows its encouragement on the manufacturers and mechanics of a foreign country in preference to its own is unwise. It discourages and enfeebles itself, and ultimately must work its own ruin. . . .[59]

Nor was this a mere abstraction to Guelph's business community. As the same newspaper remarked on another occasion, "Canada must become a manufacturing country so far as her domestic wants are concerned, and to encourage our manufactures is a duty each one owes to himself. The principle applicable to a country is equally applicable to a community. . . ."[60] Moreover, it was equally clear to them that roads and railways, however desirable, were not, themselves, capable of creating wealth. As the *Advertiser* argued in 1856:

Railways would be useless unless there were people to travel on them and produce to carry over them; railways afford facilities for business, but do not create it; railways present a cheap mode of transit for merchandise and the products of the farm, but they neither manufacture the one nor grow the other, whilst the working expenses attending them are very great.[61]

58. *Guelph Advertiser*, September 18, 1856.
59. *Guelph and Galt Advertiser*, November 9, 1847.
60. *Ibid.*, July 9, 1847.
61. *Guelph Advertiser*, May 22, 1856.

The perception that long-term commercial prosperity and economic growth depended entirely upon the development of production, and that both were strictly limited unless the "*internal* productiveness of the place"[62] could be built up, would be critical when the construction of the Toronto, Grey and Bruce and Wellington, Grey and Bruce railways threatened to undermine the commercial base of the town.

Of course, in spite of the rather pessimistic statements about Guelph's lack of manufacturing enterprises, by the 1850s, several substantial businesses had developed to serve local needs. In September 1854, the *International Journal* made these comments about Guelph:

> There are three foundries in Guelph. The establishment of Mr. John Watt gives employment to over 60 men. . . . The foundry and Tinware manufactory of Smith, Mathewson & Co., is devoted to stoves, agricultural implements, etc., and is a neat and well arranged establishment. The other Foundry is that of Mr. A. Robertson, devoted to general castings and machinery. . . . The tannery of Mr. John Harvey is very extensive and turns out annually about 5,000 side of sole leather, besides upper leather and kipskins. Mr. Gow has a large tannery inoperation, and there are three others in the suburbs of the Town, owned by Mr. Jackson, Mr. Clark and Mr. Horning. A Fanning-mill manufactory is carried on by Mr. James Mays, where fanners capable of cleaning a bushel of wheat per minute, are made and sold for $25 each.
>
> There is a Chair Factory in the town, carried on by Mr. Allen, and a number of Furniture Manufactories, some of which are aided by steam or water power, and use the most approved machinery for expeditious work. . . .
>
> In the suburbs of the Town, there is a Woolen Factory in operation, by Messrs. Campbell and Co.[63]

With the exception of John Watt's foundry (which failed in the post-1856 depression), these were all small operations employing not more than ten hands each. In additon to these "manufactories," there were two large grist mills in town, one owned by William Allan (the Guelph Mills built by the Canada Company in 1830) which consisted of a grist mill with six runs of stones, a barley mill, distillery (using 100 bushels of grain daily), malt kiln, barns for feeding 50 cattle and 250 pigs, a carding and fulling mill, dyehouse, cooperage and machine shop. Allan employed about thirty men on a regular basis.[64] Frederick George's Wellington Mills was about the same size and contained four run of stones for grain, a distillery (using 200 bushels of grain per day), a piggery for 200 pigs, a saw-mill, tannery, and a foundry building leased to tennants.[65] In 1860-61 a third, somewhat smaller mill was built by James Goldie.[66]

62. *Ibid.*, September 15, 1854.
63. Reprinted in *ibid.*, September 20, 1854.
64. *Ibid.*, June 30, 1853.
65. *Ibid.*, September 30, 1854.
66. David Allan, "The Milling Industry, Guelph and Vicinity,"*Wellington County Historical Society*, 1932, vol. 1, pp. 6-7

Several other, much smaller mills existed from time to time, but none had the capacity to compete effectively with Allan, George or Goldie.

Before the completion of the Toronto and Guelph railway, there was, in Guelph, a complete integration of manufacturing and local retail sales. Although this was obvious enough in the numerous artisanal activities (e.g., shoemaking, millinery, tailoring, baking, confectionery, saddle and harness-making, cooperage, etc.), it was equally true of even the largest manufactures, such as that of John Watt, which made a wide range of agricultural machinery and implements, steam engines and boilers, mill gearing and drives, stoves, and custom casting, boring and turning of all kinds.[67] In addition, Watt's firm erected mills on a contract basis.

The close relationship between local retail sales and manufacturing which existed in all Guelph's industries is illustrated especially well by the stove foundry that grew out of the stove retailing business owned by George Sunley. In 1856 the *Advertiser* reported:

> [In 1845], Mr. Sunley occupied but one building for his dwelling, workshop, and show and business establishment; now we find him the owner of a splendid stone store of three stories, his former establishment turned into a handsome stove shop, and his workshop thrust into a handsome stove shop, and his workshop thrust into a building in the rear. On looking over his stock a few days since, we learned that he and his partner, Mr. Melvin, have for sale about 70 different patterns and styles of stoves, to obtain which they have visited the best markets, and laid under contribution the foundries not only of Canada, but Buffalo, Albany, and other American cities, thus presenting to the purchaser as great a variety, and equal facilities for a choice, as are offered in any part of the Province. . . . And, not satisfied with what others may manufacture, Mr. Sunley has a share in the foundry business of Smith, Sunley & Co., by which means he presents to the buyer the additional choice, and the lovers of "domestic manufactures" can have, of Guelph make, as good a stove as may be found on the continent.[68]

The common characteristic of virtually all goods manufactured in Guelph before the railways were opened was that they were of the most common sort, bulky in nature as a ratio of their value and, therefore, were protected by the "tariff of bad roads." High-value goods with a limited market were generally imported and sold by the artisans and manufacturers who made the low-priced goods. Of course, the retail merchant attempted to compete at all levels and, not infrequently, as was the case with Sunley, entered into, or backed, a manufacturer to improve his own position.

The opening of the railways, combined with the reduction in American competition caused by the Civil War, opened the way for quite a differ-

67. *Guelph Advertiser*, September 20, 1854.
68. *Ibid.*, November 27, 1856.

ent sort of manufacturer. The Armstrong, McCrae & Co. (knitting and weaving, founded in 1860), Raymond Sewing Machine Company (founded, 1861), and the Bell Organ Company (founded, 1864) all produced high-quality goods for a much wider market. The largest of these, Armstrong, McCrae & Co. employed about thirty hands during that period.[69] All of these firms employed advanced technologies and, in the unsettled American Civil War period became well established in Canada—indeed, both the Raymond and Bell firms became sufficiently technically advanced to enter foreign markets. These, then, were Guelph's main industries when the construction of the Toronto, Grey and Bruce, and the Wellington Grey and Bruce forced the town's businessmen to look for expansion in Guelph's "internal productiveness."

No sooner had it become clear that the railways north of Guelph would be built, than in 1866 all the town's major businessmen came together to form a Board of Trade. A conscientious effort was made by the general membership to have the executive of the board reflect all the economic sectors of the town. Thus, in 1873, the council of the Board of Trade consisted of Robert Melvin (foundry), president; David Allan (flour mill, distillery, etc.), vice-president; and James Goldie (flour mill), George Murton (malt house), W. S. G. Knowles (auctioneer), J. Stewart (planning mill and lumber dealer), John McCrae (coal oil refinery), H. W. Peterson (lawyer), David McCrae (woollen mill), Charles Raymond (sewing machine factory), J. T. Brill (butter and pork merchant), James Massie (wholesale groceries, dry goods, etc.), Chas. Davidson (agent, Canada Permanent Building and Loan Society) and John Hogg (dry goods and clothing store).[70] The activities of the Board of Trade appear to have concentrated upon two goals: first, to overcome and to reconcile differences between its members and to democratically arrive at solutions satisfactory to all; and secondly, to promote the general economic interests of the town. In the latter function they intervened in virtually every aspect of town life from city beautification to industrial develop-ment. In the latter endeavours, one board tended to work behind the scenes through its close relationship with the town's newspapers (the owners and editors were active members of the board) and through the town and city council where members frequently held a majority. In the promotion of industrial development the Board of Trade took a leading role.

During the general depression of 1866-68, while the railways were under construction, the Board of Trade concentrated upon commercial matters. But as the lines neared completion, the board undertook two major steps towards industrial development: the creation of an agricultural implement factory and the Guelph Gas Works. In the creation of these two enterprises, the Board of Trade organized public meetings, publicized

69. "Goldie Scrapbook," Guelph Historical Society.
70. Town of Guelph Directory, 1873, p.20.

the necessity of industrial development, sold shares and organized the companies. *The Guelph Mercury*, in backing the project, made it clear what the establishment of new industries meant to Guelph at this juncture:

> At no time in the history of our town has it been more necessary for its citizens to look well to its standing than at the present moment. We are, as it were, in a state of commercial chrysalis, in a condition of transition from old to new that may prove either injurious or beneficial to our entire hereafter, according to the manner in which we mould and shape the opportunities of the moment. It cannot be denied that the railway extension so rapidly progressing to the northwest of us, while producing many beneficial results, will also bring with it accompanying changes and diversions, opening up new channels of trade and closing old ones, therefore it behooves us to prepare ourselves for the change that we may lead as many of these new channels as possible to flow hither. No step could be taken better calculated to secure this result than the establishment of manufactures upon, at least, a respectable scale, and there is no branch of industry which, at the present moment, promises so large a return as the production of agricultural implements.
>
> Possessing unrivalled and increasing methods of communication in all directions, with a rich and occupied tributary district, it will be our own fault if we do not secure for ourselves a monopoly of this trade for an area of scores of miles. Manufactories beget manufactories. Once firmly established they branch off and spring up side by side in a marvellous manner.[71]

The campaign was, apparently, successful, because within a few months, the Guelph Agricultural Works, Levi Cossitt, proprietor, began to advertise "The Farmers' Friend Gang Plough—The most successful *Plow* wherever exhibited, unsurpassed for simplicity and durability, and stands without a rival."[72] The creation of the Guelph Gas Company with a capital of $30,000 followed an identical pattern.[73] During the Franco-Prussian War boom, 1871-74, both industries appear to have prospered, but after 1874, the Agricultural Works in particular experienced hard times due to vigorous competition, and never lived up to the board's expectations.

Backed by the vigorous promotion of the board, and private capital raised from local sources, a number of smaller enterprises (some of which would eventually grow into large businesses) were established between 1868 and 1871. Of these Caleb Chase, carriages (later the Guelph Carriage Goods Company), 1868; Samuel Smith (later Beardmore) Tannery, 1869; Thomas Worswick, steam engines and machinery, 1871; and the Guelph Spring and Axel Company, 1871, were the most important.[74] All of these

71. *Guelph Mercury*, November 29, 1870.
72. See, for example, the advertisements in the *Mercury* and in the (Toronto) *Globe Weekly*.
73. *Guelph Mercury*, July 29, and August 24, 1870.
74. This list is drawn from newspaper items, directories, local histories, etc. It is likely that several industries, which did not have a long history, were missed.

industries prospered and grew rapidly until 1876 when high American tariffs and the international depression struck a heavy blow to Guelph's industries.

The importance of the large number of new industries created or fostered by the Board of Trade is illustrated by the population growth which followed their creation. During the very prosperous American Civil War years of 1861-1867, Guelph's population had grown from 5,067 to 5,357, an increase of only .92 per cent per year. In contrast, from 1867 to 1879, population increased by 4,715 to 10,072, an increase of 7.33 per cent per year.[75] So strong was Guelph's image as a prosperous, booming industrial centre that population continued to grow rapidly for several years after the industries had run into serious difficulties.

Particularly injurious to Guelph's manufacturers were the changes in American tariffs against Canadian manufactured goods imposed in 1877. Whereas in 1876 Guelph's sewing machine manufacturers had exported some $158,180 worth of machines to the United States, in the first quarter of 1877 these were reduced to only $7,003.[76] The combination of the international depression and American tariffs either bankrupted or weakened some of Guelph's oldest and largest firms. As John Horsman, a local hardware merchant and staunch Conservative, wrote Sir John A. Macdonald in 1879:

> Guelph is now the dullest place in the province and quite uninviting - David Allan's Mills & Distillery have lately passed into other hands paying about 10 cts. in the $. The Wilkie & Osborn Sewing Machine Factory all Mortgaged to the Bank of Commerce and working on half time. The Raymond factory but little better, Cositt's Agricultural works under a cloud, McBean & Co. (wholesale) Hardware called another meeting of their creditors, the Guelph Lumber Co. . . .a lot of businessmen in town have been soaped into that. . .and. . .the Ontario Bank is worrying them about to death.[77]

Here was a dilemma, indeed. Guelph needed new industry to survive and grow, but from every indication, investment in manufacturing was a risky business at best. It was time to return to the strategy that had been used in 1847 and 1851-52: when private investment was too risky for investors to undertake necessary tasks, it was time for municipal intervention.

In order to give the local government the widest possible powers to undertake its tasks, it was decided to ask for incorporation as a city. Because there was some question as to whether Guelph's population had reached the requisite 10,000 inhabitants, a special act of the Legislature was passed conferring that status in 1879.[78]

75. Population data are from the Canada Census and various city directories.

76. *Guelph Weekly Mercury*, February 22, 1877.

77. John Horsman to the Right Honourable Sir John Macdonald, Guelph, March 24, 1879; Macdonald Papers, 31293-31295, P.A.C.

78. 42 Victoria, cap. 41.

Figure 2: Guelph about 1880.
Guelph became a city in 1879 with a population slightly over 10,000.
SOURCE: George Grant, *Picturesque Canada*, vol. 2 (1882)

Immediately after city status was acquired the campaign to attract industry by municipal subsidization began in earnest. On January 17, 1880, the *Mercury* carried this editorial which set the tone and terms of the bonuses and subsidies that would be offered over the next couple of decades:

> Let us once more scan our position as a city with respect to this question. We are at a disadvantage as compared with some other towns and cities in the matter of railway competition. Several places offer bonuses to well established industries to locate in their midst. We desire to flourish in trade. We must, therefore, offer some inducement to industries as an equivalent for that railway competition which we do not possess and which other places do, and we must offer these inducements also because other places are doing the same. Is that not a reasonable conclusion?
>
> But there is more than justice on the side of the (industrial subsidization) by-law. There is a powerful argument in the shape of what it means financially and industrially to Guelph. . . . It will immediately feel the benefit of money expended among masons, carpenters, laborers and others engaged in the erection of the new factory. Those who have houses to rent and land to sell will find an extra demand for the same in consequence of the additional number of men to be employed. Merchants will realize larger sales for the same reason. These are the classes who will be indirectly benefitted.[79]

The various subsidization bylaws passed by the Guelph city council in the 1880s normally took the form of a remission of taxes and the provision of free water from the city-owned water works. There were, however, several strict provisions set, whether the subsidy was offered to a new industry to come to Guelph or to an old industry wanting to expand. Thus the bylaw usually stated that it did not apply unless an entirely new product or line of merchandise was to be manufactured; that a stated minimum number of persons had to be employed in this line for the subsidy to be retained; and provided for annual inspection of the firm's books and premises in order to ensure performance.[80]

Guelph's program of municipal subsidies showed immediate results in bringing to Guelph new industries and encouraging the expansion into new lines by established firms. Between 1880 and 1900 the following subsidy bylaws were passed and taken up: 1880, A. Murchy & Co., Malleable iron castings; 1881, David McCrae, textile fabrics, William Bell, cabinet organs, Wicks, McNaughton & Co., cutlery, and Guelph Carriage Goods Co., carriage hardware; 1882, Burr & Skinner, furni-

---

79. *Guelph Daily Mercury*, April 3, 1886.
80. The first bylaw of this type, passed June 21, 1880, stated that the new firm, A. Murchy and Company, would receive a remission of taxes for ten years provided that they built a factory to manufacture malleable iron castings, and employed an average of not less than ten workmen. Moreover, by the bylaw's terms, the company would lose its subsidy if it entered any other line of production. City of Guelph, Bylaw 29. The original is in the city clerk's office, Guelph.

ture; 1886, Burr Bros., chairs; 1889, Williams, Green & Rome, shirts, collars and cuffs, and James Hough, Jr., paper boxes and stationery; 1891, Thomas Griffin, stoves; 1893, The Curtain and Upholstering Mfg. Co., chenille curtains, etc; 1894, Rachel Crawford's Mfg. Est., shoddy yarns; 1895, Laughlin Hough Drawing Table Co., architects' and school supplies; Guelph Norway Iron and Steel Co., rolling mill; Charles Raymond, sewing machines and cash registers; 1897, Wallace and McAdoe, pork packing and bacon curing; 1898, Guelph Carpet Mills, carpets; 1899, Guelph Iron and Steel, iron goods, and A. R. Woodyatt & Co., lawnmowers, etc.[81] Similar bylaws would continue to be passed until well after the First World War.

The industrial subsidy system combined with the National Policy and the prosperity during the Canadian Pacific Railway construction period, resulted in the development of a considerable industrial base for Guelph. The *Guelph City Directory* for 1885-86 gives this description of the city's major employers:

Prominent among them may be mentioned Raymond's Sewing Machine Works, employing 200 hands; the Bell Organ Works, employing 200 hands; the Woollen Mills of McCrae and Co., employing 250 hands; the Piano Factories of Sweetnam and Hazelton, and Rainer and Son, employing 25 hands each; Gowdy's Agricultural Works, employing 35 hands; Silver Creek Brewery, employing 30 to 35 hands; the Foundry of Thos. Worswick, employing 30 hands; Crow's Iron Works, 30 hands; the Furniture Factory of Burr Bros, 85 hands; the Foundry of W. H. Mills, 21 hands; the Cigar Factory of Solomon Myers, 20 hands; the Guelph Sewing Machine and Novelty Works, 60 hands; the extensive Flouring Mills of James Goldie; the Guelph Carriage Goods Works, employing from 50 to 70 hands; the Ornamental Flower Goods Factory of W. N. Marcon, employing 25 to 30 hands, Brown's Boot and Shoe Factory, 22 hands; the Guelph Axel Works [T. Pepper & Co.]; the Oatmeal Mill of H. Murton; Tolton's Agricultural Works; the Union Foundry; Thain's Agricultural Implement Factory the Knitting Factory of Francis Smith; Clark & Thompson's Carpet Works; Eureka Paint and Color Col; and others of lesser note.[82]

There was, however, a major problem facing Guelph's business community. In spite of all the new industrial growth and the creation of its own "internal productiveness," the city's overall economy did not grow. The combination of the loss of its market area and the industrial depression from 1885 to 1900 left Guelph with little net gain for all its efforts. For example, population, which had reached 10,072 in 1879, declined to 9,890 in 1881 and remained virtually stationary thereafter, reaching 10,537 in 1891 and 11,496 in 1901.[83] Thus for the twenty-two year period

81. Originals of these bylaws are in the city clerk's office, Guelph.
82. *Guelph City Directory: 1885-1886* (Guelph, 1886), p. 12.
83. Census data are drawn from various city directories and Canada Census.

54 LEO A. JOHNSON

1879-1901 total population increased by only 1,424 or .64 per cent per year. Industrial subsidization, therefore, although it had prevented collapse, had failed to bring about the desired and expected growth. It was this failure and the growing tendency towards monopoly capitalism that ultimately turned the business community's attendion towards municipal ownership.

IV

The first major step into direct municipal ownership was made in 1886-87 with the construction of the Guelph Junction Railway. In the 1850s, one of the main arguments used to justify the share purchases and loan to the Galt and Guelph Railway was that it would provide competition to the Grand Trunk, thereby guaranteeing reduced rates. From 1867 on, however, the Grand Trunk and Great Western systems had agreed to fix rates at mutual points in order to maximize returns. One such point, named specifically in the agreement, was Guelph.[84] This situation was made permanent in 1882 when the two companies amalgamated. The results were predictable. Freight rates from Montreal to Guelph were set at 170 percent of what they were from Montreal to Galt which was served by both the amalgamated Grand Truck and the competing Credit Valley Railway (the present C.P.R. line). As the *Mercury* bitterly remarked:

These railway companies make fair promises to secure the monopoly of a paying business, but they never give one inch more than they have to, until competition compels them.'[85]

In a letter to Sir John A. Macdonald in 1886, James Goldie, Guelph's largest miller and grain exporter said:

This city is at the mercy of the Grand Trunk Railway, our Millers are subjected to unjust unfair and ruinously high rates when compared to what American millers pay. Our importers and exporters have all the same grievances and the G.T. Ry., in having the control, is not improving the trade in an otherwise energetic community.[86]

What was true for the millers was equally true for other manufacturers as well.

In 1884, a group of Guelph's leading merchants and manufacturers had obtained a federal charter to construct a rail line from Guelph to the Credit Valley Railway (opened in 1879) which ran from Toronto to

84. *Hamilton Evening Times*, October 19, 1867.
85. *Guelph Weekly Mercury*, May 8, 1884, in an article reprinted from the *Brantford Expositor*.
86. James Goldie to Sir John Macdonald, Guelph, May 29, 1886; Macdonald Papers, no. 58581-3, P.A.C.

Galt via Milton and Campbellville.[87] The projected cost of this line, some $200,000, quickly dampened the enthusiasm of the private investors and the matter was let drop. The severe depression after 1885, however, put a different face on things. Whereas monopoly freight rates had been a severe burden in 1884, they were a disaster in 1886. Faced with sure bankruptcy unless the Grand Truck's monopoly could be broken, the Board of Trade and city council agreed that a line must be built. With private funds unavailable, they concluded, municipal funds would have to be used. Thus ten businessmen paid the first call of $100 each (10 per cent of the face value of one $1,000 share) so that they might qualify as directors, while the city council paid in full $20,000 for twenty shares and agreed to lend the line $155,000.[88]

The campaign to persuade the ratepayers to sanction this transaction was opened in July 1886 (the vote was in December) by a series of advertisements in the *Mercury* placed by J. D. Williamson and Co., proprietors of the Golden Lion dry goods stores, of Guelph and Glasgow, Scotland. These advertisements display clearly the critical role that industrialization was seen to play in the economic well-being of Guelph's mercantile community:

> If we allow our manufactures to close up and move away without any effort on our part to retain them; if we see manufactures starting up in other places, which we might have had for a very little cost to the City; if we sit calmly by in indolent lassitude and allow the G.T.R to have a monopoly of our trade and the C.P.R. to cut off a large section of country, with which we at one time did a large trade [one branch of the Credit Valley Railway reached Fergus], without any effort on our part to construct a line of railroad which would break the monopoly of the G.T.R. on the one hand, and reconnect a large trade which we lost on the other. . .the result must prove disastrous. . . .
>
> The advantages of such a line would be innumerable. Some of the immediate benefits would be the reopening of Mr. Spence's mill, enlarging the trade of all our manufacturers, especially in the iron and milling line, where the freights are heavy, the cultivation of a wholesale and the increase of the retail trade of our merchants, our markets would increase and Guelph would become more of a shipping port. Manufacturers would settle here without any further inducements than those offered by the natural advantages of Guelph. We have no doubt that this railway would make Guelph boom for years to come.[89]

Another merchant, Ryan, Berkinshaw & Co., "P. S. - Remember we still sell Dry Goods," argued:

> We are today receiving as much trade from the country as we may reasonably expect for some time to come, as it is as thickly populated today as it will be in

---

87. The businessmen's names are listed in *Statutes of Canada*, 47 Victoria, cap. 79.
88. Dr. Marvin W. Farrell, *The Guelph Junction Railway, 1884-1950* O.A.C. (Guelph, 1951).
89. *Guelph Mercury*, July 2, 1886.

## Our Location and Railway Facilities

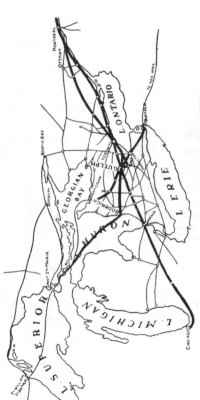

Our factory is situated at Guelph, in Central Ontario, about fifty miles from Toronto, in the centre of one of the finest agricultural sections in the country. Our City is the home of the well-known Ontario Agricultural College, which includes among its students young men from nearly every country in Europe, from India, South Africa, Japan, Australia, New Zealand, the Argentine Republic, and many sections of the United States, besides every province in Canada. We also have the famous Macdonald College, which draws its students not only from our own country, but from the Old Lands as well. Our educational facilities are the very best, and we own all our public utilities. Guelph is also the home of the Winter Fair.

For railway accommodation we can be equalled by few places. We are on the main line of the Grand Trunk Railway, going both East and West, and North and South, and of the Guelph and Goderich Railway, while the main line of the Canadian Pacific Railway is just a few miles from us, and is connected with the city by a branch line with an excellent service. This insures prompt delivery of our goods.

Figure 3:     Guelph "boosterism" in a catalogue of the Louden Machinery Company.
              SOURCE: University of Guelph Archives

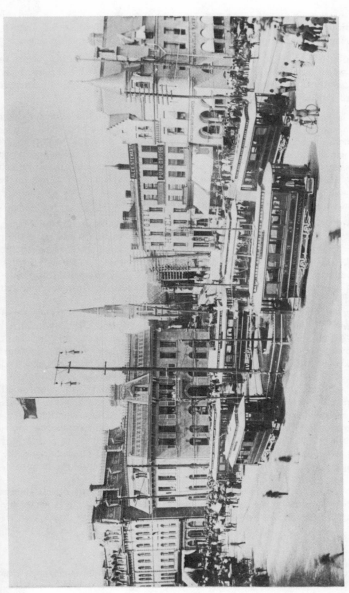

Figure 4: St. George's Square, Guelph, about 1900.
SOURCE: Guelph Civic Museum

ten years, so how can we expect the place to grow, if we do not create trade within ourselves by increasing the number of manufactures, and that cannot be done without placing Guelph on the same footing as our neighboring towns.[90]

Walter Macdonald, a third merchant, in a letter to the editor, chose to answer charges made by Grand Trunk agents that with tax remissions, free water, money bonuses, and now a railway, Guelph's manufacturers were receiving a disproportionate degree of favour:

> There should be no false jealousy of the manufacturers, it is in them that the chances of the future prosperity of Guelph chiefly lie, none of us should grudge them prosperity and think they are the only ones to be benefited. They risk their capital, they employ men and pay them, and spend the money in the place, which they get by the sale of their products far and wide. If they get rich we should rejoice and be glad, for their prosperity is ours, and will increase the value of every man's property be he rich or poor, and to those who have no property, it will give steady work and payment therefore.[91]

In spite of a vigorous campaign against the bylaw by the Grand Trunk, the voters accepted the advice of the business leaders and gave their overwhelming approval.[92] The line was completed and leased to the C.P.R. in 1888 for 40 per cent of the gross revenues.[93] Although the line lost money consistently until 1903, once the C.P.R. extended it to Goderich it became a consistent money maker, benefiting from the through traffic.

From 1902 until 1910, Guelph undertook a massive extension of municipal ownership and control. During that period the city acquired the Arkell Springs in order to extend pure water service to all users, built a sewage system, bought the Guelph Street Railway from its owners, took over the gas and electric company and became a major backer and customer of Adam Beck's Niagara Power Commission. Although the explanations and justifications for the first three undertakings continued to repeat the commercial-industrial rhetoric that marked the various bonus bylaws and Guelph Junction Railway campaigns the decision to take over the Guelph Light and Power Company and to back the "public power" campaigns of Adam Beck was based on quite a different ground. The struggle against monopoly which had characterized the campaign to build the Guelph Junction, now was extended to become a struggle against monopoly capitalism.

In order to appreciate the basis for the development of a struggle against monopoly capitalism by Guelph's capitalist class, it is necessary to say a little about the North American fuel cartels that dominated all

90. *Ibid.*
91. *Ibid.*
92. *Ibid.*, November 29 and December 2, 1886.
93. Ultimately the line cost $240,000. It received a bonus of $3,200 per mile ($42,000) from the federal government, and in 1889 the city lent it an additional $18,000.

forms of energy in the late nineteenth and early twentieth centuries. These cartels ranged from monoliths such as John D. Rockefeller's Standard Oil Company (which reached even Guelph to force the closing of local coal oil refineries in the 1880s)[94] to provincial and local "fuel dealers associations" which fixed prices and controlled fuel supply to local customers. It was the monopolistic behaviour of the fuel dealers and the determination of the local gas and electric company to maximize its profits which triggered the attack upon monopoly capitalism in Guelph.

For years, each winter the local newspapers complained about fuel shortages and skyrocketing prices. Various causes and solutions had been suggested and attempted but none seemed effective. The depth of local concern was illustrated by a section of Mayor John H. Hamilton's inaugural address to the Guelph City Council, delivered January 12, 1903:

> The Coal Situation, - Gentlemen, I should be glad to know that I am mistaken, but I am seriously of the opinion that we are rapidly nearing a period of panic regarding the coal supply. Immediate action should be taken by this council and a special committee struck off to take such action that anything of this nature should not occur again as far as this city's supply is concerned. We should have at least a year's supply of coal ahead of the period required for consumption. . . .[95]

It was in the midst of this concern about fuel shortages and rapidly rising prices that the takeover of the Guelph Light and Power Company was being discussed.

In 1870 when the Guelph Gas Company had been incorporated, it had been given a fifty-year franchise.[96] Initially it had supplied gas for street and private lighting, but in 1887 it had its franchise extended to include the generation of electricity. Four years later, it switched from water power to steam generation.[97] By 1900, the Guelph Light and Power Company (as it was now named) was the subject of a good deal of public criticism. Service was poor (one city alderman claimed that the quality of the gas was "just within a fraction of being within the law",)[98] prices were high, and profits generous to say the least. On an operation whose physical assets were assessed at $21,200, the Light and Power Company had earned, net of expenses, almost $40,000 in the three years from 1899 to 1901.[99] Although the company had, in 1902, enjoyed the

---

94. The operations of the crude oil cartel in Canada during this period are well documented in the *Monetary Times* during this period. See also my *History of Guelph.*

95. *Guelph Evening Mercury*, January 12, 1903.

96. See Bylaw 200, Town of Guelph.

97. *Guelph Weekly Mercury*, January 19, 1888 and November 24, 1891.

98. *Guelph Evening Mercury*, October 21, 1902.

99. *Ibid.*, January 2, 1902. This figure includes $12,000 dividends, $8,250 paid on debenture debt, and $19,111.68 contributed to the rest account.

most profitable year in its history it now demanded that the city pay greatly increased charges for city lighting. When the company refused to negotiate, the city council decided to expropriate the utility. The civic campaign to get ratepayer approval for a $155,000 bond issue to take over the company was, therefore, fought in the midst of the coal crisis. It was also fought in the midst of Adam Beck's campaign to bring publicly owned Niagara power to Ontario.

In January 1903, a startling revelation appeared in the Guelph newspapers. In the midst of the severe coal shortage it was revealed that eight Guelph major establishments had been the victims of a blacklist by Ontario and Guelph coal dealers. The reason for the blacklist was that they had supplied coal to their employees at cost rather than force them to pay the exorbitant prices being charged.[100] Several of these firms had had to suspend operations because of the blacklist. The *Mercury* was furious. A combination of coal magnates, they stated, held the economy of the city in its grip, and Guelph faced possible ruin. It required no great leap of imagination to connect the coal dealers' cartel with the need for public ownership of all utilities:

> If anything is hurrying on municipal ownership and socialism, when the people are hardly prepared to administer them on a large scale, it is discoveries such as these. That such a necessity as coal should be subject to trust prices, not alone at the seat of production, but provincially and locally, and that provincial and local dealers should be merely agents of the coal operators and coal carrying companies is a state of affairs that will not be borne any longer than the people can find effective means to deal with it.[101]

When it came to the Niagara power scheme, the *Mercury* was absolutely clear as to why it was needed:

> The tendency of private capital in large enterprises is to consolidate and to fix monopoly rates, and it was the danger of such a contingency that has stirred Toronto and other municipalities to aim at municipal or government ownership and control of Niagara power. . . .[102]

It was this tendency of capital, the *Mercury* argued, that had spurred Ontario's residents to use public ownership to escape corporate control:

> The growth of trusts and monopolies, and the extension of the power and control of corporations over public necessaries, has rapidly educated the people to the necessity of public ownership in larger things.[103]

Based on these arguments, the bylaw to purchase the gas and electric utility passed easily.

100. *Ibid.*, January 26, 1903.
101. *Ibid.*, January 27, 1903.
102. *Ibid.*, February 10, 1903.
103. *Ibid.*, February 5, 1903.

Between 1903 and 1906, the Guelph Board of Trade, the city council and business leaders such as J. W. Lyon, Christian Kloepfer and Lincoln Goldie were at the forefront of the provincial drive to acquire Niagara power. In 1908, the city electric utility signed to take and pay for at least 2,500 h.p. of electricity and to guarantee $347,420 worth of Hydro Electric Power Commission bonds.[104] Only Toronto and London did more to assure that project's financial success.

Guelph's vigorous program of municipal ownership quickly made it famous throughout Canada, and Guelph's businessmen were invited to speak to other Boards of Trade and Chambers of Commerce. Their speeches, however, concentrated upon the profitability of Guelph's investment, rather than upon the anti-monopoly capitalist arguments which had been used in Guelph in 1903. For example, in an interview given to the *Winnipeg Free Press* in 1909, J. W. Lyon, after describing the profitability of each aspect of Guelph's publicly owned enterprises, summarized the whole by saying:

> On the entire investment in municipal undertakings in Guelph, there is being realized 12 percent net. Gas is now furnished for all purposes for $1 per thousand feet. Power is furnished from Niagara to Guelph at one half price at which it is furnished to Buffalo, although we are more than three times as far from the source. . . .[105]

As the *Toronto Telegram* had correctly predicted in October, 1903:

> Guelph may become a centre of influence in favor of public ownership just as Hamilton has become a centre of influence in favor of corporate privilege. . . . Guelph must incidentally reap great advantage from an absolute control of every public franchise. Gas, electric light and power and street railway services are all under the control of the municipality and Guelph must grow by reason of its ability to offer immunity from the exactions of private ownership.[106]

And grow Guelph did. Between 1901 and 1911 total population increased by 3,679 (from 11,496 in 1901 to 15,175 in 1911) an increase of 3.2 per cent per year, more than four times the rate of growth from 1879 to 1901. In 1910 the *Mercury*, enthusiastic about Guelph's rapid growth, proclaimed:

> Brantford has over 21,000 population. We'll be right on your heels soon Brantford. We're striking our stride now, here in Guelph.[107]

In municipal ownership, Guelph believed that it had, at last, found the key to rapid and permanent economic growth.

It was not to be. Two factors beyond Guelph's control, the adoption

104. W.R.Plewman, *Adam Beck and Ontario Hydro* (Toronto, 1947), pp. 52-53.
105. *Guelph Evening Mercury*, September 27, 1909.
106. Quoted in *Ibid.*, October 8, 1903.
107. *Guelph Weekly Mercury*, September 29, 1910.

by competing municipalities of all of Guelph's successful strategies, and the tendency towards cartelization and centralization of industry in Canada during the monopoly capitalist period, quickly removed Guelph's advantages and slowed growth once again.

Where vigorous competition for industry existed, many errors of judgment regarding the soundness of the prospective firms were made. For example, in 1907 both the Board of Trade and the city council supported a loan of $25,000 to the Morlock Furniture Company,[108] but by 1909 that firm was bankrupt, the workers not even receiving their wages.[109] One measure of the weakness of the subsidized firms during this period of increased municipal competition can be found in a report issued in 1914 by the industrial department which had been created by city council in 1910. According to that report, in the three years of its operation, the industrial department had been instrumental in attracting sixteen industries to Guelph.[110] By 1919, however, a survey of Guelph industry showed that no less then ten of the sixteen had either closed shop or moved elsewhere.[111]

Finally, the cartelization and centralization of industry in Canada as capitalism reached its monopoly stage[112] dealt several severe blows to Guelph's industrial base. Although the tendency towards centralization of industry by railway concentration had caused industries to leave Guelph previously (the move of Inglis and Hunter, later the John Inglis Company, to Toronto in 1881 had followed this pattern),[113] the takeover and amalgamation of competing firms during the cartelization period, and the subsequent rationalization of production, cost Guelph some of its largest and oldest firms. Two of these, the Raymond Sewing Machine Co., and the Bell Organ Company, both heavily subsidized by the city, were particularly important. Of these, the sale of the Raymond company to an American firm, the White Sewing Machine Company of Cleveland, Ohio, provides a classic example of de-industrialization due to cartelization.

When it was announced, in 1916, that the Raymond firm had been sold to the White Co., the *Mercury*, basing its opinions on statements made by the White company officials, greeted the news enthusiastically:

108. *Guelph Evening Mercury*, November 10, 1906.

109. *Ibid.*, December 20, 1909.

110. *Ibid.*, January 29, 1914.

111. *Ontario Provincial Fair Review* (Guelph, 1919), p. 27. This total is arrived at by comparing the two lists.

112. The peak of the early cartelization process in Canada was reached from 1908 to 1912 when no less then 57 major amalgamations covering whole industries took place. See, for example, *Monetary Times*, January 6, 1912. I am here using the term "monopoly capitalism" in the sense that V.I. Lenin used it in *Imperialism, The Highest Stage of Capitalism* (Peking, 1965) chap. X, pp. 148-51.

113. See the *Mechanical and Milling News*, October, 1886, for a sketch of John Inglis' career. This sketch was reprinted in the *Guelph Mercury*, October 9, 1886.

One of the most important industrial transactions that has taken place in the history of Guelph was closed yesterday when the White Sewing Machine Co., of Canada, took possession of the business of the Raymond Manufacturing Co. of Guelph. . . . The coming of the White Company to the Royal City is a big win for Guelph. . . .

"I have always liked Guelph," said Mr. Chase, secretary of the White Co., of Cleveland, to the Mercury. . . . We intend to manufacture here not only for Canadian business, but for Great Britain and the British possessions. Our Russian business may also be done here, also our business with France. Our sales in Australia are increasingly large, and it may be that we will manufacture for Australia in Guelph.[114]

In conclusion, Chase stated that production of sewing machines in Guelph would be increased from thirty to three hundred per day, and that more than $200,000 would be invested to build a new plant. None of these promises were kept. In 1922 the business and equipment was transferred to the main plant in Cleveland, and the Raymond distribution network was used to sell American-made "White" machines in Canada.[115] The Bell Organ Company suffered a similar fate, being sold to a British syndicate in 1888. It continued to operate at a reduced rate until the 1930s when it was closed.

Once Guelph's municipal rivals adopted Guelph's strategies, and the results of cartelization began to be felt, growth slowed in spite of continued industrial subsidization. Between 1911 and 1921, Guelph's population increased by 2,953 for an average of 1.9 per cent per annum. Moreover, the national boom of the 1920s had virtually no effect in Guelph. From 1921 to 1931, population increased by only 2,947 (from 18,128 to 21,075) for an average increase of only 1.6 per cent per annum. Like an impoverished gentleman who had seen better days, the city's ideological leaders put on a bold face. Now, rather than emphasizing Guelph's unique economic advantages alone, they tended to concentrate upon the cultural and recreational amenities that Guelph had to offer.[116] As the results show, to prospective industrialists culture was a poor substitute for economic advantage. Such would remain the case until after the Second World War.

In their pursuit of prosperity and growth, Guelph's business and ideological leaders showed a good deal of adaptability and innovativeness. Far from exhibiting the antagonism, subjectively or objectively, which Tom Naylor claims lies at the heart of relations between the commercial and industrial capitalist classes,[117] Guelph's merchants and commercial men

114. *Guelph Evening Mercury*, April 26, 1916.
115. David Allan, *Guelph and its Early Days*, p. 22.
116. See especially the Centennial Edition of the *Guelph Evening Mercury*, July 20, 1927.
117. R. T. Naylor, "The Rise and Fall of the Third Empire of the St. Lawrence," *Capitalism and the National Question in Canada*, ed. Gary Teeple (Toronto, 1972), pp. 1-41. See especially pp. 20 and 32.

proclaimed the necessity of production as the basis of trade, and when rural production proved incapable of serving their needs, undertook to foster, by every method available to them, the industrial, "internal productiveness," of the city. They were not defeated until the cartelization aspects of monopoly capitalism brought about changes which their power over Guelph's municipal government was unable to control—and even then they ventured, for a period, into attacks upon monopoly capitalism itself. In sum, Guelph's business community had a remarkably clear perception concerning their interests as capitalists, and their ideology and strategy were, within the limits of their power, successful in protecting and enhancing the productive base of their economy. It was only when the contradictions within capitalism moved to levels beyond their reach that they failed.

# Montreal Banks and the Urban Development of Quebec, 1840-1914

RONALD RUDIN

Banks have always played an important role in urban development. N.S.B. Gras found that the establishment of financial control by banks and other financial institutions was the crowning point in the growth of a centre to metropolitan status. "In Boston it is State Street, in New York Wall Street, in Chicago LaSalle Street. Here in these districts are the banks, brokers' offices, stock exchanges and insurance offices. Here is the most sensitive spot in the metropolitan nerve centre."[1] In addition to symbolizing the maturity of the metropolitan centre, these financial institutions had an impact upon the development of towns within the hinterland of the metropolis. Particularly for a town which did not possess much capital, the accessibility to the capital mobilized by metropolitan financial interests was essential to local growth. Industrial development was the key to urban growth, and without access to this capital local industrialists could not operate successfully and entrepreneurs from outside the city would not establish facilities.[2]

Because chartered banks controlled over half of the assets held by Canadian financial intermediaries prior to the First World War, Canadian towns looked to them for the provision of services that might foster industrial growth. While banks in Canada, like those in Great Britain and the United States, shied away from granting long-term loans to industry, they provided an indispensable boost to economic develop-

1. N.S.B. Gras, *Introduction to Economic History* (New York, 1922), p. 269.
2. Allan Pred, *The Spatial Dynamics of United States Urban-Industrial Growth* (Cambridge, 1966), p. 46.

ment by making considerable short-term credit available.[3] As the economic historians Rondo Cameron and Hugh Patrick have noted,

> Characteristically, commercial banks grant short-term credit for working capital. Partly for this reason many authorities have asserted that banking has made a negligible contribution to the formation of industrial capital. Apart from overlooking the many instances in which banks departed from the rule of short-term credit, this view fails to take into account the way in which entrepreneurs use short-term credit to free their own resources for fixed investment.[4]

By the middle of the nineteenth century Montreal had reached Gras' highest stage of metropolitan development. Having already become the commercial, industrial and transportation centre of British North America, Montreal also achieved financial control through the Bank of Montreal which functioned as the government's banker. Because of the concentration of financial power in Montreal, towns in Quebec looked to it to gain access to capital for the encouragement of urban growth. As Gras suggested, capital filtered down to these towns through the establishment of branches.[5] His suggestion is confirmed by an analysis of the decision-making process regarding loans used by seven banks between 1840 and 1914, six with headquarters in Montreal and one based in Quebec City. The branch manager played a key role in assessing the desirability of granting loans. The manager's decision was rarely overturned by the directors of the bank and, even more importantly, these banks hardly ever granted loans to entrepreneurs in cities lacking a branch.[6] Accordingly, leaders of Quebec cities such as Saint-Hyacinthe who wanted to see their towns prosper, recognized the importance of attracting branch banks as absolutely necessary. "For the importer, for the speculator, for the entrepreneur, for all our merchants, for those whose businesses require considerable capital. More importantly, as long as we do not have a bank at Saint-Hyacinthe it will be difficult to count upon the establishment of industries."[7]

The purpose of this paper is to examine the degree to which Montreal banks were responsive to the needs of the growing urban centres of the

---

3. Regarding the role of banks in industrial development, see Tom Naylor, *The History of Canadian Business* (2 vols.; Toronto, 1975); Sydney Pollard, "Fixed Capital in the Industrial Revolution," *Journal of Economic History*, XXIV (1964), pp. 299-314; Lance E. Davis and Douglass C. North, *Institutional Change and American Economic Growth* (Cambridge, 1971).

4. Rondo Cameron *et al.*, *Banking in the Early Stages of Industrialization* (New York, 1967), p. 11.

5. Gras, *Economic History*, p. 259.

6. The banks under consideration are the Bank of Montreal, the Merchants' Bank, the Molson's Bank, the Banque d'Hochelaga, the Banque Jacques Cartier and the Banque Provinciale.

7. *Courrier de Saint-Hyacinthe*, 1 April 1857. This and all subsequent passages which were originally in French have been translated by the author.

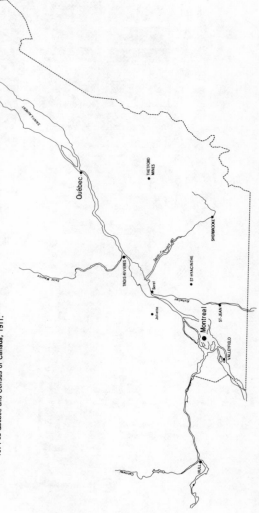

QUEBEC CITIES OF OVER 5000, 1911.*

SCALE 1:1,250,000

*not including those cities situated within a Metropolitan Census
area as defined in the 1961 Census.

SOURCES: Ministère des Terres et Forêts Tenure des Terres Forestières,
1974 du Québec and Census of Canada, 1911.

Figure 2   The Bank of Montreal (1846) in Place D'Armes. Lithographed in 1848 from a C. Krieghoff painting.

province. In searching for financial services, these towns rarely looked beyond Montreal. In 1913 only 50 of the 474 branch banks located in Quebec had been established by institutions with head offices outside of the province. Of the banks which established the other 424 branches, all but one had its headquarters in Montreal.[8] The relative absence of significant urban centres raises the question, however, of how well these Montreal institutions served the province.

The percentage of the Quebec population living in urban centres was comparable to that in Ontario throughout the period under study. While 14.9 per cent of the population of Quebec lived in urban areas in 1851, 23 per cent in 1881 and 44.5 per cent in 1911, the comparable percentages for Ontario in those years were 14 per cent, 23.8 per cent

Table 1: Montreal and Toronto, 1851-1911

|  | % of provincial urban population | | % of value of provincial industrial production | |
|---|---|---|---|---|
|  | *Montreal* | *Toronto* | *Montreal* | *Toronto* |
| 1851 | 43.5 | 23.1 | | |
| 1861 | 48.9 | 17.4 | | |
| 1871 | 45.2 | 16.8 | 46.5 | 11.9 |
| 1881 | 48.0 | 18.4 | 50.2 | 12.4 |
| 1891 | 51.6 | 24.5 | 45.9 | 18.8 |
| 1901 | 45.0 | 23.6 | 47.9 | 24.2 |
| 1911 | 52.7 | 30.1 | 47.4 | 26.6 |

SOURCE: Census of Canada, 1851-1911

Table 2: Number of Cities With Populations of over 5,000

|  | *Quebec* | *Ontario* |
|---|---|---|
| 1871 | 4 | 12 |
| 1881 | 8 | 19 |
| 1891 | 9 | 23 |
| 1901 | 10 | 28 |
| 1911 | 13 | 37 |

SOURCE: Leroy Stone, *Urban Development in Canada* (Ottawa, 1967), p. 69. Note that Stone excluded cities with populations of over 5,000 which fell within the Metropolitan Census Area as defined by the 1961 census. For instance, Lachine was not included as one of the Quebec cities of over 5,000 in 1911 because it fell within the 1961 boundaries of the Montreal metropolitan area.

8. *Monetary Times*, January 1913.

and 49.5 per cent.[9] Clear differences emerge between Quebec and Ontario, however, when the distribution of the urban population of the two provinces is considered. The Quebec urban system was distinguished by the concentration of a large percentage of the urban population of the province in a single centre, Montreal, and by the general absence of other substantial centres. By contrast, because a number of important urban centres existed in Ontario, Toronto held a much smaller percentage of the Ontario urban population than did Montreal in Quebec. Table 1 indicates the position of Montreal in the Quebec urban system and that of Toronto in the Ontario context, while Table 2 gives an idea of the number of important centres in each province. In this last regard, Louis Trotier has noted, "Towards the end of the third quarter of the nineteenth century, the Ontario urban system was already more complete than that in Quebec, not only because there were more small cities in Ontario but also because of the existence in Ontario of three cities with between 10,000 and 20,000 inhabitants. There were few medium-size cities in Quebec and those that did exist were extremely small."[10]

The problems of industrial, and accordingly of urban, development in Quebec have generally been explained by two factors: the unenterprising attitude of the francophone Quebeckers and the disadvantageous location of the province.[11] Both of these factors, however, are by themselves incapable of explaining the pattern of urban development that emerged in Quebec. The assumption that francophones tended to shy away from functioning as entrepreneurs is weakened by a closer inspection of the economic development of the lesser towns of the province. For instance, in the town of Saint-Hyacinthe, located thirty miles east of Montreal, a number of local entrepreneurs worked diligently to develop the city until it was the fourth-largest centre in Quebec in 1895.[12] These interests constantly complained, however, that their efforts were impeded by an inaccessibility to the capital controlled by the Montreal banks.[13] The distance of Quebec towns from the resources and raw materials of Ontario and the American midwest certainly played some role in discouraging Montreal financial interests from investing in these centres.[14] This point should not be overstated, however, because Quebec itself was

9. Between 1851 and 1911 an urban area was defined as an incorporated unit with a population of more than 1000.

10. Louis Trotier, "Genèse du Réseau Urbain du Québec," in Marc-André Lessard and Jean-Paul Montmigny, eds., L'Urbanisation de la Société Canadienne-française (Quebec, 1967), p. 27.

11. See René Durocher and Paul-André Linteau, eds., Le 'Retard' du Québec et l'Infériorité Economique des Canadiens-français (Trois-Rivières, 1971); Albert Faucher, Québec En Amérique (Montreal, 1973).

12. Saint-Hyacinthe would not rank as high if the industrial suburbs of Montreal such as Maisonneuve were included.

13. Courrier de Saint-Hyacinthe, 19 January 1871.

14. Faucher, Québec En Amérique, p. 221.

rich in resources and offered a market that Montreal could have dominated with no competition from centres such as Toronto. There were
also other factors which reduced the flow of capital from Montreal to
the lesser centres to a trickle. As a result, Quebec urban development
was retarded.

II

Two different types of banks existed which could be approached by
Quebec towns eager to gain facilities for the transaction of financial
affairs. On the one hand, there were the banks with headquarters in
Montreal which were controlled by anglophone interests. The most
important of these banks were the Bank of Montreal, the Merchants'
Bank and the Molson's Bank. It was rare for a francophone to serve on
the boards of these institutions. Only four of the 107 men who served as
directors of the Bank of Montreal prior to 1914 were francophones, with
not a single French Canadian serving in that capacity between 1836 and
1914. The Merchant's Bank and the Molson's Bank were also dominated by English Canadians as they were the creations of two powerful
Montreal families, the  Allans and the Molsons. These three banks
ultimately become one, as the Merchant's Bank was absorbed by the
Bank of Montreal in 1922, while the Molson's Bank experienced the
same fate three years later.

The second group of banks was dominated by francophones and,
with one notable exception, was also concentrated in Montreal. The
most important of these banks were the Banque d'Hochelaga, the
Banque Nationale, the Banque Jacques Cartier and the Banque Provinciale.[15] In 1924 the Hochelaga and the Nationale, whose head office was
in Quebec City, merged to form the Banque Canadienne Nationale.

Table 3: Bank Assets, 1871-1911 (in millions of dollars)

| Bank | Date of establishment | 1871 | 1881 | 1891 | 1901 | 1911 |
|---|---|---|---|---|---|---|
| Montreal | 1817 | 29.8 | 42.1 | 50.0 | 99.8 | 217.4 |
| Molson's | 1853 | 3.3 | 8.1 | 12.3 | 21.9 | 44.2 |
| Merchants | 1864 | 16.6 | 17.6 | 21.5 | 31.6 | 70.9 |
| Hochelaga | 1873 | | 1.5 | 3.1 | 10.7 | 23.9 |
| Nationale | 1859 | 2.6 | 4.6 | 3.9 | 7.3 | 17.7 |
| Jacques Cartier | 1861 | 3.0 | 2.1 | 2.8 | | |
| Provinciale | 1900 | | | | 3.1 | 9.9 |

SOURCE: *Monetary Times*, at start of each year listed.

15: Another anglophone bank was the Commerical Bank, while there were other francophone banks such as the Banque du Peuple and the Banque Ville Marie.

After the Jacques Cartier suspended operations in 1899 it was reorganized and emerged as the Provinciale in the following year. While there were banks in each category other than the ones listed here, these seven banks operated eighty per cent of the branches controlled by institutions in the province in 1913.

The anglophone banks were also distinguished from the francophone ones by their greater wealth. As Table 3 indicates, the assets of the smallest anglophone bank were consistently greater than those of the most important francophone institution. Accordingly, Quebec towns in search of banking facilities first directed their attention towards the anglophone institutions. It quickly became apparent, however, that these banks were not going to meet the needs of Quebec urban centres. *Le Courrier de Saint-Hyacinthe* recognized as early as 1861 that "trying to convince the Montreal bankers to open branches in this city is a useless endeavour."[16] In 1871 the Bank of Montreal had twenty-seven branches, twenty-four of which were located in Ontario. Of the remaining branches two were located in Montreal and one in Quebec City.

Table 4: Branches of the Anglophone Banks in Quebec

| Year | Total Branches | Bank of Montreal Branches in Quebec | | Quebec Branches, not in Montreal | |
|------|------|------|------|------|------|
| | | No. | % of total | No. | % of total |
| 1871 | 27 | 3 | 11 | 1 | 4 |
| 1881 | 28 | 2 | 7 | 1 | 4 |
| 1891 | 37 | 4 | 11 | 1 | 3 |
| 1901 | 47 | 4 | 9 | 1 | 2 |
| 1911 | 126 | 22 | 17 | 12 | 9 |
| | | *Molson's Bank* | | | |
| 1871 | 7 | 1 | 14 | 1 | 14 |
| 1881 | 16 | 2 | 13 | 1 | 6 |
| 1891 | 24 | 3 | 13 | 2 | 8 |
| 1901 | 40 | 9 | 22 | 7 | 18 |
| 1911 | 82 | 26 | 31 | 19 | 23 |
| | | *Merchants' Bank* | | | |
| 1871 | 24 | 2 | 8 | 2 | 8 |
| 1881 | 29 | 3 | 7 | 2 | 7 |
| 1891 | 30 | 4 | 13 | 3 | 10 |
| 1901 | 36 | 4 | 11 | 3 | 8 |
| 1911 | 151 | 17 | 11 | 12 | 8 |

SOURCE: Annual reports of banks published in *Monetary Times*.

16. *Courrier de Saint-Hyacinthe*, 4 March 1861.

Similarly, six of the seven branches of the Molson's Bank were in Ontario, as were twenty-two of the branches of the Merchant's Bank. In subsequent years some branches were located in Quebec towns other than Montreal and Quebec City, but these branches never accounted for more than a small percentage of the entire branch system of these banks.

The figures noted above take on greater interest when it is understood how these banks decided upon the towns in which branches were to be established. The banks generally granted branches only to towns which submitted petitions requesting such facilities. While the banks claimed that they ruled on these petitions solely on the basis of the prospect of making a profit from the establishment of a particular branch, the rejection of three petitions from Quebec towns suggests that other factors may have been involved. In 1880 the Merchants' Bank refused to establish a branch in Sherbrooke, while in 1889 the same bank rejected a petition from a group of Saint-Hyacinthe residents. In the latter case the bank commented that it had "already established branches in the leading centres of Ontario, Quebec and the Maritimes."[17] In 1901 a Trois-Rivières request was rejected by the Molson's Bank, ostensibly because a branch in that city would not turn a profit.[18]

Had these requests been made when the three towns were in decline then the rejections would be understandable. In fact, however, Sherbrooke was a growing industrial centre of 7,000 in 1880, while Saint-Hyacinthe was experiencing considerable industrial growth in 1889 and Trois-Rivières was a city of almost 10,000 in 1901. In all three towns industrial activity had given life to local commerce and had encouraged the productivity of farmers in the surrounding region. Accordingly, these rejections become even more curious upon taking note of certain towns that did receive branches of these banks. The Montreal anglophone banks had established branches in the primary and secondary towns of Ontario and were already opening branches in the agricultural service centres of that province by the 1880s. While the Merchants' Bank rejected Sherbrooke's request in 1880 it had opened a branch in Walkerton with a population of less than 2,600 only a few years before. Saint-Hyacinthe was turned down at approximately the same time that Mitchell with a population of 2,101 was given a branch of the Merchants' Bank, and Trois-Rivières was rejected a year after the Molson's Bank had established branches in the Ontario centres of Alvinston and Chesterville which had populations of 878 and 932 respectively in 1901. While the Montreal anglophone banks rarely located a branch anywhere in Quebec, they quite willingly established a large percentage of their Ontario branches in towns of less than 5,000 residents.

17. Public Archives of Canada (hereafter PAC), Merchants' Bank, minutes of directors' meeting, 7 April 1880; 19 June 1889.
18. PAC, Molson's Bank, minutes of directors' meeting, 29 January 1901.

Table 5: Ontario Branches of the Merchants' Bank

| | A: Ontario Branches | B: Branches in Towns of less than 5,000 | B as % of A |
|---|---|---|---|
| 1871 | 22 | 16 | 73 |
| 1881 | 24 | 15 | 63 |
| 1891 | 24 | 10 | 42 |
| 1901 | 26 | 11 | 42 |
| 1911 | 67 | 39 | 58 |

SOURCES: Annual reports of Merchants' Bank in *Monetary Times*; Census of Canada, 1871-1911

These small Ontario muncipalities where branch banks were located might have been more attractive than the larger Quebec towns had there been considerable competition from other banks in the latter towns and a relatively open field in the former. In fact, there was as much, if not more, competition in the Ontario towns as there was in the Quebec centres. For instance, when Sherbrooke was rejected in 1880 by the Merchants' Bank the city was only being served by two relatively small banks: the Banque Nationale, which had a branch in Sherbrooke, and the Eastern Townships Bank, a regional institution whose headquarters were located there.[19] In the same year Almonte with a population of approximately 2,600 was the site of a branch of the Merchants' Bank as well as an agency of the powerful Bank of Montreal. Saint-Hyacinthe was turned down by the Merchants' Bank at a time when the Molson's Bank and two rather weak institutions, the Banque Jacques Cartier and the Banque de Saint-Hyacinthe, were serving the city.[20] In 1889 Perth with less than half of the population of Saint-Hyacinthe was being served by branches of both the Merchants' Bank and the Bank of Montreal. Finally, Trois-Rivières possessed only a branch of the Banque d'Hochelaga when it was turned down by the Molson's Bank in 1901. By contrast, in the same year Owen Sound with fewer people was the site of branches of the Molson's Bank, the Merchants' Bank and the Bank of Hamilton. In 1911 Trois-Rivières, one of the fastest growing industrial centres in Canada, still had only attracted the Bank of Montreal of the three anglophone banks to locate a branch within its borders. By contrast, Alvinston, Ontario, with a population of 806 was the site of branches of all three banks by 1911.

19. The Eastern Townships Bank was established in 1859 by anglophone residents of the Sherbrooke region, who felt that the Montreal banks were not providing satisfactory service. Upon its absorption by the new Bank of Commerce in 1912, the bank had assets of $2.8 million and a chain of 99 branches that stretched across Canada.

20. La Banque de Saint-Hyacinthe was established in 1873 by local residents desirous of better banking facilities. Upon its failure in 1908, the bank had assets of $1.6 million and a chain of five branches in the immediate vicinity of Saint-Hyacinthe.

Just as the anglophone banks set different standards for the establishment of branches in Quebec from those used in Ontario, so too did they have a higher set of standards for francophone communities in the province than those applied to anglophone communities. All three of the anglophone banks concentrated a significant number of their branches upon the island of Montreal. The Bank of Montreal, for instance, located ten of its twenty-two Quebec branches in 1911 in Montreal. The distribution of the remaining Quebec branches across the province suggests, however, that the concentration of branches at Montreal was not solely a function of the size of that city. Within the province the anglophone banks located a large percentage of their branches in communities such as Montreal which had substantial anglophone populations. In 1911 nearly half of the Quebec branches of the Bank of Montreal and the Molson's Bank were in such English towns, while the Merchants' Bank had approximately 40 per cent of its Quebec branches located in these anglophone enclaves. By 1901 the Molson's Bank had refused to place a branch in Trois-Rivières and had opened branches instead in the tiny English hamlets of Knowlton and Kingsville with 760 and 1,300 residents respectively.

The establishment of branch banking facilities by one of the anglophone banks in a town did not mean that the centre would automatically take on great importance. Clearly, the location of branches in towns such as Alvinston and Knowlton did not lead to considerable industrial growth. For this growth to occur, a city also had to possess certain characteristics attractive to entrepreneurs, such as water power, good transportation facilities, proximity to raw materials and markets, and a cheap labour supply. In addition, it was necessary for local interests to actively work at drawing entrepreneurs and capital to their city. Lacking most of these attributes the Knowltons and Alvinstons of central Canada could not develop into important industrial centres even with the establishment of branch banking facilities. On the other hand, there were Quebec towns such as Saint-Hyacinthe which possesed most of these characteristics yet had difficulty in developing because of an inaccessibility to the capital controlled by the anglophone banks.[21] These banks undoubtedly established branches in a manner which seemed to them most likely to maximize their profits. It does seem clear, however, that some excellent opportunities in francophone Quebec were passed over. In pursuing profits, these banks largely restricted themselves to locating branches in both Ontario and Quebec so that anglophone interests could be served.

The anglophone banks often located branches in central Canada in growing industrial towns and did provide aid to industrial firms in these centres. Nowhere was this as clear as in the few cases in which branches

21. See Ronald Rudin, *Saint-Hyacinthe and the Development of a Regional Economy, 1840-1895* (Toronto, 1977).

were established in predominantly francophone towns. In these cases, however, the banks generally oriented their operations towards serving local anglophone interests. The Molson's Bank opened its first Quebec branch in Sorel in 1871 because of the Molson family's earlier connection with the town due to the shipbuilding industry.[22] When the Molson's were active in Sorel prior to the 1850s, there was a significant anglophone business community, and the Molson's Bank came to Sorel in 1871 to serve the remnants of that community. The only francophone firm to receive a major industrial loan through this branch was a foundry which received a $20,000 loan in 1914. As if to justify granting a loan to a francophone firm, the directors of the bank emphasized that the loan "was to enable it [the firm] to execute an order for the British govenment."[23] By contrast, between 1871 and 1914 anglophone firms received loans totalling almost $40,000.[24]

Similarly, in the largely francophone towns of Saint-Hyacinthe and Trois-Rivières francophone entrepreneurs received little assistance from the anglophone banks. The most successful entrepreneur in Saint-Hyacinthe was a shoe manufacturer who in 1865 established a factory which by 1892 was employing 240 workers. In both Canada and the United States this man was recognized as an important inventor of machinery used in the production of footwear. Nevertheless, he was rejected when he went to the manager of the Molson's Bank branch in Saint-Hyacinthe in 1888 for a $10,000 loan.[25] By contrast, an American-controlled textile mill had no difficulty in securing $45,000 in loans from the bank.[26] The Merchants' Bank which retained a branch in Saint-Hyacinthe from 1871 to 1877 consistently refused to extend credit to francophone interests, even though the prospective clients included the city's leading industrialist and its leading commercial concern.[27]

The Bank of Montreal established a branch in Trois-Rivières in the early 1850s to provide financial services in an American-controlled sawmill operating in the city. This institution did little, however, to aid in development of local industry or commerce which was largely controlled by francophones. Accordingly, a local newspaper complained that this branch bank failed "to facilitate the circulation of money which was necessary for the prosperity of the region."[28] In 1858 the sawmill was sold to a syndicate which had no interest in continuing a

22. The Molsons operated a shipyard at Sorel from 1786 to 1851.
23. PAC, Molson's Bank, minutes of directors' meetings, 17 November 1914. At the request of the Bank of Montreal, which controls the records of the Molson's Bank and the Merchants' Bank as well as those of the Bank of Montreal, the names of specific clients have been withheld.
24. Compiled from Molson's Bank, minutes of directors' meetings, 1871-1914.
25. PAC, Molson's Bank, minutes of directors' meeting, 20 July 1888.
26. Ibid., 26 February 1897.
27. PAC, Merchants' Bank, minutes of directors' meeting, 16 June 1877; 15 July 1876.
28. L'Ere Nouvelle, 13 April 1857.

working relationship with the Bank of Montreal and that bank "resolved to close its agency at Trois-Rivières there no longer being any advantage to keep it open."[29] The Bank of Montreal only returned to Trois-Rivières in 1908 after anglophone interests had established themselves in the city.

Finally, in Sherbrooke, where significant anglophone and francophone business communities existed, the anglophone banks again showed their preference for dealing with businessmen who spoke the same language. One of the few major loans made through the Sherbrooke branch of the Merchant's Bank was a $10,000 loan to a local anglophone entrepreneur who was operating a machine factory.[30] The Bank of Montreal also contributed to the economic growth of the city by means of a $373,000 loan to the Paton Manufacturing Company in 1874.[31] This loan was significant, as it was one of the few granted by these banks to a firm in a town where a branch was not located. This exception was made because George Stephen, then vice-president of the Bank of Montreal, was also a major investor in the Paton firm. For francophone entrepreneurs in Sherbrooke, however, no such assistance was forthcoming. The most successful entrepreneur in the city was O. Gendron, whose Eastern Townships Corset Factory was giving work to two hundred employees by 1890. This success was achieved without the help of the anglophone banks. In 1887 he noted, "I established this factory in 1880 with almost no capital." He was successful only because he found "in the local French Canadian population a group of capitalists with faith in the success and the future of my firm, and thanks to them our factory now has a capital of $45,000."[32]

The problems of the francophone trying to deal with one of the anglophone banks were summed up in the case of Coaticook, a town in the Eastern Townships. A branch of one of the anglophone banks had been established there to serve the anglophone minority, and not the francophone majority, of the town. Accordingly, the Montreal business journal, *Le Prix Courant*, noted: "In Coaticook, as is the case in Montreal and elsewhere, francophone merchants are at a disadvantage when competing with anglophone merchants to whom the banks provide services which are denied to us." The solution was "to establish some French-controlled banks."[33]

29. PAC, Bank of Montreal, minutes of directors' meeting, 3 September 1858.
30. PAC, Merchants' Bank, minutes of directors' meeting, 8 August 1883.
31. Paton Manufacturing Company, minutes of directors' meeting, December 1874.
32. *Le Pionnier*, 20 January 1887.
33. *Le Prix Courant*, 8 February 1889.

III

The francophone towns of the province initially looked to the anglophone banks of Montreal for financial services because of the superior resources controlled by those institutions. The francophone banks, like the towns of Quebec, did not have easy access to capital, a limitation which resulted in the failure of several of these banks prior to the First World War. La Banque du Peuple closed in 1895, followed by the Jacques Cartier and the Ville Marie in 1899. *Le Moniteur du Commerce* referred to the weakness of the francophone banks in 1883; "The majority of our banks are so weak that they are unable to assist our progress in commerce. These banks simply discount notes and are unable to play any role in the development of French-controlled industry."[34] Despite these problems the Quebec towns were forced to turn to the francophone banks. For example, in 1860 a group of Saint-Hyacinthe residents petitioned the Banque Nationale to establish a branch in their city since the anglophone banks were unwilling to provide facilities. After a great deal of debate the directors decided that "it is impossible to establish a branch at Saint-Hyacinthe now, but the directors will provide Saint-Hyacinthe with as much service as it is possible to accord."[35] None of the francophone banks had the resources during the 1860s and 1870s to establish a system of branches. Nevertheless, the response of the Banque Nationale indicated a concern for a francophone town that was absent from the deliberations of the anglophone banks.

Table 6: Growth of the Francophone Branch System

| Bank | 1871 | 1881 | 1891 | 1901 | 1911 |
|------|------|------|------|------|------|
| Nationale | 0 | 3 | 3 | 17 | 47 |
| Hochelaga | 0 | 2 | 5 | 12 | 32 |
| Jacques Cartier | 0 | 3 | 16 | | |
| Provinciale | | | | 9 | 45 |
| Total | 0 | 8 | 24 | 38 | 124 |

SOURCE: Annual reports of banks published in *Monetary Times*.

The first of the francophone banks to establish a significant system of branches was the Jacques Cartier, which expanded its system from three to sixteen branches during the 1880s. This bank experienced some difficulties during the financial crisis of the 1870s, but emerged from these hard times with the strategy that responding to the needs of Quebec towns might lead to profits instead of losses. The directors of the bank were very pleased with the results of this new policy. In 1888 they

34. *Le Moniteur du Commerce*, 5 October 1883.
35. Archives of the Banque Canadienne Nationale, minutes of directors' meeting, Banque Nationale, 27 October 1860.

reported, "The good returns obtained from the establishment of these various branches are more and more satisfying each year." The branch policy had revitalized the bank and further expansion was anticipated as "some pressing demands have been made by several important towns for the establishment of new branches."[36]

After having experienced some difficult years, the Banque Jacques Cartier was forced to close its doors in 1899. Nevertheless, the francphone branch bank system continued to expand during the 1890s and the first decade of the new century due to the activities of the Nationale, the Hochelaga and, after 1900, the Provinciale. There was considerable room for expansion in Quebec because the francophone banks appreciated the possibilities for profit that had been bypassed by the anglophone institutions. For instance, the Banque d'Hochelaga established a branch in Sherbrooke because, unlike its anglophone counterparts, it recognized that "the importance of this city in terms of industry and commerce becomes more pronounced each day."[37] Because of the opportunities that existed within the province these banks rarely established branches outside of Quebec. In 1911, 93 per cent of the 124 branches operated by the francophone banks were in Quebec, and those situated outside of the province were oriented towards serving francophone communities which, like most of Quebec, had not been well served by the anglophone banks. In 1911 the Banque Provinciale had branches in the Acadian towns of Moncton and Caraquet in New Brunswick and in Alfred, Ontario, which had a significant French-speaking population. The main orientation of these banks, however, was towards serving Quebec. The opportunities for establishing branches within the province were seen to be so great that the Provinciale and the Nationale arrived at an arrangement so that their resources would not be wasted by competing against each other in a given town. The Provinciale was to have first call on establishing a branch in any town west of Trois-Rivières, while the Nationale was to have the same right in the eastern half of the province.[38]

In addition to establishing a chain of branches in Quebec, the francophone banks attempted to serve the needs of francophone interests, often assisting entrepreneurs rejected by the anglophone banks. In Sorel, while the Molson's Bank served the anglophone business community, the Banque d'Hochelaga served local francophone entrepreneurs.[39] While the Merchant's Bank and the Bank of Montreal assisted

36. Archives of the Banque Provinciale, minutes of directors' meeting, Banque Jacques Cartier, 30 June 1888.

37. Archives of the Banque Canadienne Nationale, minutes of directors' meeting, Banque d'Hochelaga, 15 June 1898.

38. Archives of the Banque Provinciale, minutes of directors' meeting of the bank, 12 March 1908.

39. The directors of the bank approved loans of $2,500 to La Compagnie de Chaussures de Sorel (28 February 1888), $10,000 to Cyrille Labelle (10 March 1896), and $30,000 to La Compagnie Electrique de Sorel (25 September 1908).

the anglophone interests of Sherbrooke, the Gendron corset factory was
aided by a $25,000 loan from the local branch of the Banque Natio-
nale.[40] The Saint-Hyacinthe shoe manufacturer rejected by the anglophone banks received a $25,000 loan from the branch of the Banque
Jacques Cartier located in his city.[41] Finally, the oldest francophone-
controlled industrial establishment in Trois-Rivières, the Balcer Glove
Factory, was granted a $50,000 loan by the Banque d'Hochelaga in 1902
for the expansion of its facilities and an advance of another $50,000 in
1907 to help it through some hard times.[42]

Despite their best efforts, however, the francophone banks were not
able to provide the level of services that the anglophone banks could
make available. While the Bank of Montreal could establish branches in
the smallest towns in Ontario, the Banque Nationale was only capable
of erecting sub-agencies in the lesser centres of Quebec. By 1914,
seventy-three sub-agencies had been established at which deposits could
be made but from which only very small loans could be secured. Even in
the more important centres which did secure branches of the francophone banks, entrepreneurs were often refused loans or were forced to
accept loans which could not satisfy their needs. When the Banque
Nationale established a branch in Sherbrooke in 1875, francophone
entrepreneurs hungry for capital descended upon this new institution.
Due to the bank's own lack of capital, however, the manager of the
Sherbrooke branch was told to advise his clients that the bank was
unable to "provide long-term loans" and that it could only "assist local
commercial activity."[43] The sort of commitment made by the Bank of
Montreal to the Paton firm in Sherbrooke could not have been made by
the francophone banks to francophone entrepreneurs.


IV

Most Quebec towns, and particularly those dominated by francophones, were poorly served by the Montreal banks. The anglophone
institutions seemed to be generally uninterested in serving the province,
while the French-controlled banks lacked sufficient resources to ade-
quately assist economic development. This failure on the part of Montreal interests to satisfy the needs of the province is evident upon
comparing the ratio of population to branch banks in Quebec with the
ratios for the other provinces. As Table 7 indicates, the ratio for Quebec

40. Archives of the Banque Canadienne Nationale, minutes of directors' meeting, Banque
    Nationale, 15 April 1889.
41. Archives of the Banque Provinciale, minutes of directors' meeting, Banque Jacques
    Cartier, 3 September 1884.
42. Archives of the Banque Canadienne Nationale, minutes of directors' meeting, Banque
    d'Hochelaga, 24 January 1902; 25 January 1907.
43. Archives of the Banque Canadienne Nationale, minutes of directors' meeting, Banque
    Nationale, 11 February 1875.

in 1912 was only lower than those for New Brunswick and Prince Edward Island. The ratio for the francophone population of Quebec, however, was much higher than that for the province as a whole. As the anglophone banks rarely provided services to the francophone population, the 195 banks operated by the French-controlled banks were left with the task of serving the 1,600,000 francophone Quebeckers. The resulting ratio of one branch for every 8,205 francophone Quebeckers was considerably higher than that for any of the provinces.

Table 7: Bank Branches and Population, 1912

| Province | Number of Branches | Population Divided by Branches |
|---|---|---|
| Saskatchewan | 349 | 1299 |
| Alberta | 241 | 1547 |
| British Columbia | 223 | 1677 |
| Manitoba | 195 | 2332 |
| Ontario | 1068 | 2360 |
| North West Territories | 3 | 3333 |
| Nova Scotia | 111 | 4161 |
| Quebec | 440 | 4547 |
| New Brunswick | 74 | 4754 |
| Prince Edward Island | 14 | 6695 |
| Canada | 2718 | 2606 |

SOURCE: *Moniteur du Commerce,* 4 October 1912.

The key to this situation was the position taken by the wealthy anglophone banks of Montreal towards assisting the development of the province. Unfortunately, however, it is easier to describe the behaviour of the anglophone financial community than to understand the motives of these concerns for taking little interest in the development of Quebec. Future research will perhaps make this behaviour more comprehensible.[44] Nevertheless, there would seem to be two types of motives to consider. The bankers themselves always claimed that their behaviour was based solely upon economic factors.[45] If they failed to establish numerous branches in Quebec and refused to assist francophone entrepreneurs, it was only because such behaviour did not appear to be profitable. It cannot be denied that there were more opportunities for profits in Ontario and that anglophone entrepreneurs were generally

44. The author is pursuing a larger study of the operations of both the anglophone and francophone banks with headquarters in the province and their impact upon the economic development of Quebec.
45. PAC, Merchants' Bank, minutes of directors' meeting, 19 June 1889.

more successful than their francophone counterparts, but it would also be unfair to overstate the case. While the anglophone banks provided considerable resources to aid in the progress of agriculture in Ontario through the establishment of branches in small centres, they did little to aid such activity in Quebec even though the growth of agricultural production during this period paralleled that in Ontario.[46] As was noted earlier, these banks were not averse to assisting industrial development, but provided little assistance in this field in Quebec despite the cheap labour, water power and abundant resources that the province possessed. Finally, the disdain for francophone interests which the anglophone banks seemed to display when they did locate branches in Quebec seems unwarranted in the light of recent work on francophone entrepreneurship by Gerald Tulchinsky and Paul-André Linteau.[47] The success that was achieved with minimal resources by the francophone banks further suggests that it was possible to reap profits by operating in Quebec and by doing business primarily with francophone concerns.

Factors which were not entirely economic in nature would seem to have played some role in influencing the behaviour of the anglophone banks. They oriented the establishment of branches to communities that happened to be predominantly anglophone, and when they did locate agencies in francophone centres tended to cater to anglophone interests. There is no evidence of conscious discrimination on the part of these anglophone bankers. Rather, the behaviour of these financial interests may have been influenced by the amount and nature of information that they possessed about prospects in Quebec and the reliability of francophone entrepreneurs. It would seem reasonable to expect that the barriers of culture and language would have restricted their knowledge of francophone Quebec, and perhaps would have made them shy away from investing there. The geographer Allan Pred has considered the impact of the availability of information upon business decisions in the United States at the start of the nineteenth century.[48] He found that in a pre-telegraphic society the business community of each city developed a

---

46. Between 1871 and 1911 the value of agricultural production increased by 179 per cent in Quebec and by 165 per cent in Ontario. See Ronald Rudin, "The Development of Four Quebec Towns: A Study in Quebec Urban and Economic Development," (Ph.D. Dissertation, York University, 1977), pp. 255-56.

47. Gerald Tulchinsky, *The River Barons: Montreal Businessmen and the Growth of Industry and Transportation, 1837-1853* (Toronto, 1977); Paul André Linteau, "Quelques réflexions autour de la bourgeoisie Québécoise," *Revue d'Histoire de l'Amérique Francaise*, 30 (1976), pp. 55-66. See also Linteau's article on Maisonneuve in this volume and Linteau and Jean-Claude Robert, "Land Ownership and Society in Montreal: An Hypothesis," in Gilbert A. Stelter and Alan F.J. Artibise, eds., *The Canadian City: Essays in Urban History* (Toronto, 1977), 17-36.

48. Allan Pred, *Urban Growth and the Circulation of Information: The United States System of Cities, 1790-1840* (Cambridge, 1973), pp. 277-83.

distinct approach towards similar problems because each possessed different information. Due to different sorts of barriers, the anglophone and francophone financial communities possessed different information about business prospects in Quebec, and each accordingly behaved differently. Such an approach is worthy of being pursued further.

What is clear, however, is that Quebec did not have the best of financial services in the pre-World War I era. There was a striking similarity between the distribution of banking facilities across the province. As has been indicated, both francophone and anglophone banks were known to aid industrial development, but for various reasons little capital filtered down to industrial firms in towns such as Saint-Hyacinthe and Trois-Rivières, thus hindering their development. As the major Saint-Hyacinthe newspaper noted, "There can be no doubt that a factory in a town where a bank is located has a better chance to succeed than it would have in a town lacking a bank."[49] Due to a lack of satisfactory facilities, industrial development was retarded in the province outside of Montreal and few important industrial centres emerged. These problems were seen most clearly by a group of French financiers brought to Quebec by the provincial government in 1880 to induce them to establish financial institutions there. After having traveled throughout Canada they observed that "several English banks have agencies in the English part of the country while the French Canadian is abandoned to himself. . . . The resulting inequality of capital causes an observable difference in the development of the two parts of the country and places French Canada in a position of relative inferiority."[50]

49. *Courrier de Saint-Hyacinthe*, 19 January 1871.
50. *Monetary Times*, 6 August 1880.

# Staples and the New Industrialism in the Growth of Post-Confederation Halifax

L.D. McCANN

In the late nineteenth and early twentieth centuries, the course of urban development in the Maritimes diverged sharply from the path followed by central Canada. Because Maritime cities grew at slower rates and with fewer opportunities than their Ontario and Quebec counterparts, they remained much smaller in size and less diverse in their functional activities.[1] Scholars are now attempting to explain this divergence. Several have emphasized the differentiating role of the entrepreneur. T.W. Acheson, for example, has argued recently that the Maritime entrepreneurial class, comprised mostly of a long-established mercantile elite, was unable to meet the challenge of the new industrialism of the late nineteenth century. The limited industrial experience of these community-oriented businessmen and their lack of access to capital markets led eventually to the collapse of the region's urban-industrial base and to its

SOURCE: *Acadiensis: Journal of the History of the Atlantic Region,* 8 (Spring, 1979), pp. 47-79. Reprinted by permission of the author and the Department of History of the University of New Brunswick.

The author wishes to thank Peter Smith, John Weaver and David Alexander for their comments on an earlier version of this paper. Thanks are also due Geoff Lester and the Cartography Division, Department of Geography, University of Alberta, for drafting the maps. A research grant from the Humanities and Social Sciences Research Fund, Mount Allison University, supported some of the research and is gratefully acknowledged.

1. For an overview and comparative analysis of regional urbanization in Canada, see R.E. Preston, "The Evolution of Urban Canada: The Post-1867 Period," in R.M. Irving ed., *Readings in Canadian Geography* (3rd ed., Toronto, 1978), pp. 19-46.

almost complete takeover by central Canadian interests.[2] A second group of scholars has stressed the vulnerability of the region to external forces. The demise of both shipbuilding and the carrying trade as a consequence of changing shipping technologies adversely affected the economic growth of Yarmouth and Saint John.[3] At the same time, forces of continentalism in the guise of freight rate equalization policies, business reorganization and concentration, and changing market potentials all robbed industrial towns such as New Glasgow and Amherst of access to central and western Canadian markets, thereby crippling their economic base.[4] A final group, represented by economic historians such as S.A. Saunders and David Alexander, has argued that Maritime economic development has been restricted by the marginal quality and poor management of the region's resource base.[5] Although these studies focus almost exclusively on the plight of individual staples to the neglect of the urban process, nevertheless they suggest that the resource hinterland of the Maritimes was too limited in size, variety, and richness to support intensive and sustained urban development.

This essay extends these enquiries into Maritime urban development. It offers a theoretical perspective on regional urban growth in Canada and examines the industrialization of Halifax in the post-Confederation period down to the eve of World War I to show how the regional staple economy of the Maritimes and the comparative advantages of other Canadian cities adversely influenced the course of urban-industrial development in the Nova Scotian capital.

There can be little doubt that the growth of cities in the nineteenth

2. T.W. Acheson, "The National Policy and the Industrialization of the Maritimes, 1880-1910," *Acadiensis*, I (Spring 1972), pp. 3-28. For other studies focusing on this theme, see D. Frank, "The Cape Breton Coal Industry and the Rise and Fall of the British Empire Steel Corporation," *Acadiensis*, 6 (Autumn 1977), pp. 3-34; and D.A. Sutherland, "The Personnel and Policies of the Halifax Board of Trade, 1890-1914," in Lewis R. Fischer and Eric W. Sager, eds., *The Enterprising Canadians: Entrepreneurs and Economic Development in Eastern Canada*, 1820-1914 (St. John's, 1979).

3. D. Alexander, "The Port of Yarmouth, Nova Scotia, 1840-1889," in K. Matthews and G. Panting, eds., *Ships and Shipbuilding in the North Atlantic Region* (St. John's 1978); and E. McGahan, "The Port of Saint John, New Brunswick, 1867-1911," *Urban History Review*, No. 3 (1976), pp. 3-13.

4. E.R. Forbes, "Misguided Symmetry: The Destruction of Regional Transportation Policy for the Maritimes," in D.J. Bercuson, ed., *Canada and the Burden of Unity* (Toronto, 1977), and B. Archibald, "The Development of Underdevelopment in the Atlantic Provinces" (MA thesis, Dalhousie University, 1971).

5. D. Alexander, "Economic Growth in the Atlantic Region, 1880-1940," *Acadiensis* (Autumn 1978), pp. 47-76; R.E. Caves and R. Holton, *The Canadian Economy: Prospect and Retrospect* (Cambridge, Mass., 1959), pp. 140-94; and S.A. Saunders, *The Economic History of the Maritime Provinces: A Study Prepared for the Royal Commission on Dominion-Provincial Relations* (Ottawa, 1939), pp. 14-33 and 90-99.

century was influenced by staple production.[6] Across Canada, regional urban systems first emerged when foreign and inter-regional demand fostered increased staple flows and where the commodities were sufficiently bulky, weighty and perishable to require urban-based linkages in the transport, manufacturing, and service sectors. As late as World War I, nearly one-third of Canada's industrial output was based directly upon primary production, and the indirect impact of staples on the secondary and tertiary sectors must have been considerable.[7] By this time, the growth of Canadian cities was also influenced by what Harold Innis has termed the "discrepancy between the centre and the margin,"[8] that is, by continentalism – the polarizing effects of core-periphery development. In 1910, nearly 70 per cent of the country's non-primary production was concentrated in the rapidly industrializing towns and cities of Ontario and Quebec.[9] The effects of this concentration were considerable. Quite simply, the accumulated comparative advantages of heartland cities restricted urban development in the periphery.

Thus, a framework for examining the growth of Canadian cities is of necessity derived from the staple theory of regional economic growth and the core-periphery of heartland-hinterland conceptualization of regions.[10] The core supplies those factors of production (capital, labour, technology and entrepreneurship) that are used to develop the resource base of the periphery. In return, the periphery exports staple commodities (those raw materials or resource intensive goods occupying a central position in the region's exports) to the source of demand in the core. Within this context, heartland and hinterland cities function as intermediaries (see Figure 1). As intermediaries, their economic base is characterized by functions based on handling the factors of production which include trading, transportation, manufacturing, and financial and busi-

---

6. This theme has been taken up in several recent studies by Canadian geographers. See D.M. Ray, *Canada: The Urban Challenge of Growth and Change* (Ottawa, The Ministry of State for Urban Affairs, 1974), pp. 23-26; and J.W. Simmons, "The Growth of the Canadian Urban System," *Research Paper 65* (Toronto, 1974).

7. Calculated from data presented in A.G. Green, *Regional Aspects of Canada's Economic Growth* (Toronto, 1971, p. 86.

8. H.A. Innis, *The Fur Trade in Canada* (Toronto, 1957), p. 385.

9. Green, *Regional Aspects of Canada's Growth*, p. 86.

10. R.E. Caves, "Vent for Surplus Models of Trade and Growth," in R.E. Baldwin *et al.*, eds., *Trade, Growth and the Balance of Payments* (Chicago, 1965); J. Friedmann, "Regional Economic Policy in Developing Areas," *Papers and Proceedings of the Regional Science Association*, XII (1963), pp. 41-61; J.M. Gilmour, *Spatial Evolution of Manufacturing: Southern Ontario 1851-1891* (Toronto, 1972), pp. 12-25; and M.H. Watkins, "A Staple Theory of Economic Growth," *The Canadian Journal of Economics and Political Science*, 28 (1963), pp. 141-58. I have explored some of the ideas presented in this section in a related paper: L.D. McCann, "Urban Growth in a Staple Economy: The Emergence of Vancouver as a Regional Metropolis, 1886-1914," in L.J. Evenden, ed., *Vancouver: Western Metropolis* (Victoria, 1978).

ness activities. However, the degree of specialization and the composition of economic sectors differs in heartland and hinterland cities. It is unlikely that the hinterland city will have a fully diversified economic base. Depending upon the type and distribution of resources found within the periphery, cities here will function mainly as resource towns, as central places, or as break-in-bulk points. In the heartland city manufacturing is emphasized because the core's accessibility to national markets creates an initial advantage.[11] The external economies of concentrated human resources also favour heartland cities as financial and business headquarters.

Further differences between heartland and hinterland cities become evident by examining the circular and cumulative process of urban growth. The stimuli for growth are the regional economy's urban-based functions. They set in motion the multiplier effect which spawns additional basic and non-basic economic activity. Manufacturing is the leading sector in most heartland cities and the creator of employment and population growth. This sector comprises manufacturers of primary, consumer and producer goods. Primary manufacturing industries are indicative of forward linkage effects, while the consumer goods destined for industrial markets are the result of backward connections.[12] Manufacturers of consumer and producer goods are minimal in the hinterland because the heartland can meet national market demands more efficiently. The stimulus for growth in the hinterland city is the wholesale-trading complex, particularly if the regional economy depends upon the staple trade.

Ultimately, the sustained growth and development of all cities depends upon local, regional or extra-regional thresholds, that is, upon the attainment of minimum income or population levels. Historically, heartland cities have attained higher thresholds more easily than hinterland cities and therefore have grown larger in size.[13] If it is accepted that industrialization was the "engine of growth" in the late nineteenth century, why did hinterland cities fail to industrialize? The main reasons, assessed most succinctly by Pred and by Muller, are the accumulation of external economies; transport and route developments; entrepreneurial behaviour, particularly combination practices, ologopolistic competition and inhibiting mercantile traditions; and initial advantages in the

---

11. A.R. Pred, "Industrialization, Initial Advantage and American Metropolitan Growth," *Geographical Review*, 55 (1965), pp. 165-80.

12. Backward linkages relate to inducements to invest in the production of goods required by the export sector; forward linkages relate to opportunities for investment for investment in industries using the output of the export industry as an input; and final-demand linkages describe the inducement to invest in consumer goods industries producing for factors in the export sector. See Watkins, "A Staple Theory of Economic Growth," pp. 145-6.

13. A.R. Pred, *The Spatial Dynamics of U.S. Urban-Industrial Growth* (Cambridge, Mass., 1966), pp. 16-24 and 33-37.

guise of site and situation, relative accessibility, factor immobility, and labour and capital availability.[14] These conditions placed hinterland cities at a comparative disadvantage against their heartland competitors.

An additional factor is the differentiating effect of the regional resource base. This factor is of particular significance for the cities of a staple economy. The economic base of a city which is dependent upon one particular staple can be seriously eroded by anything from resource depletion without substitution, loss of competitive position through inelastic foreign demand, and adverse shifts in demand resulting from

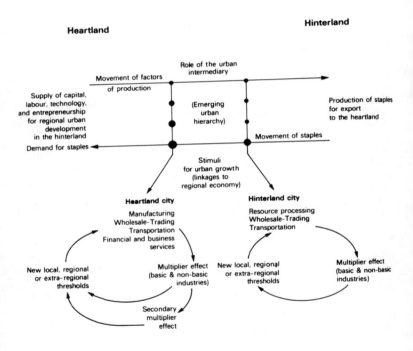

Figure 1: The Process of Urban Growth in a Heartland-Hinterland Economic System

14. *Ibid.*, pp. 46-83; and E.K. Muller, "Regional Urbanization and the Selective Growth of Towns in North American Regions," *Journal of Historical Geography,* 3 (1977), pp. 21-40.

competition from cheaper or synthetic sources of supply, to simply changes in taste. Differential urban growth is also related to the residentiary effects of individual staples.[15] Some staples, such as the cod fishery, have traditionally produced only weakly developed linkage effects.[16] Others generate social structures which inhibit supportive community development, as in the case of the company town.[17] The type and strength of linkages with other economic sectors, particularly of the final-demand variety, are affected by labour force participation and income levels. the availability of labour for residentiary industries is further weakened by competition from other resource industries and by the limited skills of a resource-oriented labour force. The extent of domestic savings derived from the resource base also influences the degree to which staple entrepreneurs invest in other sectors of the economy.[18] In each situation, the location, type and size of the resource base, together with the nature of the control associated with the staple economy, directly affects urban growth. Moreover, the number and size of urban places, and certainly the timing of urban development, is critically related to the demand for staple commodities and to the supply of scarce factors of production. These, in turn, are influenced by the social, political and economic conditions of the metropolitan economy. Urban development in the hinterland is thus governed largely by external forces.

One city adversely affected by these factors during the late nineteenth and early twentieth centuries was Halifax. Although in the 1871 to 1921 period its population almost doubled from 29,582 to 58,372, Halifax dropped from Canada's fifth ranking city to its eleventh (see Figure 2). Concurrent with the decline of Halifax are other trends: the general instability of Canada's urban system, particularly between 1881 and 1911 when there was considerable reordering of ranks; Central Canada's absolute and proportional increase of cities of 10,000 people or more; and the initial appearance and rise to prominence of cities in the western periphery. Halifax is only one example of the widening urban disparity between the eastern margin and the rest of Canada. Indeed, by the end of World War I, seven Canadian cities had won higher ranks

15. The term residentiary is used to designate secondary and tertiary industries which locate in urban areas to serve the local or regional market. See D.A. North, "Location Theory and Regional Economic Growth," *Journal of Political Economy,* 58 (1955), pp. 243-58.

16. R. Ommer, "The Cod Fishery and a Theory of Settlement Development" (unpublished paper, McGill University, Department of Geography, 1976).

17. L.D. McCann, "Canadian Resource Towns: A Heartland-Hinterland Perspective," in Richard E. Preston and Lorne H. Russwurm, eds., *Essays on Canadian Urban Process and Farm II* (Waterloo: University of Waterloo Department of Geography, 1980), pp. 209-67.

18. L.R. MacDonald, "Merchants against Industry: An Idea and its Origins," *Canadian Historical Review,* 56 (1975), pp. 263-81.

and rates of growth: Hamilton, Ottawa and London in Central Canada; and Winnipeg, Calgary, Edmonton and Vancouver in the western periphery. In the Maritimes, Halifax's downward path was not unique. Saint John and Charlottetown fared worse. The only exceptions to this pattern were Moncton, an important distribution point on the Intercolonial Railway, and Glace Bay and Sydney, the centres of coal and steel production in Cape Breton.

It is apparent that as the heartland-hinterland process emerged during the late nineteenth century, it changed regional development patterns and thereby created alternative bases for urban growth beyond those of staple production. This restructuring can be analyzed by the statistical technique of factor analysis.[19] By this procedure, measures of provincial labour force participation in 1881 and 1911 were summarized to outline the structural dimensions of the Canadian economic system (see Table 1 and Figure 3).[20] It is apparent that shortly after Confederation staple industries dominated the economy (Factor I). Agricultural production characterized Ontario and Quebec as well as Manitoba and Prince Edward Island. In British Columbia and the other Maritime provinces, fishing and mining absorbed a considerable share of the labour force. Of secondary importance to the Canadian economy, but certainly a harbinger of subsequent development, was a structural dimension of urban activity. Indeed, trading activities, government service, manufacturing and construction combined in 1881 to distinguish the industrializing core from the periphery (Factor II). Transportation comprised Factor III of the labour force analysis. It proved to be more important at the margins of the country than elsewhere. Even by 1881, therefore, the framework of the heartland-hinterland paradigm was well in place.[21] By 1911, it was an established pattern. Core-periphery contrasts, based on differences between urban-oriented activities and agricultural production, comprised the basic dimension of the economy (Factor I). Specialized secondary and tertiary activities were concentrated in Ontario and Quebec where during an earlier phase of agricultural production they

19. Michael Ray has employed the same technique to examine Canada in 1961. He concluded that core-periphery contrasts were particularly significant: "secondary manufacturing and service activity have gravitated toward the center, leaving hinterland areas reliant on primary activities which tend to play a diminishing role in national economies," "The Spatial Structure of Economic and Cultural Differences: A Factorial Ecology of Canada," *Papers and Proceedings of the Regional Science Association*, 23 (1969), p. 8.

20. To ensure comparability over time, these labour force data have been reclassified according to Canada, Dominion Bureau of Statistics, *Standard Industrial Classification Manual* (Ottawa, Queen's Printer, 1960).

21. This analysis confirms the aspatial examination of the evolution of Canada's economic system contained in O.J. Firestone, "Development of Canada's Economy, 1850-1900," in *Trends in the American Economy in the Nineteenth Century*. Studies in Income and Wealth, vol. 24 (Princeton, 1960), pp. 217-46.

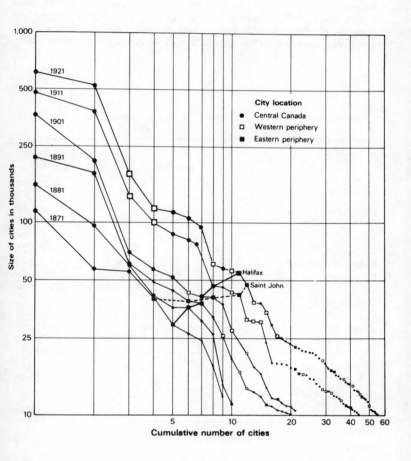

Figure 2: The Rank-Size Distribution of Canadian Cities, 1871-1921

had been merely supplemental. The agricultural frontier had shifted west to the newly created provinces of Saskatchewan and Alberta. Other forms of staple production were still important, but only on a restricted regional basis (Factor II). In the peripherally located provinces of Nova Scotia and British Columbia, mining and fishing comprised important elements of the economic base. These resource activities and government services comprise Factor III. The development of forestry, mining

Table 1:

Factor Analysis of the Canadian Space Economy, 1881 and 1911
(Variable Loadings on Varimax Rotated Factors[1])

| | 1881 | | | 1911 | | |
|---|---|---|---|---|---|---|
| | Factor | | | Factor | | |
| Labour Force Group[2] | I | II | III | I | II | III |
| Agriculture | .734 | | | -.763 | | -.578 |
| Fishing and trapping | -.905 | | | | -.949 | |
| Forestry | | .579 | | | | .585 |
| Mining | -.850 | | | | .779 | .545 |
| Manufacturing | | .742 | | .930 | | |
| Construction | -.624 | .719 | | .666 | | |
| Trade | | .938 | .927 | | | .974 |
| Personal and professional services | .944 | | | .876 | | |
| Government | | .966 | | | | .906 |
| % Total Variance | 49.9 | 25.6 | 13.2 | 56.3 | 20.8 | 12.7 |
| % Accumulated Variance | 29.9 | 75.5 | 56.3 | 77.1 | 56.3 | 89.8 |

SOURCE: Calculated by the author.

NOTES:
[1] Only variable loadings $\geq$ .500 are indicated.
[2] Percentage of provincial labour force by industrial groups.

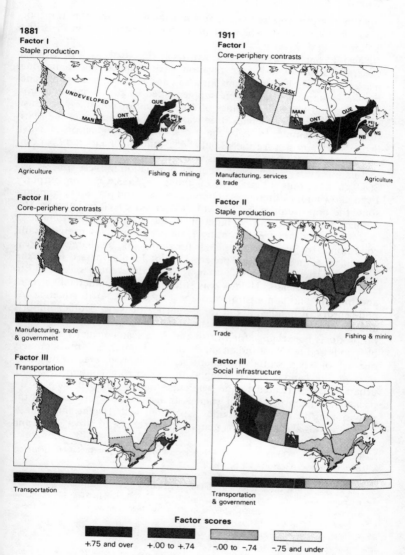

Figure 3: Changing Patterns of the Canadian Space Economy, 1881 and 1911

and other staples required extensive government investment in transportation facilities, which was most pronounced at the margin in British Columbia and Alberta.[22]

Several conclusions can be drawn from this analysis. For cities in Central Canada whose core position gave them access to national markets, the wholesaling, transportation and resource processing functions of a staple economy were supplemented by the secondary manufacturing and business activities of the new industrialism. By contrast, the opportunity for diversification was limited in the cities of the Maritimes. For example, in 1910 the gross value of production of Nova Scotia's primary sector totalled only $45.3 million, or a mere 17 per cent of Ontario's $261.9 million. Disparity in the secondary and tertiary sectors was even greater. Nova Scotia's production in these spheres was $68.9 million, only 12 per cent of Ontario's $583.6 million output.[23] Even allowing for differences in population size and per capita comparisons, the opportunities for concentrated urban development were still limited. Halifax did indeed function at the margin of development.

Until the mid-nineteenth century Halifax's growth had been hesitant, conditioned by fluctuating demands for staples and by changes in Great Britain's foreign and trading policies.[24] Saint John surpassed Halifax in population during the 1830s and in 1861 the differential was 27,317 to 25,026.[25] Nonetheless, Halifax was still a leading commercial entrepôt and military and administrative outpost in British North America. At mid-century, it handled about three-quarters of Nova Scotia's imports and about two-thirds of its exports.[26] It traded fish, forest and agricultural products in exchange for British and American consumer and durable goods and for West Indian sugar, rum and molasses. Approximately one-half of Nova Scotia's merchant marine was registered to businessmen of the capital.[27] Accordingly, the waterfront contained various urban activities related to the wholesale-trading complex such as the packaging and processing of staples, transportation services, and the provisioning and repair activities associated with freight shipment.

Halifax was also the leading manufacturing centre of Nova Scotia and

22. H.G.S. Aitken, "Government and Business in Canada: An Interpretation," *Business History Review*, 37-38 (1964), pp. 4-21.
23. Green, *Regional Aspects of Canada's Economic Growth*, p. 86.
24. These themes and the role of the Halifax merchantocracy in their development are examined in a most comprehensive study by D.A. Sutherland, "The Merchants of Halifax, 1815-1850: A Commercial Class in Pursuit of Metropolitan Status" (PhD thesis, University of Toronto, 1975).
25. Canada, *Census of Canada, 1870-71*, vol. 4, Table 1, pp. 232 and 344.
26. *Novascotian* (Halifax), extra, 14 February 1854.
27. Nova Scotia, House of Assembly, *Journals and Proceedings of the House of Assembly, 1851*, Appendix 78, Abstract of Provincial Shipping Tonnage. See Also E. W. Sager, "The Shipping Fleet of Halifax, 1820-1905" (paper presented at the Atlantic Canada Studies Conference, University of New Brunswick, April, 1978).

this activity absorbed about 22 per cent of the city's labour force in 1861 (see Table 2). Despite this pre-eminence, manufacturing was still pre-industrial in character. Since its structure was only weakly developed and its capitalization was slight, most manufacturers were small in scale, based on crafts production and catered principally to the local market.[28] This degree of underdevelopment was not due to the antagonistic attitude of the business community toward manufacturing. In fact, industrialization had been made an important platform in a programme of civic improvement. Both the Society for the Encouragement of Trade and Manufacturers, formed in 1839,[29] and its successor, the Nova Scotia Society for Developing and Encouraging Home Manufacturers, established shortly after Confederation in 1870,[30] campaigned vigorously for industrial development. But as the census data of the period indicate, their efforts brought only limited success. Large-scale manufacturing of the type found in Great Britain and the larger cities of the northeastern United States was not established in Halifax. For example, the boot and shoe industry, Halifax's largest manufacturing employer in 1871, claimed only 371 employees dispersed in 29 places of work.[31] In this respect Halifax was typical of other mid-nineteenth-century Canadian cities. They were essentially commercial in character and oriented to staple hinterlands; their complete industrial transformation awaited the last decades of the century.[32]

28. These generalizations are based on the following table compiled from data in the *Census of Nova Scotia, 1860-61*, Appendix 8, pp. 287-91. They can be compared to Pred, *Spatial Dynamics*, p. 170.

### The Structure of Manufacturing in Halifax, 1861
#### (Classified by Relationship to the Wholesale-Trading Complex)

| Function | Establishments | | Value Added | |
|---|---|---|---|---|
| | No. | % | Amount | % |
| Entrepôt | 9 | 22.5 | 36,800 | 10.5 |
| Commerce-serving | 5 | 12.5 | 39,400 | 11.2 |
| Local market | 18 | 45.0 | 69,000 | 19.7 |
| Construction | 7 | 17.5 | 25,100 | 7.2 |
| Other (gas factory) | 1 | 2.5 | 180,000 | 51.4 |
| Totals | 40 | 100.0 | 350,300 | 100.0 |

29. Society for The Encouragement of Trade and Manufactures, Halifax, N.S., *Rules and Regulations* (Halifax, 1839).

30. Nova Scotia Society for Developing and Encouraging Home Manufactures, *Address to the People of Nova Scotia. . .Constitution of the Society* (Halifax, 1870).

31. Canada, *Census of Canada*, 1870-71, vol. 3, Tables 28-54.

32. G.W. Bertram, "Economic Growth in Canadian Industry, 1870-1915: The Staple Model and the Take-Off Hypothesis," *The Canadian Journal of Economics and Political Science*, 29 (1963), pp. 159-84.

L.D. McCANN

Table 2: The Changing Economy of Halifax, 1861-1911
(Distribution of Labour Force by Industrial Groups)

| Industrial Group | 1861 | | 1881 | | 1911 | |
|---|---|---|---|---|---|---|
| | No. | % | No. | % | No. | % |
| Primary | 292 | 4.6 | 291 | 2.2 | 243 | 1.4 |
| Manufacturing | 1,379 | 21.8 | 2,242 | 17.4 | 2,750 | 15.4 |
| Construction | 802 | 12.7 | 1,088 | 8.4 | 1,034 | 5.8 |
| Trade | 819 | 12.9 | 1,232 | 9.3 | 1,297 | 7.2 |
| Transportation | 666 | 10.5 | 1,205 | 9.5 | 2,205 | 12.3 |
| Finance | 13 | .2 | 49 | .4 | 223 | 1.2 |
| Insurance and real estate | — | — | 60 | .5 | 220 | 1.2 |
| Personal and professional services | 840 | 13.3 | 3,900 | 30.2 | 3,817 | 21.3 |
| Government[2] | 141 | 2.2 | 277 | 3.1 | 1,706 | 9.5 |
| Unspecified | | | | | | |
| Commercial clerks | 312 | 4.9 | 1,251 | 9.6 | 1,676 | 9.4 |
| Labourers | 1,072 | 16.9 | 1,014 | 7.8 | 2,738 | 15.3 |
| Others | — | — | 350 | 2.6 | — | — |
| Total Labour Force | 6,338 | 100.0 | 12,959 | 100.0 | 17,909 | 100.0 |

NOTES:

SOURCES: Nova Scotia, *Census of Nova Scotia*, 1860-61, Appendix 5, pp. 190-199; Canada, *Census of Canada*, 1881, vol. 2, Table 14, pp. 232-43; and Canada *Census of Canada*, 1911, vol. 6, pp. 326-34.

[1]To ensure comparability over time, the labour force data for each year were reclassified according to: Canada, Dominion Bureau of Statistics, *Standard Industrial Classification Manual* (Ottawa, Queen's Printer, 1960).
[2]Excludes army and navy personnel of Great Britain (1861 and 1881) and Canada (1911).

The new industrialism was spurred by a combination of fiscal policies set in the guise of protective tariff schedules referred to collectively as the National Policy. Introduced in 1879, by the late 1880s these measures had encouraged the development of a number of manufacturing industries in Halifax which previously had either not existed or existed in only limited form. Several city foundries increased their capitalization to produce engines, tools, spikes, nails and construction materials for the expanding resource sector. More significantly, the mercantile community of Halifax responded favourably to these measures. Its most notable ventures were the Nova Scotia Sugar Refinery, capitalized initially in 1879 at $300,000, and the Halifax Cotton Company, formed in 1881 and also supported by locally subscribed shares of more than

$500,000.[33] Both businesses were based on existing trading patterns with British West Indian suppliers.[34] As a consequence of these and other efforts, manufacturing in Halifax advanced appreciably, nearly doubling in value between 1900 and 1910 to achieve an annual level of production of more than $12 million.[35] With the aid of the National Policy and local entrepreneurial initiative, manufacturing had maintained its leading position over the other sectors of the economy. Yet Halifax could not claim hegemony in the Maritimes as a manufacturing centre. Sydney, Amherst, New Glasgow and Trenton in Nova Scotia and Saint John, Moncton, Marysville and Chatham in New Brunswick dominated several industrial categories.[36] Nor could Halifax claim to match the sizeable advances made by the cities of Central Canada. Of the 55 Canadian cities of 10,000 people or more in 1911, an overwhelming majority, 48 in fact, experienced higher rates of industrial growth between 1890 and 1910.[37] Later, in the two years preceding the disastrous Halifax Explosion of 1917, the city's rate of increase did not reach even one per cent. By this time, its value of production had fallen to less than 5 per cent that of either Toronto or Montreal.[38] Over the next decade, as freight rates climbed and mergers took effect, the decline became even more precipitous. Between 1920 and 1926, manufacturing employment in the city dropped from 7,171 to 3,287.[39]

The process of industrialization in Halifax down to the eve of World War I thus presents a paradox. On the one hand, boosters could point to the substantial number of newly established firms and to the sizeable gains in output, and could boast of the role of manufacturing as the city's leading employer. On the other, if they cared to, they might comment on the diminished relative importance of Halifax as a national manufacturing centre. This paradox is clearly demonstrated by the location quotient technique. By this method, the functional importance of

33. Acheson, "The National Policy and the Industrialization of the Maritimes," p. 7.
34. Examination of the Mercantile Agency Reference Books of Dun, Wiman and Company and of R.G. Dun and Company for the 1879-1914 period revealed that few American and British industries set up branch plants in Halifax. Besides the mercantile community and the small industries that expanded operations, it was noted that migrants to the city from within the province, Great Britain and the United States did establish in Halifax small manufacturing industries. See, for example, the biographies of local businessmen in *Our Dominion: Halifax* (Toronto, 1887), pp. 40-117.
35. Canada, Department of Trade and Commerce, *Canada as a Field for British Branch Industries* (Ottawa, King's Printer, 1922), p. 40.
36. Acheson, "National Policy and the Industrialization of the Maritimes," p. 5.
37. Canada, Dominion Bureau of Statistics, *The Canada Year Book, 1912* (Ottawa, King's Printer, 1914), p. 224.
38. Canada, Department of Trade and Commerce, *Canada as a Field for British Branch Industries*, p. 40.
39. Canada, Dominion Bureau of Statistics, *The Canada Year Book, 1922-23* (Ottawa, King's Printer, 1924), p. 438 and *The Canada Year Book, 1928* (Ottawa, King's Printer, 1929), pp. 453-4.

any city can be accurately measured in comparison to other cities of an urban system. Using published occupation data from the 1881 and 1911 census, reclassified to ensure comparability over time, the location quotient is defined as the proportion of the urban labour force in a given industry (or occupation) divided by the proportion in that industry of some larger benchmark economy. The index takes on a value of one if the proportions are equal; a value of more than one measures an over concentration in the urban area; and a value of less than one indicates an under concentration.[40] From these measurements, it is clear that in absolute terms Halifax improved as a regional and national manufacturing centre over the 1881 to 1911 period (see Table 3). However, overshadowing this advance, cities in other regions bettered their standing relative to Halifax. For example, Halifax's national location quotient rose moderately from .79 to 1.14, but Hamilton's more than doubled from 1.21 to 2.62. This trend was shared by all the heartland cities, emphasizing the concentration of industrial activity in southern Ontario and Quebec.

Table 3: Regional and National Manufacturing Location Quotients for Halifax and Selected Canadian Cities, 1811 and 1911

| Region and City | 1881 | | 1911 | |
|---|---|---|---|---|
| | Regional | National | Regional | National |
| *Eastern Periphery* | | | | |
| Halifax | 1.04 | .79 | 1.38 | 1.14 |
| Saint John | 2.39 | .95 | 1.89 | 1.55 |
| Sydney | | | 3.59 | 2.99 |
| *Central Canada* | | | | |
| Montreal | 2.00 | .72 | 1.35 | 1.79 |
| Quebec | 2.31 | .83 | 1.45 | 1.58 |
| Sherbrooke | | | 1.36 | 1.80 |
| Toronto | 2.18 | 1.02 | 1.50 | 1.94 |
| Hamilton | 2.58 | 1.21 | 2.03 | 2.62 |
| London | 2.39 | 1.11 | 1.84 | 2.38 |
| *Western Periphery* | | | | |
| Winnipeg | | | 2.80 | .96 |
| Calgary | | | 2.45 | .83 |
| Edmonton | | | 3.16 | 1.08 |
| Vancouver | | | .94 | 1.01 |

SOURCES: The location quotients have been calculated from labour data in Canada, *Census of Canada, 1881,* vol. 2, Table 14; and Canada, *Census of Canada, 1911,* vol. 6, Table 6.

40. A.M. Isserman, "The Location Quotient Approach to Estimating Regional Economic Impacts," *Journal of the American Institute of Planners,* 43 (1977), pp. 33-41.

Canadian industrialists in the late nineteenth century could pursue two manufacturing strategies: they could process staples and they could establish secondary industries. In Vancouver, manufacturers followed a strategy of staple processing so successfully that by 1911 nearly 2,800 workers or 5.5 per cent of the city's labour force worked in the forest products sector alone.[41] In Saint John the percentage was smaller but no less significant. However, in Halifax in 1911, less than 200 workers of a total labour force of about 18,000 manufactured staple commodities.[42] The explanation for this situation lies not in the weakness of local initiative, but in the distribution and character of the regional resource base which mitigated against the location of staple processing in Halifax. Theoretically, the most critical locational consideration in processing resources is usually the weight reduction factor. Manufacturing plants locate near the source of a raw material to eliminate the higher transportation costs of shipping this material en masse to a market. Of course, there are other locational considerations: the differential in freight rates for shipping either commodities or finished products; the specific material input requirements of vertically integrated (and often multinational) corporations; the specific amount of on-site processing required to meet external market demands; the level of technology available to the industry; power requirements for processing purposes; and the spatial biases of government tariff and fiscal policies.

Examination of Nova Scotia's fishery and forestry industries confirms Halifax's inability to engage in staple processing.[43] Unlike the salmon catch of British Columbia which was usually canned before export, much of the Nova Scotia fishery was exported in either a green or dried state. Accordingly, only a limited amount of processing took place. The similar distribution pattern of the primary and secondary phases of the industry indicate that this processing was indeed located near the source of the staple itself, away from Halifax (see Figure 4). The largest quantities of fish came from the rich bank areas lying offshore from the province's southeastern counties of Yarmouth, Shelburne, Queens,

41. McCann, "Urban Growth in a Staple Economy," p. 30.
42. Canada, *Census of Canada, 1911*, vol. 6, Table 6, pp. 328-30.
43. The other major staple industries of the province, agriculture and mining, are not found immediately adjacent to Halifax, and for this reason were excluded from this analysis. Halifax did provide tertiary services for these industries.
44. Saunders, *The Economic History of the Maritime Provinces*, p. 78.
45. Nova Scotia, House of Assembly, *Journals and Proceedings of the House of Assembly of Nova Scotia, 1911*, Appendix 22, Industrial Opportunities in Nova Scotia, Canada.
46. The present concentration of fish processing in Halifax did not begin to take place until the mid-1930s when R.P. Bell established the forerunner of National Sea Foods Products. See C. Cox, *Canadian Strength* (Toronto, 1946), p. 35.
47. *McAlpine's Halifax Directory*, 1879-1914 (Halifax, 1880-1915).
48. The estimated pecuniary strength of Robin, Jones and Whitman in 1912 was about $500,000. R.G. Dun and Company, *The Mercantile Agency Reference Book, 1912*, p. 314.

Lunenburg and Halifax, and from those fishing banks situated to the east and south of Cape Breton Island. Here, with the exception of firms interested principally in the fresh fish trade, most canning, curing and freezing establishments were organized in small individual units.[44] The largest centres associated with the secondary phase were Digby, Yarmouth, Wedgeport, Lunenburg, Shelburne and Canso.[45] Few processing plants were in evidence along Halifax's waterfront. Within surrounding Halifax County, they were dispersed widely in the small fishing villages which dotted the coast. Although the fishery therefore added little to the primary manufacturing base of Halifax's economy before World War I,[46] the city's wholesale-trading complex did control an important share of the export trade and even segments of the processing sector. This indicates that business acumen was not to blame for the manufacturing deficit. At least fifteen to twenty mercantile houses dealt directly in the fishery in any one year.[47] Some of these managed branches throughout the Maritimes. A. Wilson and Son owned a freezing plant at Canso in Guysborough County. Robin, Jones and Whitman, the largest dealer early in the twentieth century, operated either collection depots or processing plants at Canso, along the south shore in Lunenburg County, along Northumberland Strait in Cumberland County, and also in the Gaspé region.[48] Other houses dealt solely in the fresh fish trade.

By contrast, the investment and leadership of Halifax in the forest industry of the province was less important. The few sawmills of the city were small in scale and served mainly the local market. At Richmond, a one-time industrial suburb of Halifax located two miles north of the city's central docks, the wharf facilities of the federal government were equipped to handle lumber exports. However, other linkages with the regional hinterland were minimal. There were few supporting firms within the business community that provided brokerage, manufacturing, transportation or other services.[49] This was surprising, because by World War I the primary phase of the industry had surpassed fishing in value of output and the making of log products was Nova Scotia's second highest valued manufacturing industry.[50]

The lack of participation in the processing phases of the forest industry by Halifax is based on three locational factors. The supply of local water to power a sawmill was limited and in the immediate hinterland the forests comprised mainly stunted growth and cut-over areas.[51] Elsewhere, access to the forest was limited because over 80 per cent of it was privately owned, chiefly in small farm woodlots or large holdings typically controlled by out-of-province interests.[52] Of most significance was

49. *McAlpine's Halifax Directory, 1879-1914.*
50. "Report of the Board of Works," *Annual Report of the City of Halifax 1891-92* (Halifax, 1892), pp. 59-86.
52. Saunders, *The Economic History of the Maritime Provinces*, pp. 81-3.

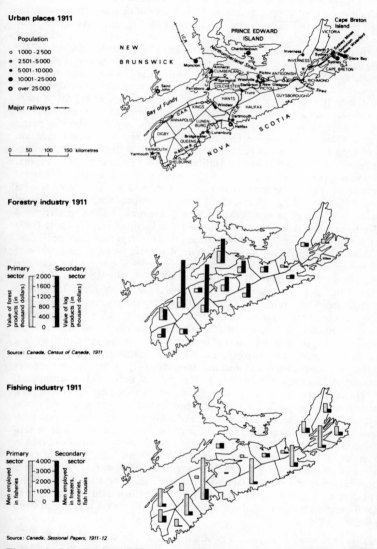

Figure 4: Distribution of Primary and Secondary Sectors of the Forestry and
Fishing Industry, 1911

the province-wide distribution and export orientation of the industry which prevented centralization at Halifax. Traditionally, most counties and many of their settlements had shared in the logging, sawmilling and exporting phases of the industry. Some of these communities had further specialized in shipbuilding, although this industry had waned by the late nineteenth century.[53] Early in the twentieth, the forest products industry was concentrated away from Halifax in the western counties of Annapolis, Digby, Yarmouth, Shelburne, Queens and Lunenburg and to the north in Hants, Colchester and Cumberland. Here, many sawmills were located along streams or small rivers at tidewater sites in order to generate power and reduce assembly and distribution costs. From a locational perspective, then, difficult lines of communication and high transportation costs made it impractical to channel raw materials over any long distance to Halifax for either processing or re-export. However, these costs did not entirely restrict the movement of finished products. This movement was usually confined to the winter season when Halifax did handle increased flows of lumber even from ice-bound areas as far away as northeastern New Brunswick.[54] The completion of the Intercolonial Railway in the mid-1870s facilitated this export trade as well as reinforcing Halifax's function as a regional entrepôt.

Hinterland cities are quite capable of growing to a substantial size either by engaging in resource processing or by servicing the resource hinterland. The growth of Vancouver before World War I is a case in point. But post-confederation Halifax was not strategically located to benefit from the industrial stimuli offered by the regional staple economy. It could, and did, act as a commercial entrepôt for staple production, channelling a sizeable share of the region's fish, forest, mineral and agricultural products to external markets; but it could not function as a processing centre for these resources. The scattered distribution, the nature of external demand and the limited supply of the region's staples restricted Halifax's accessibility to this path of industrial development.

The inability of Halifax to engage in staple processing thus provides a partial explanation for the city's more restricted course of industrial growth. Other reasons for Halifax's failure are related to its uncompetitive location at the margin. A comparative methodology, which establishes a clear picture of the structural characteristics of manufacturing in Halifax, in other towns and cities of Nova Scotia, and in major cities across Canada, isolates Halifax's shortcomings. From this analysis, it is possible to describe the external economies accruing to individual places. The changing production costs of manufacturing during the 1880 to 1910 period explain additional differentiating effects on urban-industrial growth.

53. *Ibid.*, pp. 14-22 and D. Erskine, "The Atlantic Region," in J. Warkentin, ed., *Canada: A Geographical Interpretation* (Toronto, 1968), pp. 224-6 and 253-70.
54. The major ports that competed with Halifax for the lumber export trade were Parrsboro, Amherst, Sheet Harbour, Yarmouth and Pictou.

Table 4: Manufacturing Characteristics of the Twenty Leading Towns and Cities of Nova Scotia, 1910
(Ranked by Value of Production)

| Urban Place | Population | Establishments | Employees | Fixed Capital ($000s) | Value of Products ($000s) | Value Added ($000s) | Horse Power |
|---|---|---|---|---|---|---|---|
| Halifax | 46,619 | 112 | 4,014 | 14,069 | 12,140 | 2,227 | 4,742 |
| Sydney | 17,723 | 20 | 3,890 | 24,623 | 9,395 | 1,502 | 22,002 |
| Amherst | 8,973 | 19 | 2,142 | 15,764 | 4,626 | 934 | 3,608 |
| Sydney Mines | 7,470 | 6 | 507 | 1,935 | 2,540 | 586 | 2,850 |
| Trenton | 1,749 | 3 | 1,182 | 1,853 | 2,290 | 337 | 3,435 |
| Bridgewater | 2,775 | 6 | 153 | 196 | 1,560 | 316 | 565 |
| Truro | 6,107 | 14 | 688 | 2,046 | 1,335 | 224 | 896 |
| Yarmouth | 6,600 | 34 | 714 | 1,541 | 1,198 | 387 | 757 |
| Dartmouth | 5,058 | 11 | 476 | 1,681 | 1,145 | 242 | 2,040 |
| New Glasgow | 6,383 | 21 | 776 | 440 | 1,035 | 258 | 1,160 |
| Pictou | 3,179 | 9 | 243 | 380 | 628 | 179 | 331 |
| Windsor | 3,452 | 7 | 330 | 244 | 472 | 119 | 892 |
| Kentville | 2,304 | 8 | 144 | 248 | 308 | 103 | 202 |
| Lunenburg | 2,681 | 20 | 179 | 189 | 280 | 99 | 83 |
| Canso | 1,716 | 8 | 220 | 164 | 216 | 97 | n.a. |
| Parrsboro | 2,856 | 5 | 110 | 99 | 205 | 70 | 185 |
| Stellarton | 3,910 | 7 | 89 | 649 | 201 | 102 | n.a. |
| Liverpool | 2,109 | 8 | 136 | 111 | 187 | 58 | 121 |
| North Sydney | 5,418 | 5 | 106 | 159 | 169 | 43 | n.a. |
| Glace Bay | 16,652 | 4 | 33 | 38 | 133 | 56 | n.a. |

SOURCES: Canada, *Census of Canada, 1911*, vol. 3, Tables 11 and 12; and for data on horse power, Nova Scotia, House of Assembly, *Journals and Proceedings of the House of Assembly*, Appendix 15, Factories Report.

Table 5: Structural Indices of Manufacturing in the Twenty Leading Towns and Cities of Nova Scotia, 1910 (Ranked by Value of Production)

| Urban Place | % Pop. in Mnfg. | % Increase in Output 1890-1910 | Labour | Capital ($000s) | Output ($000s) | Capital/ Labour | Capital/ Output | Output/ Labour | Labour/ Power |
|---|---|---|---|---|---|---|---|---|---|
| | | | | Average of Establishments | | | | | |
| Halifax | 8.6 | 69 | 36 | 126 | 108 | 3,505 | 6.2 | 3,025 | .85 |
| Sydney | 21.9 | 2,704 | 195 | 1,251 | 470 | 6,330 | 16.4 | 2,415 | .17 |
| Amherst | 23.9 | 526 | 113 | 830 | 243 | 7,359 | 16.9 | 2,160 | .59 |
| Sydney Mines | 6.8 | 85 | 85 | 3,225 | 423 | 3,817 | 3.3 | 5,010 | .17 |
| Trenton | 67.6 | — | 394 | 618 | 763 | 1,586 | 5.2 | 1,937 | .34 |
| Bridgewater | 5.5 | — | 26 | 32 | 260 | 1,282 | .6 | 10,198 | .27 |
| Truro | 11.3 | 58 | 49 | 146 | 95 | 2,974 | 9.1 | 1,940 | .76 |
| Yarmouth | 10.8 | -3 | 21 | 45 | 35 | 2,158 | 4.0 | 1,678 | .95 |
| Dartmouth | 9.4 | 10 | 43 | 152 | 104 | 3,532 | 6.9 | 2,406 | .15 |
| New Galsgow | 12.2 | — | 207 | 21 | 49 | 567 | 1.7 | 1,333 | .67 |
| Pictou | 7.6 | — | 27 | 42 | 70 | 1,564 | 2.1 | 2,582 | .72 |
| Windsor | 9.6 | 34 | 47 | 35 | 68 | 741 | 1.9 | 1,432 | .37 |
| Kentville | 6.4 | 79 | 18 | 31 | 39 | 1,722 | 2.4 | 2,138 | .72 |
| Lunenburg | 6.7 | -64 | 9 | 9 | 11 | 1,056 | 1.9 | 1,283 | .37 |
| Canso | 13.6 | 204 | 28 | 20 | 27 | 745 | 1.7 | 982 | n.a. |
| Parrsboro | 3.9 | -54 | 22 | 20 | 41 | 904 | .5 | 1,864 | .59 |
| Stellarton | 2.3 | — | 13 | 93 | 29 | 7,207 | 6.4 | 2,228 | n.a |
| Liverpool | 6.4 | 15 | 17 | 14 | 23 | 818 | 1.9 | 1,371 | 1.13 |
| North Sydney | 1.9 | -30 | 21 | 32 | 34 | 1,505 | 3.8 | 1,593 | n.a. |
| Glace Bay | 2.0 | — | 8 | 9 | 34 | 1,167 | .7 | 4,015 | n.a. |

SOURCES: Calculated from data in Canada, Census of Canada, 1911, vol. 3, Tables 11 and 12: and Nova Scotia, House of Assembly, Journals and Proceedings of the House of Assembly, Appendix 15, Factories Report.

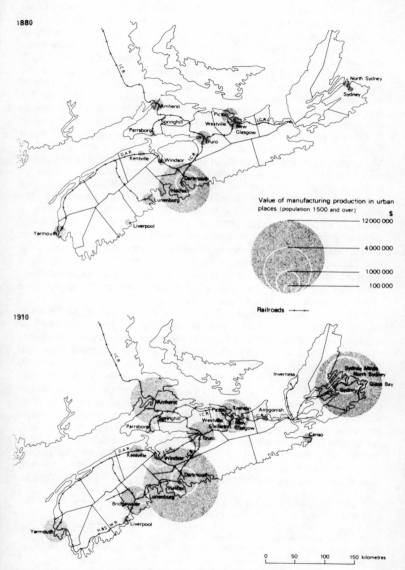

Figure 5: Changing Patterns of Manufacturing, 1880 and 1910

Most of the larger towns and cities of Nova Scotia participated in the new industrialism to such an extent that their industrial output at least doubled in the last quarter of the nineteenth century. In a number of places the advance was considerably higher. Most notable were those specialized manufacturing centres located along the Intercolonial Railway whose rates of population growth were also the highest in the province (see Tables 4 and 5 and Figure 5). Indeed, the Intercolonial was itself an important agent in the initial industrialization of the region. To create traffic, its management had established a basic freight rate structure that was between 20 and 50 per cent lower than that of Ontario's. In addition, it offered special rate concessions based upon "what the traffic would bear."[55] On traffic destined for markets west of Montreal, Nova Scotian manufacturers, including certain Halifax industries such as the sugar refineries,[56] enjoyed fixed arbitrary rates on the province to Montreal portion which gave them a stable relationship with competitors in Central Canada. The specific rate differential to the western provinces from Nova Scotia and from Ontario was about 8 cents per hundred weight.[57] Although these westbound rates did assist regional manufacturers, eastbound rates were about 12 per cent higher which meant that it was costly to import industrial materials from Central Canada. This imbalance increased the production costs of those manufacturers dependent upon external suppliers. Halifax, located far from national suppliers, was at a particular disadvantage in securing cheap materials for its industrial base.

Assisted by these freight tariffs and by other aspects of the National Policy, several of Nova Scotia's towns and cities did succeed in developing industries based on the particularistic circumstances of the region. "Busy" Amherst was led by enterprising captains of industry who capitalized on nearby coal resources to fuel their factories that produced railway cars, boots and shoes, engines, boilers, woollen goods and pianos.[58] The industrial complex of New Glasgow-Trenton-Stellarton, similarly aided by local coal deposits, developed beyond its early focus on wooden shipbuilding to establish itself as a major producer of steel products. Here, the integrated Nova Scotia Steel and Coal Company, assisted by government subsidies, became one of Canada's largest corporations with over 6,000 employees working throughout the region, spawning linked industries which manufactured railway cars, springs, boilers, tools, mining equipment, and even rifle sights.[59] In Cape Breton,

55. Forbes, "Misguided Symmnetry," pp. 60-8; and R.A.C. Henry and Associates, *Railway Freight Rates in Canada* (Ottawa, 1939), pp. 266 and 268.
56. Acheson, "National Policy and the Industrialization of the Maritimes," p. 14.
57. Nova Scotia, *Journals and Proceedings, 1911,* Appendix 22, pp. 5-6.
58. R. Lamy, "The Development and Decline of Amherst as an Industrial Centre" (BA thesis, Mount Allison University, 1930).
59. Canadian Manufacture's Association, *Evidence of the Industrial Ascendency of Nova Scotia* (Halifax, 1914),n.p.; and *The Nova Scotia Steel and Coal Company* (Halifax, 1912), pp. 5-7.

the Sydney area contained Canada's leading coal, iron and steel complex. The locational incentives for this development were the largest accessible reserves of coal in Canada and iron ore easily transported from Wabana, Newfoundland.[60] Of the cities situated away from the Intercolonial, only Yarmouth and Bridgewater ranked in the leading ten. Other towns of lesser rank were specialized in character. Some, like Parrsboro and Canso, were limited to processing staples for export; others produced only for a local market.[61]

There are also clear indications that economic factors other than accessibility to industrial materials assisted the progress of the specialized centres of Nova Scotia and placed Halifax at a comparative disadvantage. Increased economies of scale, reduced production costs and labour-saving technologies, all influenced selective urban-industrial growth. A comprehensive analysis of the 1880 to 1910 period reveals that increases in plant size and the resulting reductions in material and labour costs were most pronounced in those places sharing in rapid growth.[62] As illustrated by Table 5, Sydney, Amherst, Trenton and Sydney Mines best exemplify this pattern. The lower labour/power ratios of these places point to the ability of the specialized industries there to exploit labour-saving technologies. In addition, the centres in which manufacturing had advanced most particularly had the largest investments in physical plant and the highest capital/labour and capital/output ratios. This indicates that local industrialists successfully raised capital and used it efficiently to their advantage over competing centres. As T.W. Acheson has shown, the Nova Scotia Steel and Coal Company of New Glasgow was strongly supported by the Halifax financial community,[63] and the economies of scale associated with this corporation's policy of vertical and horizontal integration apparently were considerable.[64] This case is supported by the example of the Dominion Iron and Steel Company at Sydney. Started in the early 1890s as the Dominion Coal Company by American interests, it reorganized shortly thereafter to produce iron and steel and was subsequently controlled by Montreal and Toronto capitalists.[65] It, too, was an integrated operation, possessing, for example, a substantial shipping fleet.[66]

60. P.T. McGrath, "The Manufacture of Iron and Steel in Cape Breton," *The Engineering Journal*, 2 (1901), pp. 571-85.

61. Nova Scotia, *Journals and Proceedings, 1911*, Appendix 22.

62. The detailed tabulations of this analysis have not been reproduced here; they are available on request from the author.

63. Acheson, "National Policy and the Industrialization of the Maritimes," pp. 24-6.

64. Canadian Manufacturers Association, *Industrial Ascendency of Nova Scotia*.

65. Acheson, "National Policy and the Industrialization of the Maritimes," p. 23.

66. Part of this fleet was used to ship coal to the Montreal industrial and domestic market. Instead of returning empty to Cape Breton, these ships sometimes carried wholesale goods for distribution in Cape Breton. This practice cut into the trade of Halifax. F.W. Gray, *Mining and Transportation* (Toronto, 1909), pp. 111-28.

Table 6: Structure of Manufacturing in Halifax, 1911
(By Number and Percentage of Employees)

| Manufacturing Industry | Number | Percentage |
|---|---|---|
| Chemicals and drugs | 47 | 1.3 |
| Clothing | 752 | 20.2 |
| Food | 669 | 18.0 |
| Iron and steel | 381 | 10.2 |
| Leather and rubber goods | 177 | 4.8 |
| Liquors and beverages | 65 | 1.7 |
| Printers and engravers | 279 | 7.5 |
| Textiles | 263 | 7.1 |
| Vehicles for land | 254 | 6.8 |
| Vehicles for water | 102 | 2.7 |
| Wood products | 179 | 4.8 |
| Others | 556 | 14.9 |
| Totals | 3,724 | 100.0 |

SOURCE: Canada, *Census of Canada, 1911*, vol. 6, Table 6.

By World War I, the manufacturing structure of Halifax had deviated little from that established in the earlier phases of industrialization (see Table 6). There was no coal nearby to give an initial advantage for specialized manufacturing as elsewhere in the province. Other industrial materials were in short supply. Halifax did not even manufacture many producer goods for either the export sector or a national industrial market because of the weak linkage effects imparted by the regional staples, and also because of the sparsely populated hinterland which prevented the city from sharing in the substantial economies of production inherent in a dense network of rural and urban settlement such as in Central Canada.[67] Instead, the strength of Halifax's limited industrial structure was based on the manufacture of consumer products, such as refined sugar, confectionery goods, cotton cloth, boots and shoes and men's clothing. Some of these products entered national markets, a practise generally curtailed after the merger movement,[68] but most were consumed only by the regional market. The presence of these manufacturers was clearly a function of Halifax's entrepôt status and its nodal position within the region. As a port and rail terminus, Halifax maintained some of the advantages associated with reduced transfer costs. Its nodal position also created the advantage of relative accessibility, making it at least possible for the producers of consumers goods in Halifax to gain access to regional markets.

67. This theme is developed in J. Spelt, *Urban Development in South Central Ontario* (Toronto, 1972): and Gilmour, *Spatial Dynamics of Manufacturing.*
68. T.W. Acheson, "The Maritimes and Empire Canada," in *Canada and the Burden of Unity*, pp. 94-5.

Table 7: Manufacturing Characteristics of the Twenty Leading Cities of Canada, 1911
(Ranked by Value of Production)

| Urban Place | Population | Establishments | Employees | Fixed Capital ($000s) | Value of Products ($000s) | Value Added ($000s) |
|---|---|---|---|---|---|---|
| Montreal | 470,480 | 1,104 | 67,841 | 132,476 | 166,297 | 42,530 |
| Toronto | 376,538 | 1,100 | 65,274 | 145,799 | 154,307 | 41,302 |
| Hamilton | 81,969 | 364 | 21,149 | 58,014 | 55,126 | 15,332 |
| Winnipeg | 136,035 | 177 | 11,705 | 26,024 | 32,694 | 6,657 |
| Ottawa | 87,701 | 203 | 9,232 | 21,099 | 20,924 | 5,740 |
| Maisonneuve | 18,864 | 20 | 9,112 | 7,919 | 20,814 | 1,917 |
| Quebec | 78,710 | 175 | 8,067 | 16,488 | 17,149 | 5,305 |
| London | 46,300 | 180 | 9,413 | 15,470 | 16,274 | 4,396 |
| Brantford | 23,132 | 111 | 6,492 | 19,972 | 15,866 | 4,901 |
| Vancouver | 100,401 | 130 | 8,966 | 22,815 | 15,070 | 349 |
| Halifax | 46,619 | 112 | 4,014 | 14,069 | 12,140 | 2,277 |
| Peterborough | 18,360 | 65 | 4,029 | 6,415 | 10,633 | 857 |
| Saint John | 42,511 | 177 | 5,270 | 9,242 | 10,082 | 2,239 |
| Sydney | 17,723 | 20 | 3,890 | 24,623 | 9,395 | 1,502 |
| Berlin | 15,196 | 76 | 3,980 | 8,501 | 9,266 | 2,956 |
| Calgary | 43,704 | 46 | 2,133 | 13,082 | 7,751 | 1,502 |
| Guelph | 15,175 | 78 | 3,072 | 7,152 | 7,392 | 2,008 |
| Hull | 18,222 | 31 | 2,918 | 8,780 | 7,259 | 1,434 |
| Lachine | 10,699 | 11 | 2,239 | 7,496 | 6,296 | 1,560 |
| St. Catharines | 12,484 | 58 | 3,139 | 5,290 | 6,024 | 2,028 |

SOURCE: Canada, Census of Canada, 1911, vol. 3, Table 11.

Table 8: Structural Indices of Manufacturing in the Twenty Leading Cities of Canada, 1911. (Ranked by Value of Production within Regions)

| Region and Urban Place | % Pop. in Mnfg. | % Increase in Output 1890-1910 | Average of Establishments | | | | | Capital/ Labour |
|---|---|---|---|---|---|---|---|---|
| | | | Labour | Capital ($000s) | Output ($000s) | Capital/ Labour | Capital/ Output | |
| *Eastern Periphery* | | | | | | | | |
| Halifax | 8.6 | 69 | 36 | 126 | 108 | 3,505 | 6.2 | 3,025 |
| Saint John | 12.4 | 24 | 30 | 52 | 57 | 1,754 | 4.1 | 1,191 |
| Sydney | 23.5 | 2,704 | 195 | 1,251 | 470 | 6,330 | 14.4 | 2,415 |
| *Central Canada* | | | | | | | | |
| Montreal | 14.4 | 146 | 61 | 120 | 151 | 1,952 | 3.1 | 2,451 |
| Toronto | 17.3 | 243 | 59 | 133 | 140 | 2,234 | 3.5 | 2,364 |
| Hamilton | 25.8 | 293 | 58 | 159 | 151 | 2,743 | 3.8 | 2,606 |
| Ottawa | 10.5 | 137 | 46 | 104 | 103 | 2,285 | 3.7 | 2,266 |
| Maissoneuve | 48.3 | 469 | 456 | 396 | 1,041 | 869 | 4.1 | 2,284 |
| Quebec | 10.2 | 158 | 46 | 94 | 98 | 2,043 | 3.1 | 2,126 |
| London | 20.3 | 98 | 52 | 86 | 90 | 1,643 | 3.5 | 1,728 |
| Brantford | 28.1 | 258 | 58 | 180 | 143 | 3,076 | 4.1 | 2,443 |
| Peterborough | 21.9 | 309 | 62 | 99 | 163 | 1,592 | 7.5 | 2,639 |
| Berlin | 26.2 | 407 | 52 | 112 | 122 | 2,135 | 2.9 | 2,328 |
| Guelph | 20.2 | 149 | 39 | 92 | 95 | 2,328 | 3.6 | 2,406 |
| Hull | 16.0 | 463 | 94 | 283 | 234 | 3,009 | 6.1 | 2,576 |
| Lachine | 20.9 | 363 | 204 | 681 | 572 | 3,348 | 4.8 | 2,812 |
| St. Catharines | 21.7 | 146 | 54 | 102 | 104 | 1,886 | 2.9 | 1,919 |
| *Western Periphery* | | | | | | | | |
| Winnipeg | 8.6 | 483 | 66 | 147 | 185 | 2,223 | 3.9 | 1,847 |
| Vancouver | 8.9 | 695 | 69 | 176 | 116 | 2,545 | 7.0 | 1,159 |
| Calgary | 4.9 | 2,894 | 46 | 284 | 169 | 6,133 | 8.7 | 3,634 |

Table 9: Manufacturing Production Costs in the Twenty Leading Cities of Canada, 1880-1910 (Ranked by Value of Production in 1911)

| Urban Place | Average Size of Establishments | | | | Material Costs/Output[1] | | | | Labour Costs/Output[2] | | | | Total Costs/Output | | | |
|---|---|---|---|---|---|---|---|---|---|---|---|---|---|---|---|---|
| | 1880 | 1890 | 1900 | 1910 | 1880 | 1890 | 1900 | 1910 | 1880 | 1890 | 1900 | 1910 | 1880 | 1890 | 1900 | 1910 |
| Montreal | 23 | 22 | 48 | 61 | .62 | .56 | .51 | .53 | .17 | .19 | .25 | .20 | .79 | .75 | .76 | .73 |
| Toronto | 14 | 11 | 50 | 59 | .51 | .49 | .55 | .49 | .19 | .21 | .26 | .23 | .70 | .70 | .81 | .72 |
| Hamilton | 15 | 8 | 44 | 58 | .52 | .50 | .52 | .51 | .27 | .23 | .23 | .21 | .79 | .73 | .74 | .72 |
| Winnipeg | 9 | 8 | 31 | 66 | .56 | .54 | .58 | .56 | .24 | .20 | .21 | .23 | .80 | .74 | .79 | .79 |
| Ottawa | 14 | 12 | 33 | 46 | .64 | .59 | .49 | .47 | .19 | .21 | .31 | .23 | .83 | .80 | .80 | .70 |
| Maissoneuve | — | 49 | 227 | 456 | — | .90 | .72 | .67 | — | .04 | .15 | .23 | — | .94 | .97 | .80 |
| Quebec | 10 | 7 | 37 | 47 | .61 | .57 | .55 | .47 | .17 | .18 | .24 | .21 | .78 | .75 | .79 | .68 |
| London | 13 | 8 | 47 | 52 | .53 | .48 | .46 | .46 | .17 | .20 | .27 | .26 | .70 | .68 | .73 | .72 |
| Brantford | 9 | 11 | 82 | 58 | .58 | .44 | .50 | .47 | .22 | .24 | .28 | .22 | .80 | .68 | .78 | .69 |
| Vancouver | — | 12 | 30 | 69 | — | .45 | .54 | .51 | .22 | .29 | .26 | .26 | — | .74 | .80 | .77 |
| Halifax | 9 | 12 | 31 | 36 | .64 | .61 | .64 | .62 | .16 | .16 | .17 | .14 | .80 | .77 | .81 | .76 |
| Peterborough | 9 | 9 | 49 | 62 | .56 | .55 | .63 | .71 | .23 | .22 | .22 | .18 | .79 | .77 | .85 | .79 |
| Saint John | 13 | 8 | 25 | 30 | .62 | .56 | .52 | .54 | .18 | .21 | .24 | .22 | .80 | .77 | .76 | .76 |
| Sydney | 1 | 2 | 30 | 195 | .53 | .39 | .52 | .18 | .16 | .34 | .32 | .23 | .69 | .73 | .84 | .41 |
| Berlin | 12 | 19 | 41 | 52 | .53 | .42 | .53 | .48 | .23 | .29 | .26 | .19 | .78 | .71 | .79 | .67 |
| Calgary | — | 6 | 31 | 46 | — | .34 | .52 | .60 | — | .37 | .29 | .20 | — | .71 | .81 | .80 |
| Guelph | 10 | 12 | 32 | 39 | .58 | .57 | .52 | .52 | .21 | .23 | .23 | .20 | .79 | .80 | .75 | .72 |
| Hull | 33 | 22 | 188 | 94 | .46 | .51 | .67 | .60 | .20 | .25 | .18 | .19 | .66 | .76 | .85 | .79 |
| Lachine | 6 | 39 | 150 | 204 | .28 | .32 | .53 | .54 | .20 | .22 | .19 | .20 | .48 | .54 | .72 | .74 |
| St. Catharines | 9 | 12 | 48 | 54 | .63 | .58 | .51 | .43 | .17 | .18 | .29 | .23 | .70 | .76 | .80 | .66 |

SOURCES: Canada, *Census of Canada, 1901*, vol. 3, Table 20; and Canada, *Census of Canada, 1911*, vol. 3, Table 11.

NOTES:
[1]The term cost of materials includes the value of all material whether in the raw or partly manufactured state.
[2]Salaries and wages of employees.

Structural indices point to additional explanations for Halifax's restricted industrial development. These indices isolate the external economies which favour urban-industrial growth in heartland economies and restrict such development in peripheral regions. Manufacturing in Halifax, compared to that in the other major industrial centres of Canada, was small in scale. The average capitalization, output, and number of workers employed in its factories was considerably less than that of almost every other city (Tables 7 and 8). Of particular significance, most factories in Halifax did not increase in size over the 1880 to 1910 period to the same extent as their counterparts in Central Canada. Some firms did, such as Clayton and Son and Moirs, makers of men's clothing and confectionery products, respectively, because they were able to effect economies of scale with lower cost purchases of cotton, sugar and other materials from local companies.[69] Without the advantage of scale economies, it was difficult to compensate for the distribution costs incurred by shipping to distant markets.[70] High factor costs had generally hindered plant expansion. Even in 1880, Halifax's ratio of material costs/output was the highest of any city, and this situation did not change by 1910 (see Table 9). Most materials had to be either shipped from distant regional or national suppliers, incurring burdensome transportation costs. Also, the limited presence in Halifax of particular types of manufacturing meant that localization economies which are normally associated with bulk purchases were of no consequence. Nor could these economies be realized even by production linkages with related industrial firms because, with few exceptions, these too were minimal in the city.

To compensate for these high material costs, Halifax was forced to rely upon an efficient and productive factory system and on lower labour costs. As indicated by certain indices – capital/labour, capital/output and output/labour – Halifax did possess a comparatively efficient and productive industrial base and its labour costs were the lowest of any city.[71] But these advantages were not sufficient. Only certain labour intensive industries, such as the clothing and cotton cloth manufacturers, were in a position to take advantage of these competitive costs, principally because they employed a less-skilled and less-expensive female labour force.[72] To establish firms requiring a skilled and expensive work force would have incurred still higher costs. As it was, only three heartland cities (Maisonneuve, Peterborough and Hull) main-

69. Both firms expanded considerably during the late nineteenth century. See the biographies of these firms in Halifax Board of Trade, *The City of Halifax: Its Advantages and Facilities* (Halifax, 1909), pp. 49-50 and 66-7.

70. Pred, *Spatial Dynamics,* pp. 49-71.

71. For example, in Halifax in 1901 an electrician earned 15¢ per hour, whereas in Toronto the comparable rate was 25¢. M.C. Urquhart and K.A.H. Buckley, eds., *Historical Statistics of Canada* (Toronto, 1965), pp. 86-87.

72. Canada, *Census of Canada, 1911,* vol. 3, Table 9, pp. 230-31.

tained a higher ratio of materials and labour costs/output than Halifax.[73] The external economies of the production process bear heavily on differential urban-industrial growth, particularly the availability to cities of such factor inputs as less expensive industrial materials, labour-saving technologies, and capital financing. The comparative advantages of heartland cities, and even of some Nova Scotian centres, therefore restricted the location of secondary manufacturing in Halifax. The city was at a clear disadvantage in overcoming the high costs of obtaining industrial materials and in distributing finished products to distant markets. Without these markets, it was difficult to initiate a new round of industrial growth, and the circular and cumulative process of growth was accordingly curtailed.

The problems of the peripheral industrial location of Halifax and of the city's vulnerability to the forces of continentalism were fully recognized by local businessmen and civic officials. In 1906 the Board of Trade appraised the industrial situation and confessed "we may as well face these difficulties fairly."[74] The board then listed the critical obstacles to manufacturing development: the high cost of living, the need for cheaper power, and the distance from suppliers of materials and from markets. The high cost of rents, food and fuel was blamed for creating a shortage of skilled labour by driving away both 'provincials' and foreign immigrants.[75] There was little to compensate for Halifax's location at the margin – the tyranny of location – other than the advantage offered by "the year round commerce of our port."[76]

But many cities did attempt to compensate for locational disadvantages by offering economic inducements to entice industry. In the late nineteenth century these bonuses were of many types, including the guarantee of a bond issue, property tax exemption, and direct grants of cash, land and even water.[77] Halifax, however, could offer few of these incentives because of its limited supply of industrial land, its serious shortage of water, and its restricted tax base.[78] The latter was the princi-

---

73. "Nova Scotia is a coal producing province, yet Halifax has to pay $4.50 for a ton of coal making it higher than in almost any other place." *The Suburban*, vol. 3, No. 50 (13 January 1906), p. 6.

74. Halifax Board of Trade, *The Forty-First Annual Report of the Halifax Board of Trade for the Year 1906* (Halifax, 1907), p. 11.

75. *The Suburban* (13 January 1906), p. 6.

76. Halifax Board of Trade, *The Forty-First Annual Report*, p. 11.

77. For a general discussion of bonusing in Canada during the late nineteenth century, see T. Naylor, *The History of Canadian Business*, vol. 2 (Toronto, 1975), pp. 120-61.

78. These problems are constantly referred to in the Halifax press and also in the annual reports of the city. Despite the validity of these facts, the city was often subjected to criticism: "Inquiries regarding what inducements the City would give to new industries to locate here have been received on a number of occasions during the year, but as the City had no definite policy in this respect until recently, it has been found difficult to offset the inducements such as bonuses, exemptions, etc. that have been offered at other points." Halifax Board of Trade, *The Forty-Seventh Annual Report of the Halifax Board of Trade for the Year 1912* (Halifax, 1912), p. 24.

pal reason. Over one-third of the value of real property in the city was exempt from taxation because it was owned by various government and religious organizations.[79] This fact placed a considerable burden on the other land uses. To meet a continually rising civic debt, the city taxed not only private property, but also (despite the objections of mercantile interests) the value of goods held in local warehouses. To have granted bonuses, thereby forcing taxes to rise still further, quite obviously would have angered some local interest groups. As a consequence, bonusing was used sparingly. When it was practised, it usually aided those manufacturers aligned with the mercantile community. In this way, bonuses were given in the 1880s to the Halifax Cotton Factory, the Nova Scotia Sugar Refinery, and the Halifax Graving Dock Company.[80] After 1900, a bonus successfully enticed the Silliker Car works away from Amherst, but a long-standing search for a steel shipbuilding enterprise had failed to materialize by World War I.[81]

By examining regional urban growth in Canada within the framework of the staple theory of economic development and the heartland-hinterland conceptualization of regions, it becomes obvious why Halifax diverged from the industrial path followed by cities such as Montreal, Toronto and Hamilton. It is apparent, for example, that staple commodities have had only a marginal impact on manufacturing in Halifax. Their dispersed regional pattern, their weak endowment, and their limited processing requirements are very visible reasons for its failure to become an industrialized entrepôt. The economy of Halifax was also seriously weakened by the marketing and production problems created by the heartland-hinterland process. The metropolitan economy had little need for goods manufactured at the margin because its own industrial base produced and marketed more competitively the same products that could be produced in the Maritimes. Compounding these problems the heartland even managed to submerge and lure away vital regional industries. This centripetal force also siphoned Halifax's banking community away from its hinterland source early in the twentieth century. The forces of continentalism were difficult to overcome.

As a consequence of these problems, Halifax was forced to depend increasingly upon the government extension of the metropolitan economy for stimulating its urban economy. In the late nineteenth and early twentieth centuries, the city emerged as an important defence, trade and

79. "City Assessor's Report for 1911-12," *Annual Report of the City of Halifax 1911-12* (Halifax, 1912), p. 115.

80. "Mayor's Address," *Annual Report of the City of Halifax, 1884-85* (Halifax 1885), pp. xii-xiii and xxv; and "Mayor's Address," *Annual Report of the City of Halifax, 1885-86* (Halifax, 1886), pp. lviii-lix.

81. "Mayor's Address," *Annual Report of the City of Halifax, 1907-08* (Halifax, 1908), pp. 13-4 and 31.

transportation centre.[82] The predominantly federal responsibility for international trade, national defence, and railroad and port development fell upon this outpost of "Empire Canada." With the opening of the Canadian West, the federal government strengthened the port and the rail functions of Halifax so that it could handle the reciprocal movement of immigrants and grain. The government also strengthened the defence function. The increase was slow at first, especially after the withdrawal of the British forces in 1905, but it blossomed dramatically after the outbreak of war in 1914. The government stimulus during this period supplemented the weakening industrial base. Today, such federally supported areas as defence, transportation, research and public services are a mainstay of the local economy. Halifax has failed as an industrial city because it functions at the margin. It is only a intermediary, highly dependent upon a metropolitan economy.

[82]Regional and National Location Quotients for Selected Economic Functions of Halifax, 1881 and 1911

|                | 1881 | | 1911 | |
|                | Regional | National | Regional | National |
|----------------|------|----------|----------|----------|
| Manufacturing  | 1.04 | .79  | 1.38 | 1.14 |
| Trade          | 3.36 | 1.36 | 2.83 | 2.22 |
| Transportation | 1.50 | 1.53 | 2.12 | 1.67 |
| Finance        | 2.84 | 1.42 | 2.66 | 1.84 |
| Government     | 3.00 | 5.25 | 5.52 | 4.23 |

SOURCES: Calculated from labour force data in Canada, *Cenus of Canada, 1881*, vol. 2, Table 14; and Canada, *Census of Canada, 1911* vol. 6, Table 6.

# In Pursuit of Growth: Municipal Boosterism and Urban Development in the Canadian Prairie West, 1871-1913

ALAN F.J. ARTIBISE

Although western Canadian urban history is a rapidly expanding field, much ground remains unturned; there is, for example, still no adequate overview of the development of western cities.[1] It is the purpose of this paper to fill some of the gaps that still exist. While the emphasis in the following pages is on description, this study also attempts to move into the realm of interpretation by providing at least tentative answers to questions of significance not only to prairie cities, but to urban studies

SOURCE: This article is a revised and expanded version of "Boosterism and the Development of Prairie Cities, 1871-1913," in Alan F.J. Artibise, *Town and City: Aspects of Western Canadian Urban Development* (Regina: Canadian Plains Resource Center, 1981), pp. 209-236. This version was presented at the annual meeting of the Organization of American Historians held in Detroit in April 1981.

1. No book-length study of western urban development exists. There are, however, several articles that are of value to the student of western urban history. They include: Paul Voisey, "The Urbanization of the Canadian Prairies, 1871-1916," *Histoire sociale/Social History*, VII (May 1975), pp. 77-101; K. Lenz, "Large Urban Places in the Prairie Provinces – Their Development and Location," in R.L. Gentilcore, ed., *Canada's Changing Geography* (Toronto, 1967), pp. 199-211; L.D. McCann, "Urban Growth in Western Canada, 1881-1961," *The Albertan Geographer*, 5 (1969), pp. 65-74; J.M.S. Careless, "Aspects of Urban Life in the West, 1870-1914," in Gilbert A..Stelter and Alan F.J. Artibise, eds., *The Canadian City: Essays in Urban History* (Toronto, 1977), pp. 125-41; and Alan F.J. Artibise, "The Urban West: The Evolution of Prairie Towns and Cities to 1930," *Prairie Forum*, IV (1979), pp. 237-62. Also useful are the articles in Alan F.J. Artibise, ed., *Town and City: Aspects of Western Canadian Urban Development* (Regina, 1981).

generally. How important are the beliefs, decisions, and actions of individuals and groups in determining the shape of an urban system? In other words, did the residents of prairie urban centres play a crucial role in establishing, say, Winnipeg rather than Selkirk or Brandon as the "Metropolis of the West," or in securing for Saskatoon rather than Prince Albert the position of sub-regional metropolis? The second question has to do with what determines the growth rate of individual cities and the urbanization rate of a region. This paper, then, is a study of the urbanization process as a dependent variable; it approaches urbanization by conceiving of the city as a concrete entity whose form, structure, and growth rate are determined by a variety of variables. In describing and assessing these variables, it is possible to categorize two related sides to urban growth; that which is the produce of large-scale forces such as the settlement process, the external demand for staples, population movements, the state of agricultural and transportation technology, etc. (i.e., the development context); and the city and its residents as a force which stimulates or inhibits growth and influences the direction and location of development. One of the major tasks of urban history is the classification of the differences between, and the relationships of, these two forces. Some progress toward achieving this task has already been made, but a clear conceptual framework is only now emerging. It can be labelled the evolutionary conception of urban systems and it appears to be very promising.[2] Indeed, several researchers have already used this approach and others are in the process of doing so.[3] To date, however, no overview has been attempted.

The evolutionary view of urban development sees a continuous series of individual and collective responses to a changing development context. Cities are partially ordered collectivities which, at certain times, draw participants and, at other times, spin them off. In this situation, each round of decisions by the participants results in resource exploitation and transformation and, therefore, some greater or lesser modification in the human and natural environment. Clearly, in such a view, no city is exactly like another. Nonetheless, the interactive process of decision-making is subject to some generality. The regularities to be dealt with, however, are neither in the events themselves nor in the generalizations to be drawn from them. These are, of course, an important part of historical explanation but, obviously, events will vary from city to city, from region to region, and from one era to another. The

2. The evolutionary conception is outlined in some detail in John B. Sharples and Sam Bass Warner, Jr., "Urban History," *American Behavioural Scientist*, 21 (1977), pp. 221-44.

3. See, for example, Artibise, "Continuity and Change: Elites and Prairie Urban Development, 1914-1950," in Alan F.J. Artibise and Gilbert A. Stelter, eds., *The Usable Urban Past: Politics and Planning in the Modern Canadian City* (Toronto, 1979), pp. 130-55; and the articles by McDonald and Phillips in *Town and City*.

118                                              ALAN F.J. ARTIBISE

regularities instead lie in the concepts of process and interaction inherent to the evolutionary perspective.

To discovery these regularities, however, it is necessary to begin by applying the evolutionary conception to a particular historical setting. The development context for the prairie region in the period 1871-1913 is well known. It was a dynamic and fluid era when the rapid growth of the national economy, the settlement of low-density areas, and the initial utilization of vast resources were occurring simultaneously. In the prairie west, city-builders confronted a landscape which could be developed virtually free of the residue of earlier decision-makers.

The dynamic nature of the period is readily evident. In 1871 the prairies had no urban centres. Except for a few Hudson's Bay Company posts, there was no commercial development. The population of the region, some 73,000 persons, was entirely rural. Forty years later, in 1911, the urbanization process on the prairies was well advanced. The region could boast of twelve cities with populations of over 5,000 and one, Winnipeg, with a population exceeding 1,300,000 and over 35 per cent could be classified as urban. Furthermore, the general urban pattern of the prairies was firmly established by 1911, and five cities had emerged as dominant urban centres. These cities – Winnipeg, Regina, Calgary, Saskatoon, and Edmonton – have remained the primary urban concentrations of the region (see Tables 1 and 2).

Table 1: Populations of Prairie Cities, 1871 - 1916 (a)

| City | 1871 | 1881 | 1891 | 1901 | 1906 | 1911 | 1916 |
|---|---|---|---|---|---|---|---|
| Winnipeg | 241 | 7,985 | 25,639 | 42,340 | 90,153 | 136,035 | 163,000 |
| Calgary | — | — | 3,867 | 4,392 | 13,573 | 43,704 | 56,514 |
| Edmonton | — | — | 300(b) | 4,176 | 14,088 | 31,064 | 53,846 |
| Regina | — | 800(c) | 1,681(d) | 2,249 | 6,169 | 30,213 | 26,127 |
| Saskatoon | — | — | — | 113 | 3,011 | 12,004 | 21,048 |
| Moose Jaw | — | — | — | 1,558 | 6,249 | 13,823 | 16,934 |
| Brandon | — | — | — | 5,620 | 10,408 | 13,839 | 15,215 |
| St. Boniface | 817 | 1,283 | 1,553 | 2,019 | 5,119 | 7,483 | 11,021 |
| Lethbridge | — | — | — | 2,072 | 2,936 | 9,035 | 9,436 |
| Medicine Hat | — | — | — | 1,570 | 3,020 | 5,608 | 9,272 |
| Prince Albert | — | — | — | 1,785 | 3,005 | 6,254 | 6,436 |
| Portage la Prairie | — | — | 3,363 | 3,901 | 5,106 | 5,892 | 5,879 |

SOURCE: *Census of Prairie Provinces, 1916.*

Notes: a) Population is listed according to areas of 1916.
     b) This is an approximation taken from City of Edmonton records.
     c) The population of Regina in 1882, 1883 was between "800 and 900 souls." See E.G. Drake, *Regina: The Queen City*, p. 22.
     d) *Ibid.*, p. 71.

Table 2:

Rural and Urban  Population Growth in the Northwest Territories
and Prairie Provinces, 1871-1916
(in thousands)

| | Northwest Territories | | Manitoba | | Total Prairies | | |
| | Rural | Urban | Rural | Urban | Rural | Urban | % Urban |
|---|---|---|---|---|---|---|---|
| 1871 | 48 | — | 25 | — | 73 | — | 0 |
| 1881 | 56 | — | 52 | 10 | 108 | 10 | 8 |
| 1891 | 95 | 4 | 111 | 41 | 206 | 45 | 18 |

| | Alberta | | Saskatchewan | | | | | | |
| | Rural | Urban | Rural | Urban | | | | | |
|---|---|---|---|---|---|---|---|---|---|
| 1901 | 54 | 19 | 77 | 14 | 185 | 70 | 316 | 103 | 25 |
| 1906 | 127 | 58 | 209 | 48 | 228 | 138 | 564 | 244 | 30 |
| 1911 | 237 | 138 | 361 | 131 | 261 | 200 | 859 | 470 | 35 |
| 1916 | 307 | 189 | 471 | 176 | 313 | 241 | 1,091 | 606 | 36 |

SOURCES: *Census of the Prairie Provinces,* 1916: *Census of Canada*, 1931,
Volume 1

Note: a) Includes incorporated villages, towns, and cities.

In the years before 1911 it was by no means certain that these particular communities would become the prairies' largest cities. It is, of course, unlikely that the region could have supported many more large centres, since in this early period urban growth depended to a large degree on the resource base. The rapidity and degree of urbanization was closely tied to the agricultural development of the region. Also, as geographers and economists have pointed out, such factors as locational

advantage, agricultural and transportation technology, conscious federal policy in respect to the West, and general economic conditions
made it probable that urban development on the prairies would follow a
set pattern. Yet, even though all the ingredients for urban growth were
present, and even though they pointed toward a particular pattern, it
was, in the final analysis, only the skill and initiative of individuals and
groups that translated opportunity into reality.[4] In other words, while
the growth of prairie cities cannot be fully explained by internal factors,
such as the activities of individual and groups, neither can the growth
process be explained only by impersonal or mechanistic forces. In the
case of Winnipeg, for example, location theory does not necessarily
provide for the emergence of a single-city pattern in the Red River
Valley. A more diffuse urban pattern might well have developed.[5] The
fact that such a diffuse pattern did not develop is not easily explained,
but it can be said with some degree of certainty that one determining
factor in the particular pattern that did emerge in Manitoba, and later in
Saskatchewan and Alberta, was the initiative of civic and business
leaders.

The success of one prairie city relative to another was not determined
only by a convenient location or the impersonal forces of urbanization.
The residents of prairie cities interacted with the environment, and their
hopes, beliefs, energy, community spirit, initiative and adaptability
influenced the rate of growth, degree of prosperity, and physical form of
the cities. While the role of municipal governments and business organizations in altering the rate and pattern of urban development on the
prairies was certainly limited by outside forces, the growth, shape, and
character of the five major cities owes much to the policies devised and
vigorously applied by these bodies in response to the possibilities and
problems that emerged for their communities.

The role individuals and groups played in shaping prairie urban

4. This statement has been discussed in varying degrees of detail in the following: L. Gertler
   and R. Crowley, *Changing Canadian Cities* (Toronto, 1977), p. 152; Peter G. Goheen,
   "Industrialization and the Growth of Cities in Nineteenth-Century America," *American
   Studies*, 14 (1971), pp. 49-66; and Robert R. Dykstra, *The Cattle Towns* (New York,
   1974), pp. 3-7.

5. "The effects of the policies of the commercial elite on the urban geography of the region
   were striking. By defeating Selkirk in its bid for the railway, by winning freight rate and
   other concession from the CPR and by attaining control of the grain trade, merchant
   wholesalers and traders made possible the emergence of Winnipeg as the primate city of
   the eastern prairies. Theory does not require the emergence of a single-city pattern in the
   Red River Valley, however. In fact, it can be argued that had the railway been completed
   earlier and a more energetic immigration policy been followed, eastern whosesalers
   would have participated more vigorously in the expanded trade of the seventies and a
   diffuse urban pattern may well have emerged." Donald Kerr, "Wholesale Trade on the
   Canadian Plains in the Nineteenth Century: Winnipeg and Its Competition," in Howard
   Palmer, ed., *The Settlement of the West* (Calgary, 1977), pp. 151-52.

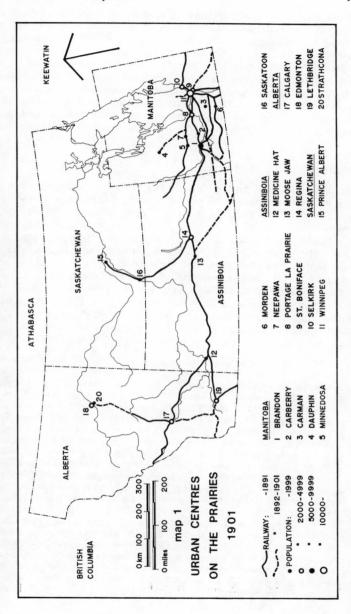

KEEWATIN

MANITOBA

SASKATCHEWAN

ASSINIBOIA

ATHABASCA

ALBERTA

BRITISH
COLUMBIA

map 1

URBAN CENTRES
ON THE PRAIRIES
1901

0 km 100   200   300
0 miles    100     200

RAILWAY:    -1891
            1892-1901
            -1999
POPULATION:  • 2000-4999
             ● 5000-9999
             ○ 10000-

MANITOBA
1 BRANDON
2 CARBERRY
3 CARMAN
4 DAUPHIN
5 MINNEDOSA

6 MORDEN
7 NEEPAWA
8 PORTAGE LA PRAIRIE
9 ST. BONIFACE
10 SELKIRK
11 WINNIPEG

ASSINIBOIA
12 MEDICINE HAT
13 MOOSE JAW
14 REGINA

SASKATCHEWAN
15 PRINCE ALBERT

16 SASKATOON

ALBERTA
17 CALGARY
18 EDMONTON
19 LETHBRIDGE
20 STRATHCONA

development has already received some attention from those interested in the city-building process in Canada. In particular, such activities as railway promotion, immigration encouragement, industry attraction, governmental reform, municipal ownership, and efforts to attain status as both provincial capital and home of the provincial university, have been studied in some depth and will only be touched on here.[6] But other topics – such as city incorporations, massive boundary extensions, huge public works programs, deficit financing, and land value taxation politicies – have received little or no attention. It is these policies, and their impact on the development of prairie cities, that will be emphasized.


MUNICIPAL BOOSTERISM

The thesis that local people played a key role in shaping the pattern of prairie urban development hinges on the ability of certain groups within western towns and cities to make and implement criticial decisions. Several case studies of prairie cities have demonstrated that these communities were controlled, in virtually all respects, by small, close-knit elites.[7] The decision-making process in prairie communities was relatively simple. From formulation to implementation, a small group controlled the process. Although several different organizations or institutions became involved along the way, membership in each was sufficiently overlapping to ensure continuity from start to finish. Thus, even though a desire for a particular policy may have originated during lunch at an exclusive club or during a game of golf, it quickly moved up the hierarchy to discussion in the press, at the Board of Trade, and finally, to

6. See, for example, Voisey, "Urbanization of the Canadian Prairies"; Alan F.J. Artibise, *Winnipeg: A Social History of Urban Growth, 1874-1914* (Montreal, 1975); E.G. Drake, *Regina: The Queen City* (Toronto, 1955); A.W. Rasporich and H.C. Klassen, eds., *Frontier Calgary: Town, City, and Region, 1875-1914* (Calgary, 1975); M. Foran, *Calgary: An Illustrated History* (Toronto, 1978); E.H. Dale, "The Role of Successive Town and City Councils in the Evolution of Edmonton, Alberta, 1892-1966," Ph.D. Thesis (University of Alberta, 1969); Jean E. Murray, "The Provincial Capital Controversy in Saskatchewan," *Saskatchewan History*, V (Autumn 1952), pp. 81-105; Jean E. Murray, "The Contest for the University of Saskatchewan," *Saskatchewan History*, XII (Winter 1959), pp. 1-22; Bruce Kilpatrick, "A Lesson in Boosterism: The Contest for the Alberta Provincial Capital, 1904-1906," *Urban History Review*, VIII (February 1980), pp. 47-109; John Gilpin, "Failed Metropolis: The City of Strathcona, 1891-1912," in Artibise, *Town and City*; and James D. Anderson, "The Municipal Reform Movement in Western Canada, 1880-1920," in Artibise and Stelter, *The Usable Urban Past*, pp. 73-111.

7. See, for example, Artibise, *Winnipeg: A Social History*; J.E. Rea, "Political Parties and Civic Power: Winnipeg," in Artibise and Stelter, *The Usable Urban Past*, pp. 155-66; Foran, *Calgary*; Rasporich and Klassen, *Calgary*; Drake, *Regina*; Lewis H. Thomas, "Saskatoon, 1883-1920: The Formative Years," in Artibise, *Town and City*; Keith Foster, "The Role of the Commercial Elite in the Development of Moose Jaw" (unpublished paper, Saskatchewan Archives, Regina, 1976); Dale, "Edmonton"; and Anderson, "The Municipal Reform Movement."

implementation by city council. Virtually the same individuals were involved at all stages. From club to church, counting house to city council, the urban elites of prairie cities formed an interlocking directorate.[8]

The decision-making process, in addition to being dominated by a small commercial elite, was also a closed process. Citizen participation is a post-1950 phenomenon; previously the commercial elites neither bothered about nor were bothered by significant opposition. A variety of factors, including a restricted franchise, plural voting, effective propaganda, and a centralized form of government assured the elites that their conceptions of desirable public policy would prevail.

The influence of the elites also stemmed from a wide range of shared characteristics. The vast majority of the elite were Anglo-Saxon Protestants of relatively humble origin who had come West from the small towns and cities of the Maritimes and Ontario. Before migrating, many had gained considerable experience in promoting growth in eastern urban centres. But the possibilities in the east were too limited for some elements, particularly, according to a British visitor, the "restless, the men who were strong and willing to dare, who saw riches around them, but beyond them, and who went West to strive, to battle, to conquer, to gain riches for themselves."[9] Within their chosen communities, the commercial elites were the men who possessed the proverbial stake in the community. Most were successful entrepreneurs who built up personal fortunes in real estate or commerce or were in the process of doing so. Most also belonged to an astonishing array of business and social organizations.

The similar backgrounds and associations of prairie urban elites meant that they were in frequent contact with each other: on the street, at meetings, in the club, at social events, and in church. Comprising a small portion of a city's residents, the elite moved in fairly tight circles. Not surprisingly, this propinquity helped to generate a general consensus about urban growth and to expedite the implementation of policies designed to encourage growth. As well, consistent plagiarism gave a unity to much rhetoric and literature about western urban development and provided a classic framework in which a city's past and future could

8. These impressions are derived from two in-depth studies of elites in prairie cities. See Artibise, *Winnipeg: A Social History*, Chapters 2 and 3; and Paul Voisey, "In Search of Wealth and Status: An Economic and Social Study of Entrepreneurs in Early Calgary," in Rasporich and Klassen, *Frontier Calgary*, pp. 221-41. Other studies also tend to confirm this view, if only in a general way. See, for example, Drake, *Regina*; J.G. MacGregor, *Edmonton: A History* (Edmonton, 1967); and D. Kerr, "Saskatoon, 1905-1913: Ideology of the Boomtime," *Saskatchewan History*, XXXII (Winter 1979), pp. 16-28. It is clear, however, that more group biographies of elites in prairie urban communities are needed.

9. J.F. Fraser, *Canada As It Is* (London, 1911), quoted in Lewis H. Thomas, "British Visitors' Perceptions of the West, 1885-1914," in A.W. Rasporich and H.C. Klassen, editors, *Prairie Perspectives 2* (Toronto, 1973), p. 192.

be contained. What the elites believed about their locale – and about successful urban centres elsewhere – shaped their notions of how they ought to live and of the values they hoped to foster in their community.

The elites shared what might be called the booster spirit, and from this mental set different ways of dealing with particular possibilities and problems were drawn. Municipal boosterism on the prairies, like its counterpart elsewhere, was both something more than a compendium of hyperbole and mindless rhetoric, and something less than a precise ideology.[10] The booster mentality was made up of a web of beliefs and attitudes, but a few stand out above the others. The most important parts of the mental baggage of the boosters were a belief in the desirability of growth and in the importance of material success.[11] To boosters, the challenge presented by the undeveloped prairies was to build there a prosperous, populous, and dynamic region as quickly as possible. They accepted the challenge eagerly, seeing themselves as energetic agents of improvement. They set no limits on their sights and were intensely optimistic, expansionary, and aggressive. Their interests lay in what has been called the "architectural aspects of society" – the broad lines of its material fabric.[12] Boosters were constantly measuring their city's progress in quantitative terms: numbers of rail lines, miles of streets, dollars of assessment, size of population, value of manufacturing output or wholesale trade, and so on.

Since progress was measured in material terms, prairie boosters directed their efforts toward encouraging rapid and sustained growth in their cities at the expense of virtually all other considerations. This meant, for example, that while all prairie boosters wanted the region to grow rapidly, they were even more concerned with the growth of their own communities. Despite the early prominence of Winnipeg, the boosters of other prairie communities optimistically envisioned their centre becoming the pre-eminent metropolis of the area. But western urban rivalry was more than a matter of community pride, although this was usually involved as well. It was a matter of self-preservation, with the losers, perhaps, dwindling to insignificance or even extinction. There was, as a result, virtually no cooperation among the five cities in their dealings with the railways, eastern industry, the federal government, or

10. Boosterism was by no means a philosphy unique to the Canadian prairie west. For an excellent discussion of boosterism in the United States, see Blaine A. Brownell, *The Urban Ethos in the South* (Baton Rouge, 1975).

11. An examination of any of the promotional literature or annual reports of the various Boards of Trade during this period will confirm this fact. A specific, published example can be found in Document III, "Winnipeg Board of Trade Report, 1905," in Alan F.J. Artibise, ed. *Gateway City: Documents on the City of Winnipeg* (Winnipeg, 1979), pp. 87-102.

12. A.R.M. Lower, *Canadians in the Making: A Social History of Canada* (Toronto, 1958), p. 364.

almost any other "outside" force.[13] Each centre zealously competed with the others for economic advantage and prestige. Commenting on this phenomenon of municipal mercantilism, one observer noted in 1915 that:

> [Western cities had not] quite reached that secure position in which a community can afford to say unpleasant things about itself. . . . If Edmonton discovers that it has unemployment, the cry is at once "Come to Calgary!" "Invest in Calgary industries!" "Buy Calgary lots!" If Calgary finds it has slum conditions, Edmonton is ready to profit by the opportunity; Regina pounces with joy on any evil spoken of Saskatoon; and the drift of capital and labour may really be affected. Therefore in each city all citizens possessing property are in a conspiracy of silence, or rather unite in so hearty a chorus on the glories of their city and district that no other sound can be heard in the neighbourhood.[14]

But while there was a noted lack of cooperation, and sometimes even goodwill among the five prairie cities,[15] there was a high degree of coordination within each city among the boosters. To prairie townsmen, not to be a booster – not to be part of the team – showed both a lack of community spirit and a lack of business sense. Boosting was essential to progress and prosperity; good citizenship and boosting were synonymous. Successful boosting demanded both collaboration and trust in the pursuit of broad, common goals and priorities on which everyone in the community could agree. An active and positive attitude toward the

---

13. One exception to this generalization was the issue of "bonusing." After many years of severe competition among prairie cities for industry, competition that usually resulted in the granting of cash bonuses or other similar benefits to prospective business firms in exchange for establishment in a city, the Western Canada Civic and Industrial League was organized at Regina in November 1912. One of the first resolutions passed stated:

    That the practice of granting money or land or bonuses of any nature to, or the guarantee of bonds of, corporations, firms, or individuals, in consideration of their establishing and operating factories, businesses, or industries, in Western Canada is not in the best interests of Western Canada and should be discouraged. (*Canadian Annual Review*, 1912, p. 626.)

    Similar resolutions were passed by the Union of Manitoba Municipalities and the Union of Alberta Municipalities. *Ibid.*, pp. 627, 624. While this move ended overt competition among prairie cities, it did not end all forms of "bonusing." In ensuing years cities used such incentives as low power rates from publicly-owned utilities as a means of attracting industry. See, for example, *Canadian Annual Review*, 1925-26, p. 400.

14. E.B. Mitchell, *In Western Canada Before the War: A Study of Communities* (London, 1915), p. 111. For another example of the same attitude, see Thomas, "British Visitors' Perceptions of the West," p. 191.

15. Each city, at one time or another, delighted in stories which indicates that their "opponents" were experiencing difficulties. In January 1913, for example, the Regina *Morning Leader* carried a story with a headline that read "GIRLS CANNOT LIVE MORALLY IN CALGARY." The story that followed stated, in no uncertain terms, that no city "the size of Calgary can compare with it in regard to girls living partly vicious lives."

community was expected and demanded of the "good citizen." As one
pioneer Saskatoon merchant put it:

> All were imbued with an optimism that Saskatoon was destined to become an
> important centre. Newcomers in business were welcomed and encouraged;
> everyone pulled together to develop and boom the town; there were no petty
> jealousies of one another. In fact there had already come to life [by 1904] what
> has rightly been called "the Saskatoon spirit."[16]

Boosters saw themselves as community builders in mushrooming cities with unbounded hopes; an environment where personal and public
growth, personal and public prosperity intermingled. Booster literature
put forth exuberant claims as to the bright destiny that awaited a particular spot among all others and asserted that great rewards awaited the
men who kept the faith with the prophecy. In such a situation there was
no room for skeptics or, as they were often called, "knockers." Indeed,
so powerful was the booster psychology of the civic and business leaders
that few in any of the cities rose to challenge their leadership. Throughout the period, the boosterism of the many overshadowed the caution or
opposition of the few.[17] But the lack of opposition to the booster ethos
also stemmed from the fact that the boosters effectively used the idea of
community solidarity as a purification ritual, a means of discrediting
opponents. An editorial in the *Saskatoon Phoenix* of May 15, 1903,
provides a typical example of this kind of pressure:

> Whenever you hear a person run down an enterprise in his town, just put him
> down for a fool, a fossil, or a malicious liar. Make no exception to this rule
> either, because there is no exception. Paste this in your hat until you have it
> indelibly impressed on your memory, and whenever you hear a local citizen
> trying to injure a home institution by talking against it . . . cut him off cold. . . .
> A cool headed and successful businessman will push his town to the front at
> every opportunity, but a fool is always working the other way.

Advocates and supporters of boosterism were intent on creating a feeling of community spirit on the basis of voluntarism, without any basic
revision of the system of economic inequity and social injustice that
existed in the cities. Booster literature and speeches were all aimed at
creating a positive image of the city in the minds of both residents and
non-residents, regardless of the complex truth. This was done by promoting the notion that all classes and groups could be united on the basis
of faith in the city, belief in its destiny, and commitment to its growth.
     The community spirit promoted by the elites of prairie cities was one
that was fashioned more out of will than out of reality, more out of

16. *Narratives of Saskatoon, 1882-1912* (Saskatoon: University Bookstore, 1927), p. 27.
17. See, for example, Voisey, "In Search of Wealth and Status"; Artibise, *Winnipeg: A Social History*; and J.H. Archer, "The History of Saskatoon," M.A. Thesis (University of Saskatchewan, 1948), chapter V and *passim*.

wishful thinking than out by experience. The distorted perceptions caused by boosterism can be seen in the universally held belief in each community's centrality, despite obvious incongruities in such a proposition.[18] The elites talked about understanding and common ties, about creating a store of wealth to benefit all citizens, but in fact were generally unconcerned about the vast majority of residents as long as they did not disrupt existing relationships. The myth of a shared sense of community was thus a valuable falsehood for the boosters since it enabled them to implement their programs with a minimum of opposition. The booster spirit was calculated not only to meet immediate needs and insure urban expansion, but also to justify social conformity and maintain the existing social and economic system since only a "united" community could prosper. Boosterism as practised in prairie cities even had a tinge of blackmail associated with it. So effective were the booster campaigns about opportunity for everyone, about wealth and greatness available for the grasping, that few residents were not involved. Nearly everyone was infected with the boom psychology, and once involved – whether in land speculation or business promotion – there was an added financial incentive to support the booster campaigns.[19]

Strong beliefs in the virtues of rapid growth, material progress, and social conformity were combined with an equally strong belief in the special role of local govenrment. The boosters shared the view that the active encouragement of growth and business enterprise was the prime concern of municipal corporation. Local government was purely functional; it was a tool serving personal and community prosperity. Unlike municipal corporations in older, established cities, the governments of the upstart prairie cities had little odour of sanctity; time-honoured precedents that might have got in the way of the uses to which government was put were non-existent. Prairie city boosters were not confronted with the problem of evading obsolete regulations or dislodging established oligarchies as were their counterparts in the east.[20] In the

---

18. An amazing example of this distortion occurred in the provincial capital dispute of 1904-1906 in Alberta. Both Edmonton and Calgary used the same map and similar population and economic figures to assert their centrality in Alberta. Both were wrong. See Kilpatrick, "Lesson in Boosterism."

19. See, for example, J.P. Dickin McGinnis, "Birth to Boom to Bust: Building in Calgary, 1875-1914," in *Frontier Calgary*, pp. 6-19; John G. Niddrie, "The Edmonton Boom of 1911-1912," *Alberta Historical Review*, XIII (Spring 1965), pp. 1-6; and Kerr, "Ideology of the Boomtime."

20. See, for example, Archer, "History of Saskatoon"; Artibise, *Winnipeg: A Social History*; Drake, *Regina*, p. 156; and Dale, "The Evolution of Edmonton." One example of the situation in the east is G. Bourassa, "The Political Elite of Montreal: From Aristocracy to Democracy," in L.D. Feldman and M.D. Goldrick, eds., *Politics and Government of Urban Canada* (Toronto, 1969), pp. 124-34. See also Anderson, "The Municipal Government Reform Movement in Western Canada, 1880-1920," in Artibise and Stelter, *The Usable Urban Past*.

west, "the dignity of government [was] much diminished, its utility greatly increased."[21] It was merely a device to be used for the benefit of the people who had managed to gain political power or influence. And in all prairie cities it was the businessmen who early gained control of government and who continued throughout the period to maintain that control.

Another element in the booster mentality was the loose degree of attachment to the gospel of Social Darwinism. It is true, of course, that boosters often sang the praises of rugged individualism and competition. But it is also true that "the reality was often at variance with the rhetoric."[22] As one study of the attitudes of Canadian businessmen during this period observes:

> Individualism, though clearly insufficient for business success, was deemed perfectly sufficient for success as a worker or farmer. Business was justified in striving mightily to overcome the laws of supply and demand and the forces of the marketplace; others should bend to them. Co-operation was the keynote of the new business philosophy; hard work and saving were to be others' code. The leaders of trade combines were business statesmen; the leaders of labour unions and farmers' organizations were parasitical demagogues. The price-cutter in business was a pirate, reckless and unethical; the scab worker was a free and honest man.[23]

The attitude of boosters toward organized labour and the poor and disadvantaged was one of scorn, and the cities the boosters dominated spent only a small fraction of their budgets on such community services as sanitation, health departments, or welfare; far less than was spent on promoting growth. In these areas, "survival of the fittest" was the norm.

The individualism of the boosters was selective, influencing only some of their activities. It often amounted to a philosophy of individualism for the poor and cooperation for the rich.[24] Boosters not only used the communities' funds for their own ends but went further to become in many instances firm believers in municipal ownership. It is also noteworthy that one of the first organizations created in all prairie cities was a board of trade, designed through cooperation to facilitate the growth

21. W.L. Morton, "A Century of Plain and Parkland," *Alberta Historical Review*, XVII (Spring 1969), p. 10.
22. R.C. Brown and R. Cook, *Canada, 1896-1921: A Nation Transformed* (Toronto, 1974), p. 145.
23. M. Bliss, *A Living Profit: Studies in the Social History of Canadian Business, 1883-1911* (Toronto, 1974), p. 141.
24. This attitude has been discussed in John Weaver, " 'Tommorow's Metropolis' Revisited: A Critical Assessment of Urban Reform in Canada, 1890-1920," in Stelter and Artibise, *The Canadian City: Essays in Urban History*, pp. 393-418. See also the comments of Max Foran, "Urban Calgary, 1884-1895," *Histoire sociale/Social History*, V (April 1972), p. 67.

of the cities.[25]

One final, general point can be made concerning the boosters' beliefs and attitudes. While the philosophy or outlook of the prairie booster was surcharged with an optimism which sometimes resulted in self-deception, it was rarely blind to the essentials of urban growth. There was a strong strain of realism in it. Boosters realized that the development of the wheat economy would result in the establishment of a large number of towns and villages linked up with a limited number of commercial cities situated at the junction points of several railway lines. They also knew that the uniform topography of the prairies placed a large number of small centres on an almost equal footing as aspirants for the coveted prize of big city status. Only a few possessed natural advantages as townsites and these were either shared with others or were not sufficiently prominent to discourage rivals. Every booster knew, almost instinctively, that the telling distinction would be the initiative and skill of his community's leaders.

## A FRAMEWORK FOR GROWTH

One characteristic that was shared by boosters in the five cities that were to dominate the prairie west by 1913 was an eagerness to attain formal city status for their communities far in advance of that title coinciding with reality. Winnipeg, unlike other prairie communities, never did go through the chrysalis stage of either village or town. When incorporated by the provincial legislature in November 1873, the approximately 1,600 residents of the small hamlet sought and obtained at the outset the clear and unambiguous title of city; a terminology, however divorced from reality, they felt would be more respected back east.[26] Regina was not quite so brash. Although it skipped the village state, it was incorporated as a town in December 1883 with the distinction of being the first

25. Winnipeg's Board of Trade was established in 1879, five years after incorporation as a city. Regina was incorporated as a town in 1883 and its Board of Trade organized in 1886. Calgary became a town in 1884 and set up a Board of Trade in 1891. Saskatoon was still a village when its Board of Trade began meeting in 1903. And Edmonton had a Board of Trade as early as 1889, three years before its incorporation as a town.

26. Artibise, *Winnipeg: A Social History*, Chapter 1. Municipal government in Manitoba was administered under four types of organization: the city, the town, the village, and the rural municipality. With the exception of Winnipeg and St. Boniface, which operated under special charters, all the municipalities were subject to the Municipal Act and the Assessment Act. From the date of incorporation until 1886 the government of Winnipeg was carried on under the powers of a special charter of incorporation granted by the provincial legislature. In 1886 this special charter was repealed and from that date until 1902 the city's affairs were administered under the provinces of the municipal and assessment acts. In 1902 the city received a special charter, which arrangement lasted until the present day.

organized centre of the territories.[27] The new town's population was around 900. Furthermore, it was not until 1903 that Regina sought and achieved incorporation as a city.[28] But this uncharacteristic tardiness in seeking city status is balanced by two significant facts. First, although legally only a town, the residents of Regina adopted the title of city in practice soon after 1883.[29] Secondly when it did apply for city status in 1903, Regina had a population of under 3,000; hardly the numbers usually associated with a city.

Calgary followed Regina's example in immediately incorporating as a town in 1884. It had, at the time, a meagre population of 506.[30] Nine years later, in 1893, with a population of less than 4,000, Calgary received its city charter, the first urban centre in the territories to acquire that status.[31] In Edmonton, members of the Board of Trade, organized in 1889, pushed for incorporation. In February 1892, with a population of only 700, Edmonton became a town. City status came in 1904 when the population was 8,350.[32] Edmonton thus had the distinction of having the largest population of any of the five prairie centres when it formally became a city, though it can be noted that as late as 1901 its population was only 2,626. Saskatoon was the last of the five communities to formally incorporate as a city; it did not attain that goal until 1906 when its population was 3,000. Saskatoon was also the only one of the five to follow the pattern of village, town, and finally city status. It had first been organized as a village in November 1901 with a population of 113. But this lowly status did not long remain in effect since in 1903 Saskatoon became a town.[33]

The manner in which all five urban centres sought and finally achieved city status was very much in keeping with the boom psychol-

27. A.N. Reid, "Informal Town Government in Regina, 1882-2," *Saskatchewan History*, VI (Autumn 1953), pp. 81-88. For general discussions of urban municipal development in the territories and later Saskatchewan and Alberta, see A.N. Reid, "Urban Municipalities in the North West Territories: Their Development and Machinery of Government," *Saskatchewan History*, IX (Spring 1956), pp. 41-62; G.F. Dawson, *The Municipal System of Saskatchewan* (Regina, 1952); E. Hanson, *Local Government in Alberta* (Toronto, 1956); and K.G. Crawford, *Canadian Municipal Government* (Toronto, 1954), pp. 43-46. In the territorial phase, municipal affairs came under the Municipal Ordinance of 1883. After 1905, Saskatchewan passed the City Act, the Town Act, and the Village Act in 1908. There were no special charters for urban municipalities in the province. In Alberta, towns, villages, local improvement districts and rural municipalities operated under general provincial laws after 1905. Cities, including Calgary and Edmonton, had special charters.

28. Drake, *Regina*, pp. 24, 116.

29. *Ibid.*, p. 102.

30. L.H. Bussard, "Early History of Calgary," M.A. Thesis (University of Alberta, 1935), pp. 64-88; and Foran, "Urban Calgary, 1884-1895."

31. Bussard, "Early History of Calgary," p. 146.

32. MacGregor, *Edmonton*, pp. 107, 132.

33. M.A. East, *Saskatoon Story, 1882-1952* (Saskatoon, 1952), pp. 43, 50.

ogy of the pre-1913 era.[34] The arguments used in each community were strikingly similar. Local pride, respect in eastern Canada, and the distinction of achieving the coveted "city" title far in advance of rivals were common refrains.[35] Municipal organization was essentially a technique of boosting – "it advertized the community and gave it a dignity which its physical appearance could not impart; . . . it was an act of faith, an expression of confidence in the future."[36] The fact that formal municipal organization was also a means to self-government was only rarely mentioned.[37] But these arguments, or the lack of them, were neither the most effective nor the most significant. The most important were that town and city status provided a broader base for borrowing funds for public works and other expenditures, and that the titles presented better opportunities for advertising and promotion.

Saskatoon's incorporation experiences are especially enlightening in this regard, although they were not much different from other prairie centres.[38] In 1902, the residents of the village of Saskatoon were aware that a group of British immigrants, the Barr Colonists, were scheduled to stop over en route to their new homes.[39] Several local businessmen immediately seized upon this anticipated event as a good reason to seek town status. They argued that to encourage immigrants to remain in the district, Saskatoon would have to carry out an extensive public works program and to do so would have to borrow large sums of money. Since villages could borrow an amount equal to 5 per cent of the total assessment while town status permitted borrowing double this amount, a campaign for incorporation as a town was hastily begun. To reach the 450 needed for town status, the organizers of the incorporation movement had to include in their census not only permanent residents but also guests in the hotels crowded with the spring immigration rush and everyone else in sight, but it was argued that the visitors would probably settle in Saskatoon anyway. In July 1903, Saskatoon was incorporated as a town.[40] Two years later similar arguments were used to achieve city

34. The five cities discussed above had many rivals. By 1913 Manitoba had four cities: Winnipeg, St. Boniface, Brandon and Portage la Prairie. Saskatchewan had seven: Regina, Saskatoon, Moose Jaw, North Battleford, Weyburn, Prince Albert and Swift Current. Alberta had six: Calgary, Edmonton, Medicine Hat, Red Deer, Lethbridge and Wetaskiwin. Strathcona received a city charter in 1907 but amalgamated with Edmonton in 1912.

35. See, for example, Artibise, *Winnipeg: A Social History*, pp. 15-19; and *Narratives of Saskatoon*, pp. 88-89.

36. Thomas, "Saskatoon."

37. Archer, "History of Saskatoon," p. 79.

38. See, for example, Drake, *Regina*, p. 116; and E. Hanson, "A Financial History of Alberta," Ph.D. Thesis (Clark University, 1952), pp. 1-158.

39. The best account of the Barr Colony is E.H. Oliver, "The Coming of the Barr Colonists," Canadian Historical Association, *Annual Report*, 1926, pp. 65-87.

40. *Narratives of Saskatoon*, pp. 88-89; East, *Saskatoon Story*, p. 43; Archer, "History of Saskatoon," pp. 79-81; and W.P. Delainey and W.A.S. Sarjeant, *Saskatoon: The Growth of a City, 1882-1960* (Saskatoon, 1974), pp. 9-10.

status, which allowed borrowing to 20 per cent of assessment. This time the increased power was needed to help in the effort to obtain both the provincial capital and the main line of the Grand Trunk Pacific Railway. The success of the campaign for incorporation as a city was never in doubt.[41]

As important as the achievement of city status was to the boosters, it alone did not satisfy their grand visions. In the years prior to the recession of 1913, all five cities underwent vast spatial expansions, bringing huge areas of land within their municipal boundaries. Winnipeg, which started off with an area of 1,920 acres in 1874, had grown to include 14,861 acres by 1913. This was accomplished by two major annexations of territory, in 1882 and 1906, and by several smaller additions in 1875, 1902, 1905, 1907 and 1913. Significantly, the city did not again expand its boundaries until 1963 except to add very small areas of land.[42] In Regina's case, the expansion was completed in one fell swoop. With an area of 1,942 acres in 1883, Regina increased its size fourfold in 1911 by adding 6,458 acres. It, too, did not find it necessary to make any further major additions until the 1950s.[43]

The town of Calgary began in 1884 with 1,600 acres and through annexations in 1901, 1903, 1906, 1907, 1910, and 1911 it reached a size of almost 26,000 acres by 1912. No further additions were made until 1951.[44] Saskatoon's expansion was more modest. When incorporated as a city in 1906 it included 2,567 acres and, prior to 1950s, made only one major addition – in 1911 it increased its territory by 5,913 acres.[45] The record for expansion clearly belongs to Edmonton. In 1892 it was a town compromising only 2,162 acres. But additions quickly followed in 1899, 1904, 1908, 1911, 1912 and 1913. Together with two further annexations in 1914 and 1917, the city grew to no less than 26,290 acres. Again, no further major additions were undertaken until the 1950s.[46]

41. The members of Saskatoon's Board of Trade led the campaign for city status. See *Narratives of Saskatoon*, pp. 70-71; Murray, "The Provincial Capital Controversy in Saskatchewan"; East, *Saskatoon Story*, p. 50; and Archer, "History of Saskatoon," pp. 95-99.

42. See Artibise, *Winnipeg: A Social History*, Chapter 8; H.A. Hossé, "The Areal Growth and Functional Development of Winnipeg from 1870 to 1913," M.A. Thesis (University of Manitoba, 1956); and "Map of the City of Winnipeg Showing Original Corporate Limits and Various Extensions Thereto" (Winnipeg, 1930; revised, 1948, 1949, and 1963).

43. See "City Growth Maps" in City of Regina, *Housing Survey Report* (Regina, 1957), p. 10.

44. See "Map of Boundaries of the City of Calgary and Metropolitan Calgary as of January 1, 1975 and Annexations to the City of Calgary, 1893-1975" (Calgary: City Planning Department, 1975). In 1923, 550 acres were removed from the city of Calgary.

45. See "Map of City of Saskatoon, Expansion of City Limits" (Saskatoon: City Planning Department, 1975).

46. See "Map of City of Edmonton, History of Annexations" (Edmonton: City Planning Department, 1974).

The "land hunger" of the booster-oriented city councils of the five prairie cities gave them a special status among Canada's urban centres. In 1921 four prairie cities – Edmonton, Calgary, Saskatoon, and Regina – had the lowest population density ratios of Canada's twenty largest urban areas (Table 3). Edmonton, for example, had only 2.2 persons per acre in 1921 compared to 19.2 for Montreal, 31.5 for Toronto, and 13.6 for Halifax. Regina's population of 34,432 was spread out over 8,429 acres; a ratio of 4.1 per acre. In contrast, Windsor, with a population of 38,591, had a land area of 2,726 acres and a population density of over 14 per acre.

Table 3: Area and Density of Population of Canada's
Twenty Largest Cities, 1921

| City | Rank By size | By density | Population | Land Area in sq. miles | Population per sq. mile |
|---|---|---|---|---|---|
| Montreal | 1 | 3 | 618,506 | 50.24 | 12,311 |
| Toronto | 2 | 1 | 521,893 | 25.89 | 20,158 |
| Winnipeg | 3 | 9 | 179,087 | 23.22 | 7,712 |
| Vancouver | 4 | 10 | 117,217 | 16.89 | 6,703 |
| Hamilton | 5 | 6 | 114,151 | 12.11 | 9,426 |
| Ottawa | 6 | 2 | 107,843 | 6.44 | 16,745 |
| Quebec | 7 | 5 | 95,193 | 8.84 | 10,768 |
| Calgary | 8 | 19 | 63,305 | 40.50 | 1,563 |
| London | 9 | 11 | 60,959 | 10.03 | 6,077 |
| Edmonton | 10 | 20 | 58,821 | 42.50 | 1,384 |
| Halifax | 11 | 8 | 58,372 | 6.72 | 8,686 |
| Saint John | 12 | 16 | 47,166 | 14.31 | 3,296 |
| Victoria | 13 | 13 | 38,727 | 7.25 | 5,341 |
| Windsor | 14 | 7 | 38,591 | 4.26 | 9,058 |
| Regina | 15 | 17 | 34,432 | 13.17 | 2,614 |
| Brantford | 16 | 12 | 29,440 | 4.93 | 5,972 |
| Saskatoon | 17 | 18 | 25,739 | 12.50 | 2,059 |
| Verdun | 18 | 4 | 25,001 | 2.22 | 11,262 |
| Hull | 19 | 15 | 24,117 | 6.25 | 3,859 |
| Sherbrooke | 20 | 14 | 23,515 | 4.85 | 4,848 |

SOURCE: *Census of Canada, 1921*, Vol. I.

The boundary extensions of prairie cities were products of the boom psychology of the period preceding 1913 and provide perhaps the best example of the inflated expectations of the boosters. Large areas were added to the cities far in advance of population pressure; not only did prairie cities have some of the lowest population densities among Canadian cities, but in every case the population was far from evenly distributed throughout the area. Indeed, vast stretches of the city's lands were completely uninhabited or unused.[47]

Since population pressure can hardly account for any but a very few of the many annexations of territory, the reasons for the physical growth must be sought elsewhere. They are not hard to find. Among the arguments used by expansionists was the fact that extended boundaries gave the cities larger total assessment values, thus increasing both tax revenues and the borrowing power of the municipal corporation.[48] Large assessment figures were also useful when cities went seeking customers for city bonds and debentures.[49] Another reason for the expansions was the argument that much of the land surrounding the cities was held by foreign (i.e., non-resident) speculators and that by bringing the land within the city's taxing authority, a source of revenue that would be of benefit to local citizens could be tapped.[50] The most important reason, however, was the pressure exerted on city councils by local real estate interests. In every city vast chunks of agricultural land were subdivided and offered for sale.[51] In Saskatoon, for example, there were only 28 real estate firms in 1908; by 1912 there were 267 operating firms with land available for sale in no less than 107 subdivisions. Despite the fact that over 5,900 acres had just been added to the city in 1911, there were, in 1913, forty square miles of agricultural land outside the city

47. See, for example, maps of spatial development in D. Ravis, *Advanced Land Acquisition by Local Government: The Saskatoon Experience* (Ottawa, 1973); Dale, "The Evolution of Edmonton"; and *The Metropolitan Development Plan* (Winnipeg: The Metropolitan Corporation of Greater Winnipeg, 1968). In 1914, Calgary had 26,763 vacant lots served with watermains and sewers. See W. Van Nus, "The Fate of City Beautiful Thought in Canada, 1893-1930," in Stelter and Artibise, *The Canadian City: Essays in Urban History*, pp. 162-185.

48. In Regina and Sasaktoon, the borrowing power of the cities was limited by provincial statute (The City Act of 1908) to 20 per cent of the total assessment. The other three cities were not thus limited in their powers by provincial statute, but their city charters did contain restrictions based on percentages of total assessment. See R.M. Haig, *The Exemption of Improvements From Taxation in Canada and the United States: A Report Prepared for the Committee on Taxation of the City of New York* (New York, 1915), p. 112 and *passim*. In terms of tax revenues, the boundary extension undertaken by Calgary in 1910 lowered the tax burden to the average taxpayer by about ten percent. See Max Foran, "Land Speculation and Urban Development in Calgary, 1875-1914," *Frontier Calgary*, p. 217.

49. See, for example, comments in the *Regina Morning Leader*, 23 January 1913.

50. Of course not all the land surrounding the cities was held by non-residents and in some cases large land owners protested the inclusion of their land within city boundaries because of the increased taxation. See, for example, Foran, "Land Speculation and Urban Development in Calgary," p. 217.

51. For an excellent description of this situation, see "Sub-Division Conditions and Land Speculation in Canada," *Canadian Annual Review*, 1913, pp. 35-42.

limits subdivided for intended urban expansion. Had the real estate bubble not burst in 1913, Saskatoon probably would have expanded its boundaries still further.[52] In Calgary, expansion was urged on by most of the city's two thousand real estate agents who, in 1912, operated out of 443 firms.[53] And in Winnipeg in 1913 there was land for sale in no less than twenty subdivisions outside the city limits. The *Canadian Annual Review* of 1913 summed up the situation:

> Through the Western Provinces . . . everybody had been speculating in real estate, every village and town had been anticipating the days when it would be a city or important centre. Nearly everyone for a time had made money out of selling properties to others in their locality, to the visitor or investor from abroad, to the speculator in another city, to syndicates which further exploited the property or combination of properties as Subdivisions, to the American sharper who bought land for a trifle miles away from the centre of a town and flooded Eastern or English newspapers with flashy advertisements of "a choice residential centre" close to such and such a progressive town, or rising centre, or seat of future railway and industrial development. . . . The whole thing was ephemeral, a natural product of exotic progress, an outgrowth of Western enthusiasm.[54]

The advantages, for owners of real estate, that came with the inclusion of their property within city boundaries were twofold. First, it was a decided benefit to be able to legitimately advertise one's property as being in a city, rather than on the outskirts, although it must be noted that this obstacle was often overcome by other means.[55] Secondly, inclusion within boundaries meant that urban services – waterworks, sewers, roads, streetcar lines, and so on – were likely to be extended to one's property and this obviously led to increased prospects for both sales and profits. Thus, as with the move to city status, inclusion of land within city boundaries had the effect of raising the value and appeal of real estate holdings. It also provided an ideal framework in which to pursue growth.

52. Ravis, *Advanced Land Acquisition*, p. 13; Delainey and Sarjeant, *Saskatoon*, pp. 22-34; and R. Rees, "The 'Magic City on the Banks of the Saskatchewan': The Saskatoon Real Estate Boom, 1910-1913," *Saskatchewan History*, XXVII (Spring 1974), pp. 51-59.
53. Foran, "Land Speculation and Urban Development in Calgary," p. 213. A similar situation existed in the other cities. See, for example, W.C. Mahon, *Real Estate Highlights, 1912-1972* (Regina, 1972); and MacGregor, *Edmonton*, pp. 163-208.
54. *Canadian Annual Review*, 1913, p. 39.
55. See, for example, Kerr, "Ideology of the Boomtime"; and *Canadian Annual Review*, 1913, p. 38.

THE BUSINESS OF URBAN PROMOTION

The fantastic spatial growth of the prairie cities, coupled with the rapid population growth that occurred in the pre-1913 period, presented a gargantuan problem for local authorities. Municipal services needed to be expanded at an ever-increasing pace. In Alberta's cities, for example, the expenditure and debenture debt increased at a startling pace in the years 1906-1913. In 1906, Alberta's cities spent only $2 million; in 1912 they spent $16.6 million. The total surged to a high of $36.5 million in 1913. The city of Edmonton alone spent 40 per cent of the 1913 total, or one and a half times the entire provincial government expenditure in that year.[56] Calgary spent $1 million on civic building in 1911 and another $3 million on street railways, waterworks, sewers, schools, electric lighting, paving and parks.[57] In the five years from 1909 to 1913 Winnipeg expended $81 million on buildings alone; Calgary, $50 million; Edmonton, $32 million; Regina, $20 million; and Saskatoon, $14 million.[58]

While much of this massive expenditure was made in a legitimate attempt to provide necessary services to a burgeoning population, there was also a good deal of unnecessary expansion during the period; expenditures that added burdens to the already over-taxed facilities and resources of the prairie cities. The irony of the situation was that in providing services to new and unpopulated subdivisions far from the city centres, councils neglected to provide adequate services to already developed areas.[59] In Regina, the municipally owned street railway system was under constant pressure by private real estate firms to build lines into areas "well past the limites of heavy settlement where little and often no housing existed."[60] One result of this policy of creating streetcar suburbs was a loss of over $60,000 on the operation of the system in 1913, in addition to the service charges on the $1.5 million debt of the public street railway company in 1913.[61] Yet the obvious lack of judicious planning apparent in the operation and expansion of the street railway system was contrasted with a situation in the city's foreign ghetto of Germantown, where 60 per cent of the houses lacked such basic services as sewer and water connections.[62]

A similar situation existed in Edmonton. In that city, the pressure

56. Hanson, "Financial History of Alberta," pp. 156-57. See also Niddrie, "Edmonton Boom," pp. 1-6.
57. McGinnis, "Building in Calgary, 1875-1914," p. 17.
58. *Canadian Annual Review*, 1913, p. 38.
59. See, for example, Artibise, *Winnipeg: A Social History*; and Delainey and Sargeant, *Saskatoon*.
60. C.K. Hatcher, *Saskatchewan's Pioneer Streetcars: The Story of the Regina Municipal Railway* (Montreal, 1971), p. 15.
61. Haig, *The Exemption of Improvements*, p. 41.
62. Drake, *Regina*, pp. 155-156.

exerted by private land developers in regard to the extension of city services nearly always had the desired effect. One writer, commenting on Edmonton city council's policies in regard to utility expansion, has stated:

> [Private land developers] bought lands on the periphery of the city but often could not develop or sell them without first having city services extended on them. Nor could they sell for a better price if the services were not first extended to the sites. Successfully they influenced the Council to extend to these areas not only the streetcar system but other utilities as well, chiefly water and sewerage, and light and power. In this way the value of their lands increased and, consequently, the harvest of profits.

Now it has to be admitted that the problem the Council faced was acute and complex. Yet, administrative authorities should not be guilty of favoritism. Moreover, the evidence of mismanagement, as revealed by the Council minutes of 1913, hints at this and suggests that their motives might have been self-interested. The evidence seems also to justify the conclusion that a well-thought out policy in regard to street railway extension was lacking. Before an extension was carried out no thought was given to the expenditure involved, the revenue to be derived that would not warrant the extension, the distance the property was from existing facilities, and the scattered development that would exist. . . . Thus the Council of 1913 may be criticized for permitting indiscriminate street railway extensions and by so doing increasing the burden of taxation. The waste was made manifest after the boom collapsed. The City was left with numerous pieces of vacant land, particularly on its periphery, serviced with utilities.[63]

The tremendous sums of money required to pay for the essential as well as the unessential expenditures of prairie cities during these years of rapid expansion came, of course, from either numicipal taxes or from the sale of city bonds and debentures.[64] While it is true that prairie cities at one time or another raised some revenue from business taxes, licence fees, poll taxes, and even income taxes, the major source of income for the municipal corporations was property taxation.[65] Furthermore, the property taxes levied by the prairie cities, particularly in the years immediately preceding the First World War, were for the most part taxes on the assessed value of land; improvements were either exempted or taxed

---

63. Dale, "The Evolution of Edmonton," pp. 99-100, 102.

64. In 1912 alone, according to an estimate of the Montreal *Financial Times*, there were 15 floatations of bonds in London, England by western Canadian cities totalling $27,855,000. *Canadian Annual Review*, 1913, p. 33.

65. Some revenue was also raised from municipally owned utilities but this was never a major contributor to municipal revenue. Indeed, in many cases, utilities actually lost money. For a good general discussion of taxation in Canada, including material on municipal taxation, see J. Harvey Perry, *Taxes, Tariffs and Subsidies: A History of Canadian Fiscal Development* (2 vols., Toronto, 1955).

at only a percentage of their value. While eastern Canadian cities continued in nearly every case to tax land and improvements at 100 per cent
of their value and to raise significant segments of their revenues from a
variety of other sources, prairie cities moved rapidly in the pre-First
World War era to an almost total dependence on land taxes. The reason
for this situation was the influence of Henry George's single tax philosophy throughout western Canada.

In its pure form the single tax is a system of taxation by which a
community raises its entire revenue from land taxes. The justification
given for the scheme is that the value of land is entirely due to the
presence and expenditure of the people and community should therefore
receive the benefit through taxation. Through placing all taxes on land,
Henry George hoped to nullify the negative effects of land speculation
and ultimately eliminate private land ownership.[66] It is ironic then that
the adoption of the single tax – or, more precisely, a modified version of
it since it was never followed in its pure form – should have been
adopted in western Canadian cities at a time when speculators were
making fortunes from land sales.[67] It is even more ironic that among the
chief proponents of the modified single tax were the real estate speculators themselves.[68] It is clear, however, that the prairie boosters who
supported the modified single tax philosophy did so for reasons that
probably would have caused Henry George to turn in his grave. For
while all the classic arguments in support of land value taxation were
bandied about, the adoption of a modified form of single tax in the five
prairie cities came about for at least three specific reasons; reasons quite
distinct from George's general philosophy.

First, a heavy tax on land values was a means whereby municipal
corporations could raise large sums from non-resident land owners and
speculators. In Edmonton, for example, the move to a single tax in 1904
came about largely as a way to shift the burdens of taxation to the
Hudson's Bay Company which, at the time, owned a tract of land
approximately a mile and a half long by a mile wide in the heart of the
city.[69] And even though resident speculators would also have to pay
similar taxes, the local businessmen involved in speculation "do not
expect to hold their land long, often not even over one tax-paying
period" and they thus "do not object strongly to shifting of the [tax]
burden to land."[70] The second reason for the adoption of a modified
version of the single tax was the belief that it stimulated development by

66. Henry George, *Progress and Poverty* (London, 1911). See also M.T. Owens, *Land Value
    Taxation in Canadian Local Government* (Westmount, 1953).
67. Rees, "The Magic City on the Banks of the Saskatchewan," p. 55.
68. Haig, *The Exemption of Improvements, passim.*
69. *Ibid.*, p. 91. See also Perry, *Taxes, Subsidies and Tariffs*, p. 131.
70. *Ibid.*, p. 275. See also S. Vineberg, "Provincial and Local Taxation in Canada," *Studies
    in History, Economics and Public Law* (Columbia University), LII (1912), p. 88.

encouraging building. An investigator of the western Canadian tax situation observed:

> [Real estate men] are eager to encourage anything which promises to assist in increasing land values and nothing seems to be more effective for this purpose than the rapid construction of buildings. They are convinced that the land tax stimulates building, and, as one of them expressed it, every real estate man "is for all the single tax he can get." They look upon the plan as a bonus to building, and they, being interested primarily in that part of land which is ripe for building, pay only part of that bonus. Part is borne by the other land owners, who had already purchased land when the tax was imposed – purchased it perhaps from these very real estate men. Thus, indirectly, they pay at least part of the bonus themselves.[71]

The third reason for the support the idea of land value taxation received was that it could be implemented in varying degrees without reducing the overall tax base. Since land values were increasing at a fantastic pace (see Table 4), and since prairie cities were continually expanding their boundaries, taxes on improvements and other forms of taxation could be reduced or removed entirely without the tax base diminishing.[72] Land assessment values in Winnipeg, for example, jumped from $11.9 million in 1900 to $53.7 million in 1906, and to $259.4 million by 1913. In Edmonton the increases were even more dramatic. Land assessment value stood at only $1.3 million in 1900; by 1913 it exceeded $188 million. In Calgary, property at the corner of 7th Avenue and 2nd Street West, just on the edge of the main business areas, was valued at $150 in 1895. By 1905 the same lot was worth

---

71. *Ibid.*, pp. 275-76. See also Owens, *Land Value Taxation*; P.R. Creighton, "Taxation in Saskatoon: A Study in Municipal Finance," M.A. Thesis (University of Saskatchewan, 1925), p. 31-34; and City of Edmonton, *Special Report on Assessment and Taxation* (Edmonton, 1921), pp. 1-24.

72. Haig, *The Exemption of Improvements*, pp. 266-67. It must be noted that the rapid increases in assessments were not based on realistic increases in land values. This is not only evident in declining assessments in the post-1913 period, but in comparisons made with other cities. Commenting on the situation in 1914, one observer made the following statement: "As evidence of the inflations of land values so as to obtain sufficient revenue on a low tax rate, Edmonton is a most conspicuous example. In 1914, with a population of 72,500 the assessed land value of that city was $209,000,000. Montreal's total assessed land value in the same year was $537,000,000. Had Montreal's land value been assessed at the same ratio to population as Edmonton's, the assessed value would have been $1,874,000,000, or $1,023,000,000 more than the total assessed value of Montreal's land and buildings. In other words, the assessed land value per capita was in 1914, Montreal $825, and Edmonton, $2,880, or over three to one." J. Hamilton Ferns, "Single Tax or Taxing of Land Values in Practice," *Canadian Municipal Journal*, XVI (April 1920), p. 119. See also A.B. Clark, "Recent Tax Developments in Western Canada," *Proceedings of the 13th Annual Conference of the National Tax Association* (New York, 1920), pp. 48-69.

$2,000; by 1912 the purchase price had jumped to $300,000.[73] In short, the move to an almost exclusive form of land value taxation was very much a product of the western boom.

Table 4: Land Assessment Values in Prairie Cities, 1895-1913 [a]
in millions of dollars

| Year | Winnipeg | Regina | Saskatoon | Calgary | Edmonton |
|------|----------|--------|-----------|---------|----------|
| 1895 | 11.7 | — | — | 1.5 [b] | 1.1 |
| 1900 | 11.9 | — | — | 1.0 [c] | 1.3 |
| 1902 | 12.6 | 1.0 | — | 1.0 [d] | 1.7 |
| 1904 | 25.1 | 2.2 | 0.4 | 2.1 | 3.9 |
| 1906 | 53.7 | 6.4 | 2.5 | 7.7 | 17.0 |
| 1908 | 62.7 | 12.4 | 8.1 | 10.3 | 22.5 |
| 1910 | 108.6 | 12.1 | 8.6 | 22.4 | 30.1 |
| 1912 | 151.7 | 65.5 | 35.4 | 102.2 | 123.4 |
| 1913 | 259.4 | 82.5 | 56.3 | 133.0 | 188.5 |

SOURCES: R.M. Haig, *The Exemption of Improvements from Taxation in Canada and the United States* (New York, 1951); Province of Manitoba, *Report of the Assessment and Taxation Commission* (Winnipeg, 1919); City of Saskatoon, *1975 Municipal Manual*; P.R. Creighton, "Taxation in Saskatoon: A Study in Municipal Finance" (M.A. thesis, 1925); Henry Howard, *The Western Cities: Their Borrowings and Their Assets* (London, England, 1914).

Notes: (a) The figures are net values; i.e., exempted property is excluded.
(b) 80% of net value.
(c) Includes property owned by city and exemptions.
(d) 90% of net value.

The actual extent to which prairie cities adopted a modified version of the single tax varies considerably, although no western centre was immune from the intoxicating arguments put forward by single tax supporters. Of the five major cities, Winnipeg was affected least, perhaps because of the relative maturity of its leading businessmen and politicians and because of its dominant position in the urban hierarchy. In 1909 buildings, which until then had been taxed at 100 per cent of their assessed value, began to be taxed at only two-thirds of their value.[74] In both Regina and Saskatoon taxes on improvements were

73. R.P. Bains, *Calgary: An Urban Study* (Toronto, 1973), p. 22.
74. H. Carl Goldenberg, *et al.*, Report of the Royal Commission the Municipal Finances and Administration of the City of Winnipeg (Winnipeg, 1939), p. 325.

reduced by a 1908 provincial statute to 60 per cent, but the move toward the single tax did not stop here. Regina further reduced the improvement tax to 45 per cent. In 1912 it became 35 per cent and in 1913, 25 per cent.[75] Calgary reduced its improvement tax from 100 per cent to 80 per cent in 1909 and then followed these reductions with two further ones in 1911 and 1912. In 1911 the tax was set at 50 percent; in 1912 it was reduced to 25 per cent.[76] It was in Edmonton, however, that the single tax philosophy went furthest. In 1904 the improvement tax was eliminated; a situation which lasted until 1918. Several other taxes were also abandoned; the poll and income taxes in 1910 and the business tax in 1911.[77]

The effects these tax policies had upon the development of prairie cities were mixed. Even though the adoption of a modified version of the single tax occasioned numerous reports and studies, the widsom of the policy remained a matter for debate. Supporters and detractors of land value taxation agreed on only one thing – the modified single tax did not stop land speculation. Single tax proponents argued that the tax did in fact stimulate development; its detractors labelled land value taxation a complete failure and held it responsible for the financial difficulties prairie cities faced in the decades after the First World War.[78] Although there is no doubt that taxation policies did contribute to postwar problems, in the fluid prewar period they undoubtedly served the purpose of promoting urban growth.

## THE "LOSERS"

While the promotion strategies pursued by Winnipeg, Edmonton, Calgary, Regina and Saskatoon were successful in so far as these cities were the largest on the prairies by 1913, there is evidence to support the contention that other centres – including several with greater initial

75. Haig, *The Exemption of Improvements*, pp. 40-69; Owens, *Land Value Taxation*, passim; and Creighton, "Taxation in Saskatoon," *passim*.

76. Haig, *The Exemption of Improvements*, pp. 108-28; Owens, *Land Value Taxation*, passim; and Vineberg, "Provincial and Local Taxation in Canada," p. 83; Calgary also abandoned the income tax in 1911. See Perry, *Taxes, Subsidies and Tariffs*, pp. 134-135.

77. City of Edmonton, *Report on Assessment and Taxation*.

78. See, for example, Ferns, "Single Tax"; J. Loutet, "Taxing of Land Values in Theory and Practice," *Canadian Municipal Journal*, XVI (May 1920), pp. 144-145; Haig, *The Exemption of Improvements*; Owen, *Land Value Taxation*; City of Edmonton, *Report on Assessment and Taxation*; Rees, "The Magic City on the Banks of the Saskatchewan," p. 55; Alan F.J. Artibise, "Continuity and Change: Elites and Prairie Urban Development, 1914-1950," in Artibise and Stelter, *The Usable Urban Past*, pp. 130-54; and H.L. Macleod, "Properties, Investors and Taxes: A Study of Calgary Real Estate Investment, Municipal Finances, and Property Tax Arrears, 1911-1919," M.A. Thesis (University of Calgary, 1977).

advantages than the eventual "winners" – lost out because their leaders' initiatives and policies were not up to those of the competition. And in the absence of sufficient initiative and skill, natural advantages were incapable of sustaining growth for very long. In general, leaders in the "losing" cities were more divided and complacent, and often confined their activities to broadsides in the press.[79]

Four examples of losers may be taken as fairly typical. The announcement, in 1874, that the transcontinental railway west of Lake Superior would cross the Red River at Selkirk rather than Winnipeg seemed to spell the doom of the latter. The choice of Selkirk as a crossing point made sense to Sanford Fleming, engineer-in-chief of the transcontinental project. The route was not only shorter than one through Winnipeg, it had obvious engineering and economic advantages. A railway bridge at Selkirk would have the benefit of more stable banks than existed at Winnipeg and, even more important, both bridge and railway would be safe from flooding. The site of Winnipeg had been flooded seven times since 1812 while Selkirk, situated on a high ridge of land, had never been flooded. Here, then, was a case of Selkirk having clear and considerable advantages over Winnipeg. yet Winnipeg's elite marshalled their forces and set out to win the coveted prize. Promising the Canadian Pacific syndicate major concessions in a variety of areas, the Winnipeg group succeeded. Selkirk, unwilling to match Winnipeg's generosity, soon languished. By 1901, Selkirk's population was barely 2,000 compared to over 40,000 in Winnipeg.

The rivalry between Moose Jaw and Regina is more complex. From the early 1880s competition between the two cities was constant and determined, but the "Queen City" of Regina continued to maintain a slight edge. Since both cities were in close proximity, there was little question of natural advantage for either. Indeed, for many years the race between the two communities was a close one and it was not until after 1906 that Regina moved ahead of Moose Jaw in terms of size. Yet even before 1906, Regina's elite gave evidence of an ability to move both more quickly and more effectively in grasping for opportunities. In 1883, for example, Regina became the first organized town in the North West Territories. Moose Jaw followed Regina's example in 1884, but it had already lost an important psychological advantage. It was Regina, rather than Moose Jaw, which gained the bulk of the attention of easterners eager to invest in or migrate to the west.[81]

79. In addition to the material cited below, many of the articles in Artibise, *Town and City*, support this view.
80. The Selkirk-Winnipeg rivalry is detailed in Artibise, *Winnipeg: A Social History*.
81. See K.A. Foster, "Moose Jaw: The First Decade, 1882-1892," M.A. Thesis (University of Regina, 1978); and G.R. Andrews, "The National Policy and the Settlement of Moose Jaw, Saskatchewan, 1882-1914," M.A. Thesis (Bemidji State University, 1977).

More important, however, was Moose Jaw's loss of two critical prizes between 1905 and 1909: the provincial capital and the provincial university. While four Saskatchewan cities vied for these important designations, only the two eventual winners regarded the issues of sufficient importance to conduct all-out campaigns. While Moose Jaw gave up on its capital chances early and tried to claim only the university, both Regina and Saskatoon continued throughout to attempt to gain both prizes. Ultimately, Regina did become Saskatchewan's capital and in the following few years bypassed Moose Jaw as the leading city in the province.

There were other indications that Moose Jaw's elite did not possess the same drive and ability to pursue growth as did their Regina counterparts. As late as 1913 Moose Jaw's streets were in such poor condition that one observer noted that "no self-respecting citizen could feel pride in [such] a town."[82] Other observers found Moose Jaw lagging behind Regina and other Saskatchewan cities in their move toward full land value taxation; a condition that indicates the conservatism of the city's elite.[83] While later developments would suggest that Moose Jaw probably adopted the wiser course in terms of taxation philosphy, it was not a wise course to follow in the fluid days of the boom era when a city's progressive image was an essential ingredient for success. In short, in a wide variety of areas, Moose Jaw's elite was out manoeuvred by their Regina counterparts.

The case of Prince Albert resembles that of Moose Jaw in many respects. It, too, lost out in the crucial fight for the capital and the university. The loss of the capital might have been expected; the loss of the university was not. The fact was that Prince Albert adopted a complacent attitude and simply did not conduct a strong or effective campaign.[84] The city's efforts in respect to both railways and power development also lacked the skill and drive of its competitors. In the case of power development, for example, the city's plan to bypass other Saskatchewan communities as an industrial centre through the utilization of its considerable water power resources was daring, but the plans were poorly prepared and executed and ultimately left Prince Albert with a huge debt. One historian has described the episode as "one of the greatest disasters ever to befall a western city."[85] Indeed, Prince Albert's elite, in comparison with the elites of either Saskatoon or Regina, lacked the skills necessary to take advantage of the opportunities open to them. Local government moved at a sluggish pace and despite lavish expenditures throughout the boom era, public works programs were ill-planned

---

82. Haig, *The Exemption of Improvements*, p. 38.
83. A. Stalker, "Taxation of Land Values in Western Canada," *McGill University Publications*, Series VI, No. 4 (1914), pp. 39-45.
84. Gary Abrams, *Prince Albert, 1866-1966* (Saskatoon, 1966), pp. 155-56.
85. The power project is discussed at length in chapters 12 and 13 of *ibid.*

and carried out in a slipshod manner. And in the years following the boom, Prince Albert entered a period when its civic affairs were conducted in a mood of anger, bitterness, and petty striving which was not to leave the city for fifty years. Prince Albert lacked that element of community spirit – at least among the members of the commercial elite – that so effectively overcame petty jealousies within elites in Regina and Saskatoon.

In Alberta, Medicine Hat's development consistently lagged considerably behind that of Calgary, its closest major rival. Although Medicine Hat was from the early 1880s a divisional point on the Canadian Pacific Railway, the leaders of this community were unable to use this or other considerable resources as stepping stones to big-city status. Compared to Calgary, which was organized as a town in 1884 and as a city (the first in the North West Territories) in 1894, Medicine Hat did not become a village until the latter date and was not incorporated as a city until 1906. Given the psychological and economic benefits of formal incorporation, Medicine Hat's leisurely pace was a fatal mistake. In other areas, municipal organization was also slow to get under way; as late as 1913, for example, the city's streets were still in poor condition.[86] Another indicator of a lack of drive is the city's slow adoption of the single tax philosophy. Medicine Hat moved to land value taxation in 1912, something that had been begun in Calgary three years earlier. Medicine Hat, like Selkirk, Moose Jaw, Prince Albert and a host of other prairie communities, lacked that special combination of initiative and skill that enabled the cities of Winnipeg, Regina, Saskatoon, Calgary and Edmonton to take advantage of the fluid conditions of the time.

CONCLUSION

A precise determination of the effect of municipal boosterism on prairie urban development must obviously include far more analysis than has been presented here, including a good deal more comparative work. But, even at this point, several conclusions, however tentative, can be made. It is clear, first of all, that in every aspiring urban centre efforts were made by civic and commercial elite to facilitate growth. These policies took a variety of forms, including early city incorporations, massive boundary extensions, huge public works programmes, deficit financing, special tax policies, railway promotion, immigration encouragement, industry attraction, municipal ownership, and efforts to attain status as both provincial capital and home of the provincial university. While booster policies differed in degree and kind from centre to centre, no prairie community was immune to boosterism.

The policies implemented by the municipal boosters did have an

86. J.G. MacGregor, *A History of Alberta* (Edmonton, 1972), p. 158.

impact on prairie urban development. The decisions made by urban elites were important factors in the determination of the pattern of urban development; in the absence of these efforts a more diffuse or different urban pattern would probably have emerged. Instead, in the relatively short span of four decades five urban centres had bypassed their regional competitors (see Table 1) and all had gained a prominent place within the ranks of Canada's largest cities (Table 5). Moreover, these five cities together contained 65 per cent of the region's urban population (Tables 1 and 2). Booster policies also affected the region's growth rate. In gross terms, urban growth in the prairie region in the decade preceding the First World War surpassed that in all other regions of the country. The urban population of Manitoba jumped from 28 per cent in 1901 to 43 per cent in 1911, an increase of 15 per cent. In Saskatchewan, the increase was 11 per cent (16 per cent to 27 per cent); in Alberta, 12 per cent (25 per cent to 37 per cent). During the same period urban population growth in the Maritimes was only 6 per cent (22 per cent to 28 per cent); in Quebec, 8 per cent (40 per cent to 48 per cent); in Ontario, 10 per cent (43 per cent to 53 per cent); and in British Columbia, 2 per cent (50 per cent to 52 per cent). Of course, the region's relative newness together with the development of the wheat economy of the prairies had much to do with this rapid rate of urbanization. Still, in the absence of such massive efforts on the part of urban elites throughout the prairies, it is possible that many of the immigrants and some of the capital investment that did come to the area might have gone elsewhere.

The final point that can be made about the impact of boosterism on the development of prairie cities in the prewar era is, perhaps the most important. By 1913 the five major prairie cities had a firmly established framework for future development. This framework included a variety of elements ranging from particular patterns of physical development to special tax policies and government structures. The key element, however, was the attitude of the decision-makers. The pre-1913 experience confirmed in the minds of the elites in all five cities that a booster mentality was essential to continued growth. More than ever before, the elites subscribed to the belief that "cities are made by the initiative and entreprise of its citizens." Not one of the tenets of the booster philosophy had been dislodged by past experience. In terms of attitudes towards railways, immigration and industrial encouragement, inter-city rivalries, labour, and the role of government, the commercial elites of the cities entered the post-war era intent on following past practices. It was, in many respects, a serious mistake. In the prewar period, a marked fluidity existed when the opportunities for recruiting the acknowledged attributes of city status – population, major transportation facilities, capital investment – were open to all. In this situation, the leaders of the five major prairie cities acted effectively and were rewarded with rapid growth and big city status. In the subsequent era, from 1914 to the early

Table 5: Rank of Selected Canadian Cities by Size, 1901-1921

| Rank | 1901 | 1911 | 1921 |
|------|------|------|------|
| 1 | Montreal | Montreal | Montreal |
| 2 | Toronto | Toronto | Toronto |
| 3 | Quebec | **Winnipeg** | **Winnipeg** |
| 4 | Ottawa | Vancouver | Vancouver |
| 5 | Hamilton | Ottawa | Hamilton |
| 6 | **Winnipeg** | Hamilton | Ottawa |
| 7 | Halifax | Quebec | Quebec |
| 8 | Saint John | Halifax | **Calgary** |
| 9 | London | London | London |
| 10 | Vancouver | **Calgary** | **Edmonton** |
| 11 | Victoria | Saint John | Halifax |
| 12 | Kingston | Victoria | Saint John |
| 13 | Brantford | **Regina** | Victoria |
| 14 | Hull | **Edmonton** | Windsor |
| 15 | Windsor | Brantford | **Regina** |
| 16 | Sherbrooke | Kingston | Brantford |
| 17 | Guelph | Peterborough | **Saskatoon** |
| 18 | Charlottetown | Hull | Verdun |
| 19 | Trois-Rivieres | Windsor | Hull |
| — | — | — | — |
| 36 | — | **Saskatoon** | — |
| — | — | — | — |
| 73 | **Calgary** | — | — |
| — | — | — | — |
| 77 | **Edmonton** | — | — |
| — | — | — | — |
| 97 | **Regina** | — | — |
| — | — | — | — |
| 110 | **Saskatoon** | — | — |

SOURCE: *Census of Canada, 1931*, Volume I.

1940s, many of the structures, ideas, and routines that had in the past worked to fuel growth were not longer adequate. Indeed, in many cases they became, in themselves, obstacles to continued or renewed growth. The three decades following 1913 were taken up with repeated efforts to alter or overcome the negative effects of earlier decisions. All five cities struggled to regain the initiative they realized was so important to continued success. It was true, of course, that the problems of the interwar era were aggravated by a long period of depression, but it was also true that the earlier framework of growth could be redefined or reordered to deal with new problems only with great difficulty. By the early 1940s, some of the problems had been overcome and the five centres looked forward to better times. Yet the ability of the cities to take advantage of postwar prosperity was limited; although some changes had been made and some problems overcome, the framework remained relatively rigid. The fact was that by the 1950s prairie cities were in most respects still tied to the structures, ideas, and routines of a bygone era, and no amount of simple manoeuvring could alter this legacy in a major way. Growth in the postwar era occured almost in spite of the framework and the efforts of urban elites rather than because of them.

This overview of the early history of prairie urban development does not sustain the view that local leaders through their actions created their cities, or were alone responsible for their patterns of growth and development. Nor does it reveal the unfolding of a predetermined design whereby a number of prairie hamlets emerged as the major cities of a region. It shows instead the influence of boosterism on urban development. The scope, character, and direction of community policy – conditioned by a complex framework of geographic and economic influences – had profound effects on the rise of prairie cities. It does not propose that "men, not chance" made prairie cities, but rather that the element of urban leadership must be an intergral part of any explanation of urban growth.

# Physical Expansion and Socio-Cultural Segregation in Quebec City, 1765-1840

MARC LAFRANCE AND DAVID-THIERRY RUDDEL

Rapid economic growth, military fortifications and an inefficient administration were the most significant factors influencing the development
of Quebec City in the late eighteenth and early nineteenth centuries.
Between 1790 and 1840, the British timber trade was responsible for
rapid economic, demographic and physical expansion. Parallel to this
development was the growth of an imposing military presence; a presence which limited urban growth and influenced the social composition
of the city. The partial administration of the Commission of the Peace, a
body dominated by anglophone merchants, played a significant role in
the division of the city into two areas: one for the ruling class and the
other for labouring groups. Although divisions existed during the
French regime, they were not as distinct as they became during British
rule.

In addition to the abusive use of power by administrators and military officials, the existence of a tacit agreement between anglophone
and francophone elites led to increasing differences between areas of the
city. Much has been written about the merchants' attempt to control
colonial politics, but little attention has been paid to their domination of
local affairs. Yet the merchants' mastery of the urban administration
(Commission of the Peace) played an important role in the city's
development.

Without disputing the traditional interpretation of political competi-

SOURCE:

This article is based on D.T. Ruddel, "Quebec City, 1765-1831," Ph.D. Thesis (Laval
University, 1981); and on a monograph by André Charbonneau, Yves Desloges and Marc
Lafrance, *Québec, ville fortifiée, du XVIIe au XIXe siècles* (Québec: Parks Canada and Pelican
Press, 1982). The authors also wish to thank Nancie Ensoleil, Jean-Pierre Hardy, Peter
Rider, and the editors of this volume for their comments on an earlier draft of this article.

tion between British merchants and Canadien\* professionals, this study draws attention to the agreement which existed between these groups concerning urban conditions. This is one of the reasons that complaints by residents of St. Roch were neglected – social and economic leaders did little to improve working-class areas. Consequently, conditions in these areas continued to deteriorate and social segregation became common.

## PHYSICAL EXPANSION AND URBAN DIVISIONS

The social segregation of the urban landscape did not occur until after two centuries of growth, but Quebec's functional separation came earlier.[1] By 1660, Quebec had been divided into two areas, the Lower Town, a commercial and residential centre, and the Upper Town, the administrative, military and religious centre. As the town[2] expanded, suburbs and agricultural settlements in the surrounding area also developed. During the seventeenth and eighteenth centuries, expansion was limited mainly to the Upper and Lower Towns. Between 1795 and 1895, the timber trade stimulated expansion, increasing building by over 16

---

\*The word "Canadien" is employed throughout to designate French-speaking inhabitants. This term was used by both linguistic groups in Quebec when referring to members of the majority. Only a small number of anglophones identified themselves as Canadians. Most preferred to be known as British citizens.

---

1. The legal boundaries of Quebec City changed little between 1791 and 1840. In 1791 the limits of the town of Quebec included the extensive "tract or promontory of land. . . between the Rivers St. Lawrence and St. Charles, bounded in the rear by a right line running along the easterly front of the Convent called the General Hospital, and continued from River to River." See the "Proclamation dividing the Province of Lower Canada into counties and electoral districts," May 7, 1792, in Arthur G. Doughty and D.A. McArthur, *Documents Relating to the Constitutional History of Canada, 1791-1818* (Ottawa: C.H. Parmelee, 1914). In 1831, the act for the incorporation of the city redefined the quarters but did not change existing boundaries. Between 1831 and the mid-nineteenth century, these quarters were modified for political reasons, but did not alter the physical limits as defined in 1792. The 1831 divisions are described in the act of incorporation given in full in *Le Canadien*, June 13, 1792. For more information, see Ruddel, "Quebec city," pp. 51 and 52. The evolution of Quebec's legal boundaries in mid-century closely resembled those of Montreal. For further information, see Jean-Claude Robert's doctoral thesis, "Montréal, 1821-1871, aspects de l'urbanisation" (Ecole des Hautes Etudes en Sciences Sociales, Paris, 1977), pp. 150-58.
2. The distinction between "town" and "city" made throughout the text, is based on the definition of the latter as a centre with 8,000 or more inhabitants. If one uses this definition, Quebec did not become a city before the early nineteenth century.

per cent in the suburbs (see Table 1). Following this period, the building rate diminished but construction in the suburbs continued to outstrip that of the Old City (Upper and Lower towns) until 1832, by which time the number of buildings in the suburbs had surpassed those of the Old City. Thus, although the Upper and Lower Towns included 70 per cent of Quebec's buildings in 1795, by 1842 the suburbs possessed 66 per cent.

Table 1: Building Expansion in Quebec City, 1795-1842
(yearly increase expressed as a percentage)

| Quarters | 1795-1805 | 1806-1819 | 1820-1831 | 1832-1842 |
|---|---|---|---|---|
| Upper Town | 1.2 | .26 | 1.2 | 3.1 |
| Lower Town | 3.03 | .09 | 3.2 | 2.6 |
| St. Roch | 16.3 | 5.8 | 6.5 | 2.4 |
| St. John | 16.6 | 1.6 | 6.7 | 1.8 |

SOURCES: Based on an analysis of the *curé's* 1795 and 1805 returns, the *Quebec Gazette*, Feb. 11, 1819, and the census of 1831 and 1842. The 1795 and 1805 *dénombrements* are found in Joseph-Octave Plessis, "Les dénombrements de Québec. . .," *Rapport de l'archiviste de la Province de Québec, 1948-1949.*

Physical growth was accompanied by population growth, and the pattern of suburban development was clear. Whereas in 1795 the Upper and Lower Towns accounted for 75 per cent of the population, by 1818, 56 per cent of Quebec's residents lived in the suburbs (see Table 2). This suburban growth continued, so that while the Old City was doubling its population between 1795 and 1842, the suburbs were increasing tenfold. The suburbs attracted artisans and labourers from Lower Town and farmers from surrounding villages working as labourers, artisans and carters.[3] Instead of travelling to Quebec to work and then returning to their neighbouring villages, many of these men decided to move their families to the suburbs. Expensive property in the Old City discouraged these people from attempting to settle there.

*Lower Town*

By 1660 Lower Town already possessed numerous buildings and streets and although they were not well aligned, the urban grid was well established. Vacant space still existed, but the crowded building patterns

---

3. Jean-Pierre Hardy and David-Thierry Ruddel, *Les apprentis artisans à Québec, 1660-1818* (Montreal: Les presses de l'université du Québec, 1977), p. 151.

Table 2: Growth of the Population by Area, Quebec City, 1795-1842

| Quarters | 1795 | 1805 | 1818 | 1831 | 1842 |
|---|---|---|---|---|---|
| Upper Town | 2,813 | 2,973 | 3,730 | 4,496 | 5,017 |
| Lower Town | 2,649 | 2,628 | 3,402 | 4,963 | 7,284 |
| St. Roch | 829 | 1,497 | 5,217 | 7,983 | 10,760 |
| St. John | 1,008 | 1,764 | 4,070 | 8,502 | 8,686 |

SOURCES: Figures for the years 1795, 1805 are from J.P. Plessis, "Les dénombrements de Québec"; 1831 and 1842, from census returns. The 1818 figures include the Ursulines, Seminary and Hotel-Dieu. The General Hospital and the garrison are excluded and the figures for St. Louis have been added to those for St. John. See Joseph Signay, *Recensement de la ville de Québec en 1818* (Quebec: Société historique de Québec, 1976), p. 279. Since the curés underestimated the Protestant and military communities and both the *dénombrements* and census failed to consider the floating population, these figures can only be considered as approximate.

now typical of Quebec were already evident. After a major fire in 1682, colonial administrators pleaded with the court to extend the port area.[4] Surrounded by water on one side and by the cliffs of the Quebec promonotory on the other, Lower Town residents, and particularly merchants, had no choice but to expand vertically and over the St. Lawrence River. French engineers, such as Chaussegros de Léry, planned the extension of the Lower Town into the river by land fills and the construction of wharves and fortifications. Although ordinances prescribed the dumping of rubbish in the Lower Town and the port captain dredged certain areas of the harbour, extension was limited.[5] The development of the Cul-de-Sac shipyards[6] and the construction of artillery batteries along the waterfront only slightly altered the Lower Town shoreline.

It was not until the early nineteenth century that greater port activity and extensive wharf construction led to significant expansion of the port. Increased activity and larger ships led to the extension of wharves. Merchants' demands for concessions of government land along the shoreline, from the St. Charles River to the Anse-des-Mères, multiplied at a rapid pace in the early 1800s. British shipbuilders and timber merchants added their requests to those involved in dry goods and competition became stiff and, occasionally, bitter. In 1800 the government,

4. On the urban development of Quebec in the seventeenth century, see, Rémi Chénier, "L'urbanisation de la ville de Québec, 1660-1690" (Québec: Parcs Canada, unpublished report, 1979.)
5. P.G. Roy, "Les capitaines de port à Québec," in *Bulletin des Recherches Historiques*, Vol. XXII (January 1926), pp. 3—12.
6. Jacque Mathieu, *La construction navale royale à Québec, 1739-1759*, Cahiers d'Histoire, no. 23 (Quebec: La Société historique de Québec, 1971), pp. 21-26.

realizing the increasing value of the waterfront area, decided to reserve some of it for military purposes, while leasing other parts of it to shipbuilders and merchants.[7] In the scramble for land west of Lower Town, below Cape Diamond, merchants first occupied vacant spaces and then requested grants of the land. Some failed to request grants while others petitioned for the same space. In their hurry to occupy as much land as possible, merchants built on land reserved for streets and public passages.[8] Coves and beach land at western and northwestern extremities were occupied by timber merchants. This fervent activity increased the number of wharves from eleven in 1785, to twenty-one in 1804, and thirty-seven in 1829. By 1830 the wharves extended in an almost unbroken line from the Dorchester Bridge on the St. Charles River to the Anse-des-Mères on the St. Lawrence.

At the same time as they were extending wharves over the river, merchants were also building vertically; second, third and even fourth stories were added to existing structures. Huge warehouses were built and impressive structures, such as the Customs Building, the Quebec Exchange and banks, dominated the centre of Lower Town and its major artery, St. Pierre Street.[9] Wharf construction between St. Pierre Street and the river was so considerable that the limits of some wharves were over 500 feet from the street.[10]

Besides extending over the St. Lawrence, Lower Town was also expanding along the narrow shoreline at both its extremities. As early as 1785, an important wharf was situated at Anse-des-Mères, half a mile from the nearest wharf. Construction at this place was an indication of the difficulty of finding inexpensive and appropriate land.[11] Between 1785 and 1804, Champlain Street extended to Munn's shipyard at Anse-des-Mères. Près de ville, the area between Munn's shipyard and the King's Landing in Lower Town, was inhabited mainly by labourers and their families.

Expansion caused by the timber trade led to the development of large timber coves west of Diamond Harbour. These coves developed before the access roads to them; thus, although dozens of small wooden houses, some no larger than 15 x 15 ft. shacks, and large shipping docks, blacksmith's shops, a tobacco manufacturer and a brewery, were con-

---

7. This information is based on an analysis of petitions and map descriptions in the Public Archives, especially in PAC, RG1L 3L. A summary of these sources is found in E. Dahl, et al., Inventaire des cartes de la ville de Québec (Ottawa: National Museums of Canada, 1975), pp. 61-171.

8. See Joseph Bouchette, "Plan of part of the Lower Town beach," 1804, Public Archives of Canada (PAC).

9. Geneviève Bastien, et al., Inventaire des marchés de construction des archives civiles de Québec, 1800—1870, (Ottawa: Parks Canada, 1975), pp. 226, 394 and 738.

10. Joseph Bouchette, [Plan of the limits of the lots of John Caldwell, John Goudie and Wm. Burns], 1818, PAC.

11. Anonymous, "Plan of the Town and citadel of Quebec," 1785, PAC.

structed beneath the cliffs of the Plains of Abraham, no direct road access to the city existed.[12]

Lower Town was also extending towards St. Roch. Under the French regime, it had been virtually impossible to travel between these two areas without going via Upper Town. The opening of St. Paul Street in 1816 and the development of the Palace Harbour wharves permitted easy communication between St. Roch and the Lower Town and led to the opening of new streets and the St. Paul Street market.[13] In less than fifty years, the surface of Lower Town had doubled.[14]

In order to maintain a healthy environment around their homes in the crowded, dirty, damp and often noisy area of Lower Town, merchants requested paved roads, road maintenance, drains, and police and fire protection. Most of these requests were granted, but limited to commercial streets inhabited by merchants. Streets occupied by labouring groups were neglected. Thus, contrasts between streets or small areas, typical of Quebec in the late eighteenth century, continued to exist in the Lower Town during the first part of the nineteenth century.

## St. Roch Suburb

The development of St. Roch suburb at the end of the French regime can be partly attributed to its proximity to the Intendant's Place and to the King's shipyards, but other factors were equally important. In the early eithteenth century a few people built houses along St. Vallier Street, one of the main routes to Quebec's hinterland, but settlement on the plains area was not important until the 1740s. The land, part of the King's *censive* (seignoirial fief), had been acquired in 1720 by Henri Hichè, clerk of the King's Stores, who did little with his property until 1751-1753 when he sold seventy-five lots.[15] The construction of fortifications from 1745 onward, the establishment of a military zone (a *non*

12. This information is based on an analysis of plans at the PAC and AVQ ("Archives de l'Hotel de Ville de Québec" – Quebec City Hall Archives). See, for example, "Plan of part of the beach at Wolfe's Cove . . ." (showing the existence of small houses used by labourers in the timber trade), by John Adams and Benjamin Ecuyer, 1830, PAC; John Adams, "Map of Quebec and its environs . . . ," 1826, PAC; Jean-Baptiste Larue, "Plan figuratif de la rue ou chemin nommé Chaplain," 1826, AVQ; Alfred Hawkins, *Plan of the City of Quebec*, 1835, PAC; and A. Larue, *Plan of the City of Quebec*, 1842, PAC. In 1826, sixty houses existed on the quarter-mile stretch of road between Munn's shipyard and city limits. The area beyond the city limits was far more populous than the above-mentioned stretch. This housing is indicated on Jean-Baptiste Larue's plan of Champlain Street cited above.
13. Road Committee Minutes, 1844, AVQ.
14. Raoul Blanchard, *L'est du Canada français, "Province de Québec"* (Montréal: Beauchemin, 1935), II, p. 273.
15. Louise Dechêne, "La rente du faubourg Saint Roch à Québec, 1750-1850," *RHAF*, Vol. XXXIV (March 1981), pp. 571-72.

*aedificandi* area in which civil construction was forbidden) in St. John suburb, and the laying-out in streets and squares of the undeveloped land in Upper Town, were factors which contributed to the development of St. Roch in the early 1750s. Chaussegros de Léry saw this suburb as an area which could relieve population pressure in Upper Town. In 1752, when the engineer was planning military projects for Lower Town, he viewed with satisfaction the movement to St. Roch,"... where a large part of Quebec's craftsmen live, a development which freed the rest of the Town."[16]

In the late eighteenth and early nineteenth centuries, St. Roch became less an overflow area from the upper plateau and more an extension of the Lower Town. As St. John's suburb was expanding rapidly in the late eighteenth century, St. Roch was undergoing the reconstruction of buildings destroyed in the fighting of 1759 and 1775-1776. After 1795, the timber industry and shipbuilding stimulated the expansion of the suburb: between 1795 and 1805 housing construction increased by about 16.3 percent annually (see Table 1). By 1815, St. Roch's role as a suburban residence and workplace was established; new streets were created, old ones extended and a local church built.[17] In short, "Grant's seigniory," named after a large landowner, was filling up and developers began looking at the old Jesuit property, called "La Vacherie," to the west of St. Roch. Plans for the development of this land were prepared as early as 1822, but only a few beach lots were conceded to shipbuilders in 1827.[18] Building lots for residential purposes were not readily available until the 1830s.

The rapid expansion of St. Roch between 1795 and 1840 occurred with less planning and supervision than that happening elsewhere. Speculators, such as William Grant, and shipbuilders, like John Goudie, benefited most from this kind of expansion. Grant and Goudie incorporated public passages into their enterprises along the St. Charles River and the former objected to the establishment of 30 ft. wide roads in St. Roch because it would reduce the size of his building lots.[19] Merchants and colonial authorities considered St. Roch, along with parts of the

---

16. Chaussegros de Léry, "Project pour fortifier la Basse-ville de Québec," AN (Archives nationales, France), Outremer, D.F.C., order no. 428. The French text reads, "... ou une grande partie des artisans de la ville se sont logés, ce qui à debarrassé la ville."

17. Road Committee Minutes, 1844, AVQ. Land for St. Roch's Church was donated by William Grant in 1811 to encourage settlement. The building contract can be found in the notarial file of Felix Têtu, June 14, 1811, ANQ.

18. On the management and development of the "Vacherie," see PAC, RG1, L3L, XIX, pp. 8788—89, 9123-24, 9149 and 9325-26. See also R.C. Dalton, *The Jesuits' Estates Question, 1760-1888* (Toronto: University of Toronto Press, 1968), pp. 70 and 76; and Ruddel, "Quebec City," p. 558.

19. Grant's obstruction of public passages and objection to street sizes are described in more detail in Ruddel, "Quebec City," pp. 512, 543 and 558. For a discussion of Grant's usurpation of the title of the seignior of St. Roch, see Louise Dechêne, "Saint Roch," p. 579-85.

Lower Town, as part of the city's industrial complex, which, consequently, did not need good environmental conditions to fulfill its functions. This attitude was expressed by a local doctor who terminated his evaluation of the conditions in St. Roch in the following terms: "The streets of the suburbs will be gradually paved; though it is probable that the process will be slow, as the necessity does not seem so urgent for ordinary purposes of life. . . ."[20] In fact, the only street paved in St. Roch prior to 1820 was the major commercial thoroughfare linking country roads with the port. Fewer street alignments were made in St. Roch than elsewhere[21] and drainage was almost non-existent. The results of such neglect were constantly brought to the attention of the justices by the Road Inspector, who maintained that:

> It is impossible to describe the state of the roads and the urgency of repairing at least 2/3 of them. Communication is interrupted because of blocked drains, damaged bridges, and the practice of the poor who use wood from bridges for firewood. Flooding has caused soil erosion and left gaps everywhere and carts have created huge ruts, several over 4 feet deep. [22]

Low marshy ground, increased traffic and industrial activity, the absence of most services and failure to enforce regulations, all contributed to making St. Roch the city's most unwholesome suburb.[23]

## Upper Town

The growth and organization of the Upper Town was conditioned at the outset by the policy of the French government in granting large concessions to private individuals and religious orders. The latter were particularly significant, not only because of their initial size, but also because of the long-term effect such institutions would have on the city's evolution. An attempt made by Governor Montmagny in 1636 to control and organize the town's expansion by the establishment of a town plan was seriously handicapped by these large concessions as well as by the irregular topography of the Upper Town. Thus, by the late seventeenth

---

20. William Kelly, "On the Medical Statistics of Lower Canada," *Transactions of the Literary and Historical Society of Quebec*, Vol. III (1837), pp. 193-221.

21. Statement based on an analysis of street alignments found in the "Procès-verbaux d'alignement, 1818-1831, Bureau de l'inspecteur des chemins," AVQ.

22. Authors' translation of part of a report which was typical of others made by the Road Inspector, Jean-Baptiste Larue, between 1817 and 1839. The source for the quotation is: "Rapports de l'inspecteur des chemins," May 13, 1822, AVQ.

23. Sources for this conclusion are numerous. See for example, the *Journals of the House of Assembly of Lower Canada*, 1809, petition of the inhabitants of Saint Roch, p. 227; 1816, petition of the inhabitants of Saint Roch, p. 207; 1819, petition of the inhabitants of Saint Roch, p. 35; 1823-24, petition of the inhabitants of Saint Roch, p. 63; 1824, report of a committee of the House on the Sain Roch petitions, pp. 334-40. See also chapter VI of Ruddel, "Quebec City."

century, the Upper Town was dominated by a few large private land holdings and religious concessions. The latter included cultural and social services, such as churches, convents, schools, gardens, a cemetery and hospital.[24] Land owned by private parties was used first for agriculture and later for commercial and residential purposes.

The most serious obstacle to the expansion of the Upper Town was military fortifications. Although Champlain erected a fort in the Upper Town in the early 1660s, large scale fortifications were not built until the late seventeenth century. One of the first military obstacles to expansion was the construction in 1695 of a fortification wall which limited the development of the western part of the Upper Town for half a century. This wall was replaced, between 1745 and 1752, by new fortifications, built by Chaussegros de Léry. The new walls were situated approximately 1,200 feet to the west of the 1693 fortifications, thus enlarging the Upper Town. This additional land was then laid out in streets and squares for future expansion. De Léry also established a military reserve outside the walls by demolishing the houses of the small St. John suburb. The engineer made street alignments in St. Roch to encourage settlement in that area. The military saw the low lying suburb of St. Roch as an area for future expansion from the Upper Town. These developments not only made military considerations an essential part of urban life, they also conditioned the future evolution of the town.

In establishing his project, Chaussegros de Léry had evaluated the town's past and future growth. Any reduction in the growth capacity of the Upper Town as planned by the engineer would endanger the balance he sought between urban utility and defence requirements. In other respects, the establishment of St. Roch as an overflow area and the walling-in of the Upper Town, destined the latter to become a quarter for the elite. The advent of the British regime confirmed this scheme and ensured that the Upper Town would enjoy the status of a military reserve. The British decision to build a citadel on vacant land in the Upper Town ruined Chaussegros de Léry's plan, which projected the use of this land for settlement and contributed to making this area a reserve for an elite capable of paying for increasingly rare, and therefore, valuable land.

As early as 1763, British military officials reserved land (535 acres), for the construction of a citadel.[26] Haunted by the spectre of a French

24. Allana Reid, "The Development and Importance of the Town of Quebec 1608-1760," Ph.D. thesis (McGill University, 1950) pp. 307-19; Marcel Trudel, *La Terrier du Saint-Laurent en 1663* (Ottawa: Editions de l'Université d'Ottawa, 1973) pp. 157-93; and Trudel, *Les débuts du régime seigneurial au Canada*, (Fides: Montréal, 1974) pp. 79-89.

25. This question is discussed at length in André Charbonneau, Yvon Desloges, and Marc Lafrance, *Québec, ville fortifiée*, pp. 360-64 and 424-30.

26. As indicated in de Léry's 1752 project, this sector could have accommodated fourteen regular and five irregular blocks, including about three hundred houses. See *ibid.*, pp. 425 and 426.

attack or a revolt by Canadiens, a citadel was judged essential. The reservation of this land meant that almost half the land de Léry had reserved for urban expansion was no longer available. The slow growth of Quebec in the late eighteenth century and the absorption of the surplus population by the suburbs in the first two decades of the nineteenth century, mitigated the problem of available land in the Upper Town. After 1830 land became scarce. Like their counterparts in Lower Town, proprietors in Upper Town added new wings and third and fourth stories to existing houses and filled their courtyards. By 1842 the Upper Town had almost reached its maximum capacity in terms of construction units; fifty years later only forty-five more units had been added.[27]

In spite of limited space for construction, the Upper Town possessed the most agreeable environment of the city. Encircled and enclosed by fortifications, graced with monumental structures and controlled by the religious communities, military officials and wealthy landholders, the Upper Town was largely reserved for the city's elite.

## St. John Suburb

The obvious place for expansion on the Quebec plateau was to St. John's suburb. Up to 1745 this area had been used mainly for farming and roads to the town. Although a few houses were built along these roads, settlement was sparse. By 1745 Chaussegros de Léry had this small suburb almost completely demolished in order to create a military zone in front of the military fortifications. When the British decided to build a citadel, the military zone was abandoned. St. John suburb was henceforth given over to subdivision and construction.[28]

For many people, especially artisans, this suburb was a more attractive place in which to live than the higher priced land in Upper Town or the marshy areas of St. Roch. Moreover, shipbuilding, which was the main attraction in St. Roch, had been stagnant between 1759 and 1790. Finally, more direct communication was possible between St. John suburb and Upper Town than between the latter and St. Roch. The wealthy clientele of Upper Town was, therefore, more accessible to artisans living in St. John than to those of St. Roch. Despite slow population growth between 1760 and 1800 and the destruction of the suburb during the 1775-76 American blockade, it expanded quicker after 1780 than other sectors of the town. Like new suburbs elsewhere, St. John expanded along the main artery between the town and the country. As

27. Sixty-one per cent of building contracts between 1820 and 1842 were for three-storey houses. This percentage and the figure for the 1890s were obtained by studying construction contracts listed in the *Inventaire des marchés de . . . 1800-1870*.
28. André Charbonneau, *et al.*, *Québec, ville fortifiée*, pp. 403-405.

the population increased, settlement spread out on both sides of this route, including the area known as St. Louis suburb.[29]

Following the blockade of Quebec by the Americans, the military re-evaluated the fortifications. By the beginning of the nineteenth century, the expansion of St. John suburb was viewed as a threat to the security of the fortress. Thus all land in St. John suburb deemed necessary for defence requirements began to be appropriated. Between 1811 and 1822 the military acquired 119 acres of land in St. John and by 1850 owned or controlled 35 per cent of the land in the suburb.[30]

After 1822 the expansion of the suburb was limited by military appropriations of land. The large holdings of Jean Guillet and Vallières de Saint Réal on Coteau St. Geneviève and those of Berthelots, south of St. John road, were subdivided into streets. Expansion southward was prevented by the Panets, Lotbinières and Stewarts who were keeping their large parcels of land intact with the hope of selling them to the military for high prices. By threatening the military general staff that they would subdivide their land, they were trying to raise its value. The opposition by the Lotbinière heirs to the Road Inspector's plan to extend St. Augustin and St. Eustache streets onto their land was revealing in this regard.[31]

St. John suburb also suffered from the absence of a comprehensive urban plan. Streets were traced according to partial plans, such as the one established for the Hotel Dieu in 1783 on the land between St. John Road and the Coteau St. Geneviève. The Road Inspector constantly complained about manoeuvres by petty speculators: they laid out streets to their own advantage, granted irregular plots of land, blocked road extension projects, and even subdivided land which had been reserved for streets and public squares.[32] The suburb's situation began to improve in the second half of the nineteenth century when defence requirements changed. But, for at least fifty years, military intervention in various forms had shaped the nature of the suburb's expansion.

In addition to retarding expansion, the military presence in St. John also created traffic problems. Fortifications, gates (especially when they were closed or in repair) and large tracts of land, made communication difficult within the suburb and between different areas of the town. In

29. St. Louis suburb was an extension of S. Louis Street which began beyond the gate separating the Upper Town from the rest of the plateau. Because of its limited size and proximity to St. John's suburb, St. Louis was often considered an extension of the latter.

30. Claudette Lacelle, *La propriété militaire dans la ville de Québec, 1760-1871,* Manuscript report No. 253 (Ottawa: Parks Canada, 1978) pp. 6-26.

31. Antonio Drolet, *La ville de Québec, histoire municipale,* Cahiers d'histoire no. 17 (Québec: La Société historique de Québec, 1965) II, pp. 86-87; AVQ, Bas-Canada, Juges de Paix, sessions spéciales de la paix, vol. A, no. 527, 16 juin 1819; vol. B(2) no. 1010, 16 avril 1823, no. 206, 26 mai 1824.

32. AVQ, Bas-Canada, Bureau de l'inspecteur des chemins, rapports de J.B. Larue fol. 59, 10 nov. 1818; fol. 74, 4 mai 1819, fol. 197, 16 décembre 1823, fol. 239, 8 juin 1825, fol. 225, 15 mai 1826.

spite of these problems, St. John remained a better place to live than St. Roch. The former's location on the plateau facilitated drainage and access to the services of the Upper Town.

## URBAN GEOGRAPHY AND SOCIAL DISTINCTIONS

In the late eighteenth century, socio-cultural distinctions existed but different occupational and ethnic groups shared similar areas. Streets were inhabited according to social categories: professionals and merchants, merchants and artisans, artisans and labourers. Social differentiation was consequently evident, as was interaction between members of different groups. The growth of the suburbs in the early nineteenth century accentuated existing divisions.

Both the Upper and Lower Towns continued to include members of different social and ethnic groups, but the former retained its elitist characteristics. The Upper Town was divided between the artisans and labourers living in the northwestern area, and the professionals and merchants inhabiting the southeast. According to Jenkin Jones, an agent of the Phoenix Insurance Company, the Upper Town could be divided in two by tracing a line along St. John and Fabrique streets and down St. Famille Street, with the upper section possessing the best buildings and streets and the lower half including narrow streets and crowded and "less respectably inhabited houses."[33] If one excludes soldiers, professionals included about 50 per cent of the adult work force in 1818 and 1842 (see Table 3). This high proportion of professionals gave the Upper Town a distinctively administrative character and one which changed little during the eighteenth and nineteenth centuries.

In the Lower Town occupational groups were more mixed, although even here segregation existed. Merchants preferred to live close to their wharves, and their businesses were usually located on a few major arteries such as St. Pierre Street, while artisans and labourers shared houses on streets beneath the cliffs and west of the city limits. Few professionals lived in Lower Town. Sailors excluded, the heads of family were represented in 1818 in the following way: craftsmen and transport workers, 55 per cent; merchants, 27 per cent; labourers and domestics, 6 per cent; and professionals, 5 per cent (see Table 3).

After the 1832 and 1834 epidemics, merchants from Lower Town and St. Roch began migrating to the St. Louis and St. John Suburbs. As the cholera-stricken Irish arrived by the thousands and the number of sailors and raftsmen continued to grow, Lower Town became increasingly dominated by labouring groups. An analysis of the 1818 and 1842 fig-

33. Jenkin Jones to Matthew Wilson, Montreal, 25 Aug. 1808, M6, 24, D11, PAC.

Table 3: Occupational Structure of Quebec Quarters in 1818 and 1842, according to percentages of Heads of Family

| Occupations | Upper Town | | Lower Town | | St. John | | St. Louis | | St. Roch | |
|---|---|---|---|---|---|---|---|---|---|---|
| | 1818 | 1842 | 1818 | 1842 | 1818 | 1842 | 1818 | 1842 | 1818 | 1842 |
| Commerce | 15 | 16 | 27 | 23 | 7 | 8 | 2 | 34 | 6 | 5 |
| Professionals incl. clergy, religious orders, & military officers | 49 | 49 | 5 | 1 | 3 | 4 | 7 | 19 | 2 | 4 |
| Craftsmen & transport | 30 | 20 | 55 | 33 | 67 | 54 | 61 | 19 | 58 | 66 |
| Labourers & domestics | 6 | .2 | 13 | 33 | 23 | 31 | 30 | .9 | 34 | 25 |
| Unknown | | 15 | | 10 | | 3 | | 27 | | .4 |

SOURCE: "Heads of family" include adults (except housewives, about which little information is provided), whose occupations are given in the 1818 *dénombrement* and 1842 census. As mentioned in the source for Table 2, this information, and especially that for 1818, is incomplete. Initial work on these sources was done by summer students (especially Nicole Casteran), revised by John Hare and Jean-Pierre Hardy, and classified by Ruddel.

ures[35] shows a decline in the percentage of merchants and professionals and a significant increase in the number of labourers (from 6 to 33 per cent of the Lower Town population).

Most of the merchants and professionals leaving the Lower Town moved to the upper plateau, settling in the suburbs, on the main arteries to the Upper Town – St. John's Road and Grande Allée (or St. Louis Road). This movement of merchants and professionals changed the social and ethnic composition of the population of the suburbs. In 1818

34. Because the curé underestimated the Protestant population, groups such as merchants and craftsmen are under-represented. Soldiers and sailors are also missing from these sources. Individuals for the General Hospital were excluded by the curé in 1818, but included in 1842 (by which time the city had expanded to include it). Clergy, female orders and military officers have been included in the professional category. Althought this classification is questionable (it could be argued, for example, that some members of the orders were not professionals and a number of officers were closer to the landed aristrocracy than to professionals), it provides a preliminary portrait of social groups in Quebec City. More work is needed to complete the 1818 figures and refine the categories. Such a revision would help explain the increased percentage of professionals in St. Roch in 1842, which was probably a result of the addition of the professionals from the General Hospital. The decrease in labouring groups in Upper Town can be partly attributed to the "unknowns" (a large number of which were probably workers) in the 1842 census. Additional information on this subject can be found in Ruddel, "Quebec City," pp. 666-67.

35. For the figures from which these percentages are calculated, see *ibid*.

only three anglophone merchants were noted in St. John's Suburb, but in 1842 they numbered fifty-two, almost as many as their Canadien counterparts (sixty-nine). The change in St. Louis was even more striking, where the anglophone business community jumped from zero to eighty-eight, outnumbering the smaller Canadien group of twenty-four.[36]

By 1818 a number of tradesmen had set up shops in the suburbs, but they were mainly inhabited by labouring groups of Canadien origin. In 1818, 98 per cent of the population in the three suburbs was Canadien.[37] The increase of anglophones, especially in St. John's Suburb, altered this situation, so that the percentage of Canadien workers in the suburbs decreased by 17 per cent. Anglophone artisans and labourers in St. John's Suburb had increased from less than 1 per cent in 1818 to 22 per cent in 1842.[38] The presence of anglophone workers in St. John's Suburb corresponded to the arrival of an increasing number of British merchants and immigrants, and the construction of a citadel. St. Roch remained an almost exclusively Canadien suburb – in 1842 over 90 per cent of the population was French-speaking.[39]

Although the ethnic composition of the suburbs changed slightly, the labouring characteristics of the largest quarters, St. John and St. Roch, remained constant. In 1818 and 1842, approximately 85 to 90 per cent of the heads of families in the quarters of St. John and St. Roch were artisans or labourers. If one can rely on information furnished by the *curé* in 1818 and by census takers in 1842, the composition of the labouring groups in the two quarters changed slightly. Whereas the percentage of artisans in St. John decreased between 1818 and 1842, the number of labourers increased. The reverse was true of St. Roch where the percentage of artisans increased and the number of labourers decreased. Movement between these neighbouring areas was one of the

36. *Ibid.*
37. 1,247 Canadiens and 24 anglophones were listed as suburban residents in the 1818 *dénombrement*. The *curé,s* tendency to neglect Protestants was hardly responsible for this large difference between the two ethnic groups.
38. In 1818, 2 anglophone artisans and 2 labourers were listed in St. John's suburb. In 1842, the total had increased to 366.
39. These percentages were taken from an analysis of ethnic groups given in tables 7, 8 and 9 in Ruddel, "Quebec City," pp. 662-64.

reasons for the change mentioned above, but other factors were also important.[40]

A significant change in the social composition of a suburb happened in St. Louis. In 1818, 91 per cent of this largely uninhabited area included heads of labouring families. Between 1819 and 1842, the movement of professionals, but especially merchants, from Lower Town and St. Roch transformed St. Louis into a new area for elites. Although the large number of residents whose occupation is unknown in 1842 makes an accurate evaluation difficult, it is obvious that merchants and professionals dominated the suburb: while the percentage of workers declined, elites increased from 9 per cent in 1818 to 53 per cent in 1842. By this latter date, the suburb was 70 per cent English-speaking.[41] Thus, by 1842 members of the Old City elite had created a residential area which possessed a healthier atmosphere than Lower Town and more accessible land than Upper Town.

Increasing segregation and differentitation helped maintain poor conditions in the suburbs, and especially in St. Roch. Differences in urban conditions intensified in the nineteenth century, reaching a peak in the 1840s. Merchants in Lower Town were able to maintain a healthy environment around their homes, but workers living in this area experienced some of the most crowded, unhealthy and noisy conditions in the city. The inhabitants of St. Roch lived in damp, dilapidated housing on streets that were not properly maintained. These two areas would ultimately be united in one large working-class district. Social and cultural segregation characteristics of large industrial centres was already evident in Quebec in the 1820s.

Additional work is necessary to study social and ethnic segregation. At this point, it appears that differences in street and housing conditions *within* the same area in the eighteenth century were transformed in the second quarter of the century into differences *between* urban districts. For example, differences noted by the Phoenix Insurance Company

40. Explanations for this modification in the composition of the work force are too numerous to explore at length. Obviously, an important factor was the change in the people taking the census. Although the *curé* made mathematical errors and occasionally missed streets, he was quite accurate in identifying parishioners. The reverse was almost the case with individuals taking the 1842 census. They were more accurate in their calculations, but failed to identify many individuals. Furthermore, men in St. Roch, especially those working in the shipyards, were probably more prone to identify themselves as artisans than as labourers, in spite of their skills. Classification of some occupations in the artisan category, such as sawyers, chairmakers and leatherworkers, also contributed to the number of craftsmen noted in St. Roch. Since some of these occupations resulted from specialization made possible by increasing industrialization, their identification with traditional crafts was becoming tenuous. More work is needed on this question. In any case, the factors noted above, combined with a slight movement in the population, contributed to the change in the two suburbs mentioned in the text.

41. See Ruddel, "Quebec City."

agent in 1808 between upper and lower sections of Upper Town become less noticeable than those between the Upper Town and St. Roch. The expansion of the homogenous working-class area in St. Roch in the first part of the nineteenth century indicates that once it began, the process of segregation took place during a relatively short period of time. Although Lower Town was gradually united with St. Roch, the former maintained a more heterogeneous population for most of the century. In this instance, the traditional residential character of differentiation was still evident. The growth of new Canadien working-class areas around St. Roch in the mid-nineteenth century, with inferior housing and street conditions, is an indication that segregation continued. The settlement of a number of Irish immigrants in Quebec did not alter ethnic segregation. The gradual departure of anglophones and the transfer of large financial enterprises to Montreal, probably diminished existing socio-cultural distinctions, but conditions in working-class areas did not improve; if anything, they got worse.

## EVALUATION OF THE COMMISSION'S ADMINISTRATION

The Commission of the Peace was a body of men nominated by the Governor's office which governed municipal affairs from 1765 to 1833 and from 1835 to 1840. Quebec City's Commission was dominated by Canadien and British merchants, and especially the latter. Most of the Canadien seigniors and notaries named to the Commission lived outside the city and, consequently, had little to do with municipal affairs. In 1802, for example, over 80 per cent of the justices of the peace in the city were merchants.[42] Urban administration by these men was conducted in the interests of the city's commercial and professional elite. Merchants, seigniors and colonial officers were too concerned for their own gain to be satisfactory administrators of public funds and urban regulations. Justices, and especially merchants, were interested in using their offices to advance their commercial enterprises. They achieved this end by making special concessions to other merchants and opposing changes considered prejudicial to their interests. This kind of administration often led to the expenditure of the largest part of the public money in the Old City, and services in this area, and especially in the Upper Town, outstripped those of the suburbs.[43]

Enforcement of urban regulations was a difficult task. Not only was it obstructed by the very men who were supposed to uphold the law, it was further complicated by confusion concerning the application of Road Acts and by the lack of authority and inexperience of the Road Inspec-

---

42. Information on the background of the Commission and the social and ethnic origins of its members can be found in Ruddel, "Quebec City," pp. 435-51.
43. Explained in greater detail in *ibid.*, pp. 515-17.

tors. Because the acceptance of an urban plan, which was an essential part of the Road Acts, had not occurred, proprietors questioned the Inspector's authority. According to the 1799 Road Act, the Inspector could not enforce regulations concerning street signs, boundaries and alignments until an urban plan was ratified by the Court of King's Bench. Although the Road Inspector made a plan and submitted it to the justices of the peace sitting in quarterly sessions, the latter refused to forward it to the higher court. The objection of the officials of the religious communities (especially the Hotel Dieu), private landholders and members of the military establishment, obstructed the use of a comprehensive plan for over thirty years. Since this period (1799-1833) was one of the most significant in the growth of the city, the lack of a plan contributed to chaotic development.[44]

The absence of a legal plan also led to confusing interpretations concerning city regulations. Experts contradicted each other and the Road Inspector. When the latter complained he was being prosecuted by proprietors and requested guidance from the justices, the magistrates not only failed to support him, but they also rejected cases he brought before the court, thus rendering ineffectual a costly and time-consuming part of enforcement. Since no comprehensive plan existed, Road Inspectors found themselves in the difficult situation of having to ratify individual street plans proposed by landowners, which often failed to respect basic regulations, such as street sizes and alignments.

Complaints addressed to the Commission of the Peace and city council by the Road Inspector included more infractions by anglophones than by francophones, even in areas such as St. Roch where the number of English-speaking people was small.[45] Professionals, and especially businessmen, knew what channels to take to avoid obligations and redress their grievances. Even after their colleagues, the magistrates, had ceased governing the city, the merchants continued to address themselves directly to city council.[46] If city regulations limited their interests, these men refused to obey the rules and defied the Road Inspector to prosecute them. The latter rarely took such action as he was aware of the merchants' financial resources and influence in judicial and administrative circles. Infractions of city regulations by merchants were mentioned often by the inspector: between 1833 and 1835, Larue reported twenty-eight instances of such cases in the Lower Town, including

44. For more information on this subject, see *ibid.*, pp. 510-14.
45. Information based on an analysis of the Road Inspector's reports (1817-1839) and on a study of the Road Committee minutes from 1833 to 1835, AVQ. These are the only years for which quantifiable documentation exists.
46. Statement based on an analysis of council minutes, 1833-1835, AVQ. See also Marcel Plouffe, "Quelques particularités sociales et politiques de la charte, du système administratif et du personnel politique de la cité de Québec, 1833-1867," M.A. thesis (Laval University, 1971), pp. 122-26.

twenty-four by British merchants.[47] Merchants obviously felt secure in abusing their power in Quebec's commercial centre.

Merchants felt confident in disobeying municipal regulations for a number of reasons. Prior to the acceptance of a city plan in 1833-1834, they knew the Road Inspector's authority could be challenged in some areas without fear of the law. Equally important was the knowledge that even if the Road Inspector followed established aspects of the law, merchant-justices sitting in quarter sessions often failed to uphold his complaints. A tacit agreement existed between members of Quebec's professional, commercial, military and religious elites concerning urban policies and priorities. Justices and most members of the elite agreed that streets in the Old City should be paved and repaired before those in the suburbs. Magistrates and members of the English-speaking elite accorded military projects a high priority and when officers opposed urban planning, the justices usually submitted to their wishes. The colonial government and most of the religious orders (especially officials of the Ursulines, Hotel Dieu and General Hospital) granted or leased land to merchants, allowing the latter to develop the property according to their wishes. This kind of collaboration between members of the elite conditioned the shape of Quebec.

Although often holding different political views, anglophone and francophone elites agreed on the necessity of maintaining a high quality of environmental conditions in areas where they lived. This consensus and its results were most evident in the Upper Town and St. Roch. Few Canadien justices or members of the Canadien elite attempted to improve conditions in St. Roch. Even when citizens from the suburbs petitioned the House of Assembly about their grievances, Canadien representatives did little to improve the formers' circumstances.

CONCLUSION

The military, religious communities and merchants were the three principal landowners responsible for shaping Quebec City's urban development. These groups correspond to the original functions of the town as a fort, missionary centre and port. Military intervention was particularly evident on the plateau where the citadel occupied 42 per cent of prime land in Upper Town and 35 per cent of the land in St. John's Suburb. Other military factors which conditioned the town's evolution included the fortifications of Upper Town and use of St. Roch as a residential area for the working class. The decision of the French regime to grant large land tracts to religious communities also had an important impact on Upper Town. Although their role in the development of other parts of Quebec needs further study, the ownership of land by

47. Road Committee minutes.

religious orders in St. Roch ultimately retarded expansion. The third group, the merchants, played an important role in stimulating urban growth and providing employment during periods of economic expansion. The price of such expansion, however, was shared by the inhabitants who saw their traditional means of access to the water blocked by port developments.

Increased differentiation and social and ethnic segregation occurred during this period of rapid expansion. In early nineteenth-century Quebec two movements were happening which would ultimately characterize industrialized centres later in the century. First, labouring groups, along with British sailors, were beginning to concentrate in increasing numbers throughout Lower Town and St. Roch. By 1820 socio-cultural segregation was so pronounced in St. Roch that it can be identified as one of the country's first ethnic and working-class quarters. Secondly, as more sailors, labourers and Irish immigrants frequented the Lower Town, merchants began migrating to the upper plateau, ultimately abandoning the Lower Town as a residential centre. In the early 1830s Quebec merchants had already made the distinction, which would become commonplace later, between residential and commercial areas. Although these men viewed both the Lower Town and St. Roch as undesirable living quarters, they considered the area an appropriate one for labouring groups.

Local government by the Commission of the Peace gave merchant-justices the opportunity of further influencing the city-building process. One of the merchant's most important roles in local government was the promotion of commercial growth. This was achieved in numerous ways, including the use of public funds to improve the port area and pave and maintain commercial avenues. Equally important was the task of overcoming obstacles to unbridled commercial growth, such as regulations concerning street sizes, alignments and public passages. The success of merchants, religious communities and military officials in opposing a comprehensive plan of development for Quebec facilitated the aims of these groups, while depriving the majority of the city's inhabitants of essential urban conditions. With a few notable exceptions (such as Jean-Baptiste Larue, Quebec's Road Inspector from 1817-1839), Canadien professionals did little to improve the urban conditions of Canadien workers. While benefiting from superior conditions in the Upper Town and on the plateau, most professionals turned a blind eye to the poor environment of areas such as St. Roch.

The combination of these influences resulted in the continued existence of poor urban conditions for the city's ethnic majority. Whereas members of founding groups in most North American cities enjoyed better quarters than     transient workers and immigrants, Canadien labouring families in Quebec continued to experience some of the worst urban conditions.[48]

48. See the brief comparison of urban conditions discussed in Ruddel, "Quebec City," pp. 619ff.

Figure 1: This view of Quebec, published in Bacqueville de la Potherie's *Histoire de l'Amérique septentrionale*, shows the two major divisions of Quebec in 1722. Notice the narrow beach area existing on both sides of Lower Town. Source: Public Archives of Canada.

Figure 2: An aquatint of J.P. Cockburn in the early 1830s which shows the bustling commercial centre of Quebec. Notice the government wharf at the Cul-de-Sac with its buildings, ships and soldiers and people waiting for the arrival of a dignitary. Source: Public Archives of Canada.

Figure 3: This extract of a plan of Quebec in 1804 clearly shows the development of wharves in the Lower Town and the beginnings of the St. Roch suburbs. The King's wood yard and William Grant's sawmill and shipyard were the major users of the beach at this time. Source: Public Archives of Canada.

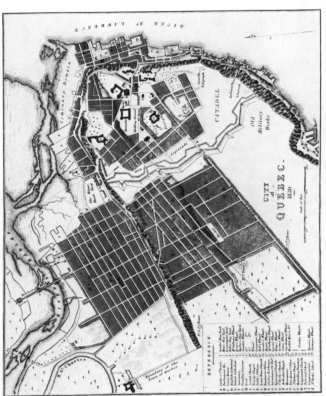

Figure 4: J.C. Waler's *City of Quebec* shows the extent of the old and new City in 1830. Evident are, the effects of land fill in Lower Town; land reserved for institutions in Upper Town; military outworks in the suburbs; and the boundaries of the City. Notice how the surveyor omitted the western part of the shoreline. Source: Public Archives of Canada.

Figure 5: On the order of the Minister of Marine, Chaussegros de Léry prepared this plan in 1752. Besides the new *enceinte* still under construction, the plan shows the urban extensions proposed by the engineer in Upper and Lower Town as well as the military *(non aedificandi)* zone established in St. John suburb. The land proposed for extension in the Upper Town will be reserved after 1760 for the construction of a Citadel. Source: Public Archives of Canada.

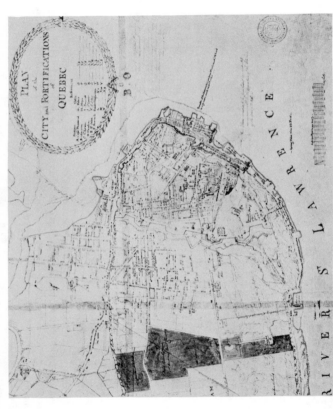

Figure 6: Though almost illegible, this detail of Dunford's 1821 "Plan of the City and fortifications of Quebec," is useful in that it shows how military property (shaded area) blocked the City's westward expansion on the plateau. Source: Public Archives of Canada.

# Speculation and the Physical Development of Mid-Nineteenth Century Hamilton

## MICHAEL DOUCET

Over the past fifteen years historians and historical geographers have begun to explore the mechanisms underlying the growth of cities and their suburbs during the nineteenth century.* In spite of this activity, there is, as yet, no very good sense of the way that the process of land development, defined here as the conversion of raw land into urban uses, operated in the Victorian city. The most comprehensive studies, such as those by H.J. Dyos and Sam Warner, have provided a useful beginning to this area of analysis, but many questions remained unanswered about both those who were involved in the land-development process and the ramifications of the process on the morphology of the city.[1] Moreover, many studies of urban development in the nineteenth century have been largely impressionistic and often anecdotal in their treatment of these questions. Two contradictory themes stand out in this body of literature. In the first place, several authors have stressed the importance of land speculators in the development of nineteenth-century cities, a theme, I might add, that persists in the "manipulated city" hypothesis that has become common in many recent analyses of current urban development. Writing of the 1880s, Mel Scott, for example, has argued that

*I wish to thank Michael Katz, Jim Lemon, Mark Stern, Brenda Torpy, Gil Stelter and Jim Simmons for their interest and comments.

1. H.J. Dyos, *Victorian Suburb: A Study of the Growth of Camberwell* (Leicester: Leicester University Press, 1961), and Sam Bass Warner, Jr., *Streetcar Suburbs: The Process of Growth in Boston, 1870-1900* (Cambridge: Harvard University Press, 1962).

almost everywhere the upsurge in population had touched off orgies of land platting and selling. . . . The subdivision of land long before home sites were really needed was such a common practice that few voices were raised against the speculators.[2]

I am not trying to suggest that there was no speculation in nineteenth-century urban North America. Such an assertion would be ridiculous, for there is considerable documentation that speculation existed and that it was widespread. But it is one thing to state that speculators were present, and quite another to assess their impact on that land market and their financial success. I am arguing, then, that we need fewer statements that speculators existed and more analyses of their role in urban development. The problem is that too many scholars have just examined the initial acquisition of property by people that they assumed to be speculators. As I will show, speculators were only one element in the process of land development in the Victorian city. We need to know a great deal more about what happened after the speculative phase had run its course.

The second theme in the impressionistic literature on nineteenth-century urban development is related to the notion that North America was a land of golden opportunity for those who came to settle there during the nineteenth century. Travel accounts from this period often stressed the relative ease with which settlers should have been able to acquire property on this continent.[3] Some historians, however, appear to have taken these comments too literally. Indeed, very few historians or historical geographers have examined the distribution of property ownership at more than a superficial level. This has led to some rather questionable generalizations about participation in the nineteenth-century land market. Ontario historian J.K. Johnson, for example, in writing of the career of mid-nineteenth-century London businessman William Warren Street, has stated that

if most of the earliest settlers in the province came with the intention of making a living from farming the land, many others, from at least as early a date, saw land primarily as a commodity to be bought and sold at a profit. . . . The buying and selling of land for possible profit, whether on a very small or a very

2. Mel Scott, *American City Planning Since 1890* (Berkeley: University of California Press, 1969), p. 2. For a critical analysis of the "manipulated city" hypothesis, see S.T. Roweis and A.J. Scott, *The Urban Land Question*, Papers on Planning and Design, No. 10 (Toronto: Department of Urban and Regional Planning, University of Toronto, 1976), pp. 19-30.

3. See, for example, Samuel Strickland, *Twenty-Seven Years in Canada West: Or the Experience of an Early Settler* (Edmonton: M. Hurtig, 1970[1853]), pp. 265-66, and William Catermole, *Emigration: The Advantages of Emigration to Canada* (London: Simpkin and Marshall, 1831), pp. 114-15. For a discussion of this issue, see David Gagan, "The Prose of Life: Literary Reflections of the Family, Individual Experience and Social Structure in Nineteenth-Century Canada," *Journal of Social History*, 9 (Spring 1976), pp. 367-81.

large scale, was almost a universal preoccupation, almost, it might be argued,
the Upper Canadian national game.[4]

This statement is enchantingly evocative, but it really does not tell us a
great deal about the ultimate role of land speculation in nineteenth-
century Ontario cities. Who were the speculators and where did they
come from? What proportion of the population was active in the vacant
land market? How profitable was land speculation in the nineteenth-
century city? What was the impact of speculation on land prices? What
bearing did the presence of speculators have on patterns of urban development? These are the issues that I intend to explore in this paper.

Land speculation is usually associated with the ownership of some
relatively large and arbitrarily determined amount of real estate which,
it is assumed, the would-be speculator hopes to sell one day for a profit.
Speculation, however, can only be successfully carried out under certain
conditions. It clearly works best, for example, whenever a monopoly or
oligopoly exists whereby a small group of individuals exercise a firm
control over the supply of land. This, however, rarely occurs. Boom
periods in the economy, on the other hand, can make speculation profitable for those who are lucky enough to have their money and land in the
right place at the right time. The apparent success of the early arrivals
will often persuade others to invest in real property. As the early
twentieth-century land economist W.N. Loucks noted,

increments in land values have always been a big factor in attracting investments in land. Tales of fabulous fortunes made from holding land for an
increase in value have been uncritically accepted in popular opinion of land
investments. Particularly in fast-growing cities . . . the possibility of pocketing
relatively effortless gains in land values has lured many investors beyond the
boundaries of caution.[5]

It is dangerous to assume that everyone who invested in land in
nineteenth-century cities made money. Yet it is only through a detailed
empirical study that we can hope to evaluate land speculation and its
impact on urban development. This is not the first such investigation,
for it was a topic of interest to land economists in the 1920s and 1930s,

4. J.K. Johnson, "The Businessman as Hero: The Case of William Warren Street," *Ontario
History*, 65 (September 1973), pp. 127-28. Johnson, however, does correctly note that
merchants had to invest in land during this period. On the distinction between the use-
value and the exchange-value of land see Michael B. Katz, Michael J. Doucet, and Mark
J. Stern, *The Social Impact of Early Industrial Capitalism: Themes in the North American
Experience* (Cambridge: Harvard University Press, 1982), chapter 4.
5. W.N. Loucks, "Increments in Land Values in Philadelphia," *Journal of Land and Public
Utility Economics*, 1 (October 1925), pp. 469.

but it is the first study to consider the problem of speculation and development in this manner for a Canadian city.[6]

Speculation has long served as a convenient scapegoat for the seemingly irrational development patterns of the Victorian city. I intend to show, however, that it was only one element in the system of land allocation during that era. Make no mistake, speculators were present in every North American city, but everyone was not speculating in land. I will demonstrate that, in urban areas at least, speculation was dominated by members of the elite. Moreover, and perhaps of even greater importance, the impact of speculators diminished quite rapidly as time passed.

This paper derives from a detailed investigation into the process of land development in Hamilton, Ontario, between the years 1847 and 1881.[7] Three data sources relate to the themes to be discussed here. First, use was made of the assessment rolls for the City of Hamilton for 1852 and 1881 to evaluate the degree of concentration of vacant lot ownership in the city at two dates. Next, registered plans are utilized to assess changes in the supply of vacant land during the years in question. Finally, registered instruments from the Hamilton Registry Office are employed to determine the identity of those who bought, sold and developed property in fourteen selected subdivisions that together contained 1,668 building lots. In all, 7,154 legal instruments were registered for these properties between 1847 and 1881.

During the early settlement phase of most nineteenth-century North American cities the ownership of vacant land was probably fairly concentrated. Those who arrived on the scene first were often able to purchase large areas of land very cheaply, and at a later date, these lands would be converted into building lots. This was clearly the case in Hamilton. Just after 1800, the thirty farm lots that would one day constitute the entire area of the city at mid-century rested in the hands of nine individuals. Over the years, the ownership of vacant land remained fairly concentrated, but the number of people involved expanded rather dramatically.

Speculation in Hamilton's future development sites was neither a universal preoccupation for the residents of the city, nor was it restricted to a few people. Although there was a great degree of concentration of ownership in the vacant land market, it was not sufficient to permit absolute control over land prices, the tempo of development, or subse-

---

6. Important studies include Homer Hoyt, *One Hundred Years of Land Values in Chicago* (Chicago: University of Chicago Press, 1933); Helen C. Monchow, *Seventy Years of Real Estate Subdivision in the Region of Chicago* (Evanston: Northwestern University Press, 1939); and G.B.L. Arner, "Land Values in New York City," *Quarterly Journal of Economics*, 36 (August 1922), pp. 545-80.

7. Michael J. Doucet, "Building the Victorian City: The Process of Land Development in Hamilton, Ontario, 1847 to 1881," Ph.D. thesis (University of Toronto, 1977).

Table 1: Concentration of Vacant Lot Ownership,
Hamilton, Ontario, 1852[1]

| No. of Lots Owned | Assessed population | | | | Total lots controlled[1] | | |
|---|---|---|---|---|---|---|---|
| | No. | % | Cumulative[2] % | | No. | % | Cumulative % |
| 0 | 2391 | 89.1 | 89.1 | | 0 | 0.0 | 0.0 |
| 1 | 170 | 6.4 (58.2) | 95.5 | (58.2) | 170 | 11.2 | 11.2 |
| 2-5 | 87 | 3.3 (29.8) | 98.8 | (88.0) | 238 | 15.8 | 27.0 |
| 6-10 | 17 | 0.6 (5.8) | 99.4 | (93.8) | 130 | 8.6 | 35.6 |
| 11-25 | 9 | 0.3 (3.1) | 99.7 | (96.9) | 145 | 9.5 | 45.1 |
| 26+ | 9 | 0.3 (3.1) | 100.0 | (100.0) | 836 | 54.9 | 100.0 |
| Totals | 2683 | | | | 1519 | | |

SOURCE: Compiled from the City of Hamilton Assessment Rolls, 1852, MSS.
NOTES:
1. Includes only lots owned by Hamiltonians. In 1852 there were 2,525 vacant parcels in the city, with 39.9 per cent controlled by non-residents.
2. Figures in parentheses relate to the proportion of all Hamilton lot owners (N = 292).

quent land uses. By 1852 almost three hundred people possessed vacant land in Hamilton (Table 1). There were, of course, some very large owners within this group. Thirty-five people possessed 73 per cent of the lots controlled by Hamilton residents, with just nine individuals owning 55 per cent of these properties. For the most part, these larger owners were long-time residents of the city who had subdivided raw land into building lots at some time during the late 1840s and early 1850s. They still owned relatively large numbers of lots because they had not been able to sell their holdings as quickly as they had anticipated. Hamilton, then, had its share of land speculators in 1852. Indeed, no clearer evidence of this can be seen then in the fact that 40 per cent of the city's vacant lots were owned by non-residents at that date. Not all of them, however, would profit from their adventures.

One hallmark of the nineteenth-century urban land market, as Homer Hoyt and Helen Monchow have shown, was the premature nature of many of the subdivisions of this period.[8] As a result, there was nearly always a surplus of building lots over and above the needs for immediate development in the Victorian city. Hamilton was no exception, for

8. Hoyt, *One Hundred Years*, p. 109, and Helen C. Monchow, "Population and Subdividing Activity in the Region of Chicago: 1871-1930," *Journal of Land and Public Utility Economics,* 9 (May 1933), pp. 192-206. See also Philip H. Cornick, *Premature Subdivision and Its Consequences* (New York: Columbia University Press, 1938).

in 1852 there were enough building lots on the market to house almost 17,000 more people at a time when the citizens of the city barely numbered 14,000. For land speculation to be profitable, then, the speculators would need to gain a tighter control over the supply of lots, or they would have to be assisted by an upswing in the economy that would induce people to invest in land. The boom years that preceded the international panic of 1857 provided just such an aid to some of those who had invested in land in Hamilton. This was fortunate for the early investors since the ownership of vacant land continued to disperse as the years passed. True oligopolistic control over the land market would await a much later day, and even then it could sometimes prove difficult to exercise any real measure of market power.[9]

Throughout the period under investigation there is very little evidence of any source of collusion on the part of the large landowners in Hamilton. New subdivisions came on the market with a rhythm that appears to have been more closely related to cycles in the general economy than to any attempt to exercise control over the supply of land.[10] Between 1850 and 1880 almost 4,800 new building lots were surveyed, registered and offered for sale within Hamilton (Table 2). Moreover, several subdivision plans were also registered for areas that lay just beyond the limits of the city. Advertisements for these lots often touted the fact that both lot prices and property taxes were lower in these areas. An excess of vacant properties continued to characterize the local land market. Many people must have found the subdivision process attractive during this period, for 94 different names were associated with the 104 plans registered during the years studied. While many of these people had similar backgrounds in terms of their occupations, religious affiliations and levels of participation in other areas of community life, the sheer numbers involved in this process mitigated against the existence of any tight control over the land market. Finally, building lots were generally offered for sale very soon after the plan of subdivision had been executed; sometimes lots were sold prior to the actual registration of the plan. In other words, there is very little to suggest that conscious attempts were being made to keep lots off the market. Newspaper advertisements of the period always noted that the entire survey would be offered for sale. As a result, there was still a sizable excess of building

9. For a discussion of this issue, see James R. Markusen and David T. Scheffman, *Ownership Concentration and Market Power in Urban Land Markets*, Discussion Paper 003 (London: Department of Economics, University of Western Ontario, 1975).
10. Note, for example, that 28.7 per cent of all lots were registered between 1853 and 1857 and 27.3 per cent were registered between 1868 and 1873. Only 10.8 per cent of the properties, however, were registered between 1858 and 1867. For a discussion of the growing literature on nineteenth-century building cycles see Doucet,"Urban Land Development in Nineteenth-Century North America: Themes in the Literature," *Journal of Urban History*, 8 (May 1982).

Table 2: Summary of Registered Plans, Hamilton, 1850-1880

| Year | Plans registered No. | % | Lots registered No. | % | Lots per survey |
|------|------|------|------|------|------|
| 1850 | 3 | 2.9 | 73 | 1.5 | 24 |
| 1851 | 1 | 1.0 | 146 | 3.1 | 146 |
| 1852 | 1 | 1.0 | 109 | 2.3 | 109 |
| 1853 | 4 | 3.8 | 463 | 9.7 | 116 |
| 1854 | 7 | 6.7 | 348 | 7.3 | 50 |
| 1855 | 5 | 4.8 | 186 | 3.9 | 37 |
| 1856 | 6 | 5.8 | 155 | 3.2 | 26 |
| 1857 | 3 | 2.9 | 219 | 4.6 | 73 |
| 1858 | 1 | 1.0 | 98 | 2.1 | 98 |
| 1859 | 1 | 1.0 | 136 | 2.8 | 136 |
| 1860 | 1 | 1.0 | 157 | 3.3 | 157 |
| 1861 | 1 | 1.0 | 32 | 0.7 | 32 |
| 1862 | 0 | - | 0 | - | 0 |
| 1863 | 0 | - | 0 | - | 0 |
| 1864 | 2 | 1.9 | 18 | 0.4 | 9 |
| 1865 | 1 | 1.0 | 37 | 0.8 | 37 |
| 1866 | 1 | 1.0 | 17 | 0.3 | 16 |
| 1867 | 2 | 1.9 | 18 | 0.4 | 9 |
| 1868 | 6 | 5.8 | 358 | 7.5 | 60 |
| 1869 | 2 | 1.9 | 358 | 0.2 | 6 |
| 1870 | 2 | 1.9 | 34 | 0.7 | 17 |
| 1871 | 3 | 2.9 | 61 | 1.3 | 20 |
| 1872 | 11 | 10.6 | 295 | 6.2 | 27 |
| 1873 | 10 | 9.6 | 544 | 11.4 | 54 |
| 1874 | 3 | 2.9 | 95 | 2.0 | 32 |
| 1875 | 6 | 5.8 | 136 | 2.8 | 23 |
| 1876 | 1 | 1.0 | 40 | 0.8 | 40 |
| 1877 | 3 | 2.9 | 170 | 3.6 | 57 |
| 1878 | 9 | 8.7 | 378 | 7.9 | 42 |
| 1879 | 4 | 3.8 | 321 | 6.7 | 80 |
| 1880 | 4 | 3.8 | 122 | 2.6 | 31 |
| Total | 104 | | 4776 | | |

SOURCE: Compiled from copies of the Registered Plans of the City of Hamilton, Map Library, McMaster University.

lots in Hamilton in 1881. Enough, in fact, to take care of population growth until some time after 1901.

Although vacant lots were still owned by a minority of Hamiltonians in 1881, the number of people involved had doubled and the overall importance of small-scale owners had increased markedly over the three

Table 3: Concentration of Vacant Lot Ownership,
Hamilton, Ontario, 1881[1]

| No. of lots owned | Assessed Population | | | | Total lots controlled[1] | | |
| | No. | % | Cumulative[3] % | | No. | % | Cumulative % |
|---|---|---|---|---|---|---|---|
| 0 | 6880 | 92.0 | 92.0 | | 0 | 0.0 | 0.0 |
| 1 | 318 | 4.3 (53.3) | 96.3 | (53.3) | 318 | 15.8 | 15.8 |
| 2-5 | 213 | 2.8 (35.7) | 99.1 | (89.0) | 587 | 29.1 | 44.9 |
| 6-10 | 46 | 0.6 (7.7) | 99.7 | (96.7) | 333 | 16.5 | 61.4 |
| 11-25 | 13 | 0.2 (2.2) | 99.9 | (98.9) | 180 | 8.9 | 70.3 |
| 26+ | 7 | 0.1 (1.1) | 100.0 (100.0) | | 598 | 29.7 | 100.0 |
| Totals | 7477 | | | | 2016 | | |

SOURCE: [1]Compiled from the City of Hamilton Assessment Rolls, 1881, MSS.
NOTES:
[1]Includes only lots by Hamiltonians. In 1881 there were 2,166 vacant parcels in the city, with 6.9 per cent controlled by non-residents.
[2]Figures in parentheses relate to the proportion of all Hamilton lot owners (N = 597).

decades under investigation (Table 3). In fact, in 1881 more than three-fifths of the building lots were owned by people who controlled no more than ten properties, in contrast to less than two-fifths in 1852. Then, too, the ownership of lots by non-residents had declined rather dramatically from forty to seven per cent. Patterns of ownership had clearly changed in thirty years. To examine these trends in more detail, it is necessary to provide an intensive analysis of activities in selected Hamilton subdivisions.

The fourteen surveys selected for detailed analysis were all placed on the market prior to 1860, and each one contained at least fifty building lots. Those who laid out these subdivisions were very definitely members of Hamilton's elite. All of them held high status occupations, some were involved in local and provincial politics, and even more served as directors of railways, banks and other important enterprises (Table 4). In a nutshell, Hamilton's subdividers were intimately involved in promoting the growth and development of their city. More often than not they were able to profit from their forays into the local land market.[11] This is not surprising given the fact that they were engaged in the conversion of raw

11. Ibid., Data were available for ten of the fourteen surveys. Money was lost on only two of these subdivisions. Over the first ten years after registration, the annual rate of return for the profitable subdivisions ranged from 7.1 to 54.3 per cent of the original purchase price of the raw land.

| Survey key No. | Subdivider(s) | Occupation | Elected office | Directorships Rail | Directorships Others | Date of plan | No. of city Blocks[2] | No. of lots[3] | Modal frontage (feet) | Sales prior to registration |
|---|---|---|---|---|---|---|---|---|---|---|
| 1 | STINSON, T. | Banker | No | - | - | 3/58 | 8 | 98 | 47 | No |
| 2 | HAMILTON, P.H. | Gentleman | No | 1 | 1 | 10/48 | 20 | 220 | 63 | No |
| 3 | FERGUSON, P. | Gentleman | No | - | - | 1/54 | 12 | 106 | 56 | Yes |
| 4 | MACNAB, A.N. | Gentleman | MPP | 3 | 2 | 10/57 | 10 | 129 | 66 | Yes |
| 5 | BEASLEY, R.G. | Merchant | No | - | - | ?/54 | 6 | 100 | 40 | No |
| 5 | BEASLEY, R.G. | Gentleman | No | - | - | 8/48 | 9 | 109 | 50 | Yes |
| 6 | TISDALE, V.H. | Merchant | No | - | - | ?/54 | 6 | 100 | 40 | No |
| 7 | WILSON, H.B. | Gentleman | CC | 2 | - | 10/47 | 12 | 206 | 52 | No |
| 8 | WILSON, H.B. | Gentleman | - | - | - | 10/47 | 8 | 164 | 52 | No |
| 9 | CAMERON, J.H. | Gentleman | MPP | - | - | 3/53 | 2 | 47 | 66 | Yes |
| 10 | PRINGLE, J.D. | Lawyer | CC | 2 | 1 | | | | | |
| | LOGIE, A. | Lawyer | No | - | - | 1/55 | 4 | 55 | 30 | Yes |
| | GRIFFIN, W. | Crown Clerk | No | - | - | | | | | |
| 11 | BILLINGS, W.L. | Physician | CC | 1 | 1 | | | | | |
| | LISTER, J. | Merchant | CC | 1 | - | 6/55 | 2 | 49 | 4 | No |
| 12 | MACNAB, A.N. | Merchant | | | | ?/57 | 3 | 68 | 50 | Yes |
| 13 | MOORE, E. | Lumber | No | - | - | | | | | |
| | MOORE, J.F. | Merchants | CC | - | 2 | ?/47 | 4 | 83 | 57 | Yes |
| 14 | KERR, A. | Merchant | No | 1 | 4 | | | | | |
| | MCLAREN, W.P. | Merchant | No | 3 | 4 | 11/54 | 15 | 234 | 50 | Yes |
| | STREET, R.P. | Actuary | No | 2 | 1 | | | | | |

SOURCE: Compiled from Hamilton Registry Office records, Minutes of the Hamilton City Council, and the extant charters of Hamilton companies and institutions.

NOTES:
[1] See Figure 1 for the exact location of each survey.
[2] Includes all blocks contained within each survey, even if only partially platted.
[3] N = 1,668.

and often unused land into urban building lots, a process that greatly increased the potential intensity at which the land could be utilized. Poor timing and bad luck, however, could still ruin the schemes of even ingenious subdividers, especially if they were under-financed and heavily mortgaged. But profits awaited most of those who subdivided raw acreage into urban building lots in the nineteenth-century city. If we can apply the term speculator to such individuals, then speculators appear to have profited in many cases. Unfortunately, few scholars have ventured beyond this phase in the development process. This is a critical shortcoming in the study of urban growth, for as we shall soon see, the subdividers were not often involved in the development of urban property, nor in most cases were the people who purchased land from them. If we wish to study the land-development process in the Victorian city we must look far beyond the role of the subdividers. They set the tone for future development by laying out streets and lots, but other people further down the development chain made the actual decisions to build.

What happened to urban building lots after the subdividers had disposed of them? We should note first that few subdividers were able to sell all of their lots overnight. Even ten years after the property first had been offered for sale, most subdividers still had some building lots to sell (Table 5). Of course, this is a reflection of the premature nature of many of the surveys. Marketing was not always an easy task. The demand for lots, even by would-be speculators, was seldom great enough to ensure the rapid sale of a subdivision. Obviously, this reduced the profitability of these projects, since some not insignificant carrying charges, such as property taxes and mortgage interest, had to be met each year.

The properties in our selected subdivisions changed hands an average of 4.3 times each during the period under investigation. Over the years examined, the owners of the lots changed not only by name but also in terms of the number of people involved and their characteristics. The 456 people who made directly from the subdividers, for example, were primarily individuals who had invested in land for purposes other than development (Table 6). Almost one-quarter of them were non-residents of the city, but this did not mean that they were necessarily strangers to Hamilton. Some had lived in the city at an earlier date, several had invested heavily in Hamilton's railways and industrial enterprises, and about one-third of them lived in towns, villages, and townships that lay within twenty-five miles of the city. Most of the non-resident land purchasers held elite occupational titles (Table 7). In fact, this pattern was true for the entire group of first purchasers, for most of those who bought land from the subdividers were involved in promoting the growth of the economy. For example, the servicers of the development industry (lawyers, bankers, agents, and the like), members of the merchant industrial bourgeoisie, and the landed gentry (gentlemen) together comprised 68 per cent of all first purchasers. Few of these people, as we shall see, were interested in property development. Further evidence of

Table 5: The Success of the Selected Subdividers in Selling Their Land

| Survey Number | Purchase date | Date of first sale[1] | No. of lots | Unsold After | | | | | | | | | | % still Unsold | Estimated[2] taxes | Best year for sales |
|---|---|---|---|---|---|---|---|---|---|---|---|---|---|---|---|---|
| | | | | 1 yr | 2 yr | 3 yr | 4 yr | 5 yr | 6 yr | 7 yr | 8 yr | 9 yr | 10 yr | | | |
| 1 | 02/50 | | 98 | 98 | 97 | 97 | 95 | 91 | 63 | 53 | 49 | 41 | 40 | 41[5] | $1086 | 1855-6 |
| 2[3] | 09/23 | 11/48 | 220 | 183 | 180 | 172 | 113 | 96 | - | - | - | - | - | - | 1116 | 1851-2 |
| 3 | 12/02 | 09/53 | 106 | 67 | 55 | 42 | 36 | 30 | 28 | 24 | 24 | 22 | 20 | 19 | 522 | 1853-4 |
| 4 | 12/32 | 05/45 | 129 | 49 | 37 | 34 | 29 | 26 | 24 | 20 | 18 | 12 | 12 | 9 | 392 | 1845 |
| 5 | 01/41 | 10/48 | 109 | 108 | 107 | 106 | 20 | 12 | 12 | 12 | 3 | 1 | - | 1 | 573 | 1851-2 |
| 6[4] | 03/54 | | 100 | 100 | 77 | 77 | 77 | 77 | 77 | 60 | 60 | - | - | 1 | 908 | 1855-6 |
| 7 | 07/48 | | 206 | 203 | 186 | 155 | 148 | 146 | 107 | 93 | 47 | 10 | 9 | 4 | 1656 | 1855-6 |
| 8 | 08/48 | | 164 | 0 | 0 | 0 | 0 | 0 | 0 | 0 | 0 | 0 | 0 | 0 | 0 | 1848 |
| 9 | 07/51 | | 47 | 47 | 2 | 2 | 2 | 2 | 0 | 0 | 0 | 0 | 0 | 0 | 83 | 1852-3 |
| 10 | 06/53 | | 55 | 26 | 14 | 12 | 8 | 6 | 5 | 5 | 5 | 5 | 5 | 9 | 137 | 1853-4 |
| 11 | 06/55 | | 49 | 49 | 49 | 46 | 43 | 33 | 27 | 27 | 8 | 4 | 4 | 8 | 435 | 1862-3 |
| 12 | 12/32 | 10/48 | 68 | 17 | 14 | 12 | 8 | 8 | 4 | 4 | 4 | 4 | 4 | 6 | 113 | 1848-9 |
| 13 | 10/46 | | 83 | 68 | 68 | 62 | 50 | 31 | 31 | 14 | 5 | 3 | 1 | 1 | 500 | 1850-1 |
| 14 | 07/53 | | 234 | 144 | 19 | 19 | 16 | 16 | 16 | 16 | 16 | 16 | 16 | 7 | 441 | 1854-5 |

SOURCE: Compiled from the records of the Hamilton Registry Office and from City of Hamilton assessment rolls, 1847-1881.

NOTES:
[1] For surveys that were purchased long before they were offered for sale, the date of the first lot sale as the base point.
[2] Taxes based on average of $1.50 per unsold lot per year.
[3] Peter Hamilton died in 1854. The unsold lots were divided among his heirs.
[4] Valentine Tisdale lost 60 of his lots through foreclosure in 1862.
[5] Stinson built 41 cheap houses on these lots and rented them to members of the working class. These dwellings were still owned by the family in 1881.

Table 6: A Comparison of the Initial and Final Lot Owners

| Variables | First Purchasers | Final Owners |
|---|---|---|
| *Occupational Group* | | |
| Development Industry | 25.6% | 19.1% |
| a) Builders | 12.3 | 12.0 |
| b) Servicers | 13.3 | 7.1 |
| Government Officers | 3.9 | 1.2 |
| Merchant/Industrial Bourgeoisie | 27.7 | 19.4 |
| White Collar | 1.4 | 7.9 |
| Skilled Workers | 4.8 | 15.8 |
| Unskilled and Semi-Skilled | 1.5 | 6.7 |
| Professionals | 1.4 | 2.0 |
| Landed Gentry | 26.6 | 9.6 |
| Women | 3.0 | 13.0 |
| Unclassifiable | 4.1 | 5.4 |
| *Place of Residence* | | |
| Hamilton | 75.5% | 91.6% |
| Nonresidents | 24.5 | 8.4 |
| *Number of Lots* | 1686 | 1811 |
| *Number of Owners* | 456 | 937 |

SOURCE: Compiled from Hamilton Registry Office records, 1847-1881.

Table 7: Occuaptions of Hamilton Land Buyers by Place of Residence

| Occupational Group | Residents | Non residents | All |
|---|---|---|---|
| Development industry | 21.2% | 5.6% | 19.0% |
| a) Builders | 16.2 | 2.0 | 14.2 |
| b) Servicers | 5.0 | 3.6 | 4.8 |
| Government officers | 1.8 | 1.3 | 1.7 |
| Merchant/industrial bourgeoisie | 18.6 | 12.5 | 17.7 |
| White collar | 4.1 | 2.3 | 3.9 |
| Skilled workers | 15.2 | 5.9 | 13.9 |
| Unskilled and semi-skilled | 7.9 | 2.0 | 7.1 |
| Professionals | 2.3 | 4.0 | 2.6 |
| Landed gentry | 10.9 | 40.3 | 15.0 |
| Women | 13.8 | 12.5 | 13.6 |
| Unclassifiable | 4.3 | 13.5 | 5.6 |
| Totals | 1880 | 303 | 2183 |

SOURCE: Compiled from Hamilton Registry Office records, 1847-1881.

Table 8: Changes in the Distribution of Lot Ownership: A Comparison of the Initial and Final Owners

| No. of lots owned | Initial Owners | | | | | | Final Owners | | | | | |
|---|---|---|---|---|---|---|---|---|---|---|---|---|
| | Individuals Involved | | | Lots controlled | | | Individuals Involved | | | Lots controlled | | |
| | No. | % | Cum. % | No. | % | Cum. % | No. | % | Cum. % | No. | % | Cum. % |
| 1 | 259 | 56.8 | 56.8 | 259 | 15.4 | 15.4 | 585 | 62.4 | 62.4 | 585 | 32.3 | 32.3 |
| 2 | 83 | 18.2 | 75.0 | 166 | 9.8 | 25.2 | 198 | 21.1 | 83.5 | 396 | 21.9 | 54.2 |
| 3-5 | 61 | 13.4 | 88.4 | 224 | 13.3 | 38.5 | 110 | 11.7 | 95.2 | 385 | 21.3 | 75.5 |
| 6-9 | 20 | 4.4 | 92.8 | 137 | 8.1 | 46.6 | 29 | 3.1 | 98.3 | 204 | 11.3 | 86.8 |
| 10-19 | 7 | 1.5 | 99.1 | 177 | 10.5 | 74.9 | 2 | 0.3 | 100.0 | 192 | 10.6 | 97.4 |
| 20-49 | 7 | 1.5 | 99.1 | 177 | 10.5 | 74.9 | 2 | 0.3 | 100.0 | 49 | 2.6 | 100.0 |
| 50+ | 4 | 0.9 | 100.0 | 423 | 25.1 | 100.0 | | | | | | |
| Totals | 456 | | | 1686 | | | 937 | | | 1811 | | |
| | | | | | | | 937 | | | 1811 | | |

SOURCE: Compiled from Hamilton Registry Office records, 1847-1881.

the speculative nature of these first purchases can be seen in the distribution of the lots after the initial round of transactions (Table 8). Fifty-one people had each bought at least six properties in our selected subdivisions, and together they controlled about 63 per cent of the building lots. Only about one-fifth of the lots were developed by the initial buyers. Speculation, then, was still occurring, but even at this stage it was being carried out by a well-defined class of investors. Taken together, the three occupational groups that were most prominent in buying land from our subdividers constituted a rather small fraction of Hamilton's workforce, about 12 per cent in both 1852 and 1861 (Table 9). On the other hand, white-collar, skilled, unskilled, and semi-skilled workers, who together made up more than half of Hamilton's labour force, were involved in less than 8 per cent of the purchases from the subdividers.

Table 9: Occupational Structure, Hamilton, 1852-1881

| Occupational Category | 1852 | 1861 | 1872 | 1881 |
|---|---|---|---|---|
| Development industry | 19.3% | 15.1% | 14.5% | 13.5% |
| a) Builders | 17.7 | 13.1 | 13.0 | 11.5 |
| b) Servicers | 1.5 | 2.0 | 1.5 | 2.0 |
| Government officers | 1.6 | 1.6 | 0.8 | 0.9 |
| Merchant/industrial bourgeoisie | 6.2 | 7.1 | 6.5 | 7.0 |
| White collar | 8.9 | 8.5 | 10.4 | 10.7 |
| Skilled workers | 24.3 | 18.8 | 29.2 | 29.1 |
| Unskilled and semi-skilled | 25.8 | 26.3 | 24.5 | 20.4 |
| Professionals | 2.4 | 2.3 | 1.3 | 1.6 |
| Land gentry | 4.0 | 3.0 | 3.1 | 3.1 |
| Women | 5.2 | 11.7 | 8.3 | 9.8 |
| Unclassifiable | 2.4 | 5.5 | 1.5 | 3.7 |
| Total number | 2412 | 3324 | 6517 | 7477 |

SOURCE: Calculated from the City of Hamilton Assessment Rolls, 1852, 1861, 1872, and 1881, MSS.

By 1881, 78 per cent of the properties under scrutiny became developed. On average, these parcels had changed hands 3.2 times prior to development. Under current land market conditions, we would expect such activity to drive land prices increasingly higher. Such, however, was not always the case in nineteenth-century Hamilton (Figure 2). To be sure, land prices were rising rapidly during the years preceding the depression that followed the panic of 1857, but vacant lot prices were much lower in the 1860s and 1870s than they had been in the early 1850s. There were two basic reasons for this pattern. First, the depression had a devastating effect on the North American urban economy. Many of those who had invested heavily in real estate were ruined

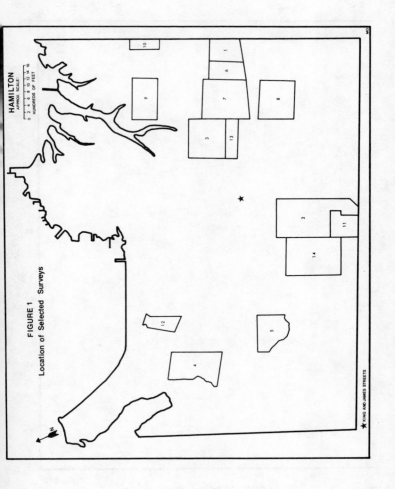

FIGURE 1

Location of Selected Surveys

HAMILTON
APPROX. SCALE!
0  2  4  6  8  10  12  14  16
HUNDREDS OF FEET

★ KING AND JAMES STREETS

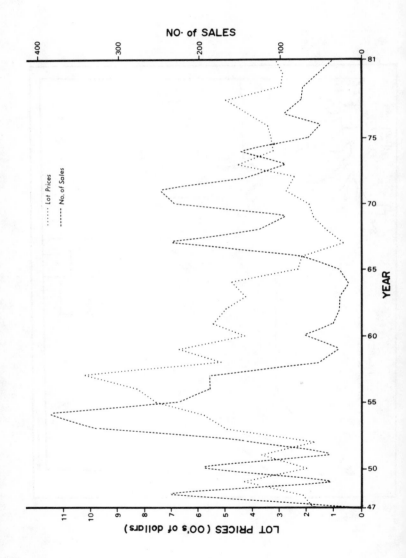

financially and were no longer even able to pay the relatively modest property taxes that were levied at this time. Many lots were seized by local authorities and sold for the amount of the back taxes. In 1865 alone almost 1,300 building lots were auctioned for this reason in Hamilton, and from this date, the tax sale was an annual fall event. Naturally, the dumping of these cheap lots (most sold for less than twenty dollars) on the open market each year would mitigate against the steady advance of lot prices. The second factor that operated to curb inflation in land prices was the continual increase in the supply of building lots throughout most of the years under study (Table 2). Indeed, in the 1870s some low-priced subdivisions were placed on the market in which it was possible to purchase building lots for as little as fifty dollars (Figure 3). Under these conditions it would be very difficult to justify continually escalating prices in the sale of lots in already established subdivisions. While many of the earlier surveys were located close to the centre of the city, Hamilton was so small during this period that relative proximity, by itself, would not be enough to sustain great price differentials, especially for residential lots (Figure 4). This same phenomenon was probably not as true in larger cities where the land-value gradients were more complex than was the case in compact cities like Hamilton.

The trends discussed above seem to suggest that investments in land could not always be expected to pay off in the Victorian city. Yet we must also stress that not every transaction that was recorded for a given parcel of land represented a real transfer of property ownership. There were other reasons to dispose of land besides the profit motive. For example, many transactions were at less than arm's length since they were concluded between individuals who were related to each other. In still other deals, ownership switches were completed between people who were friends and/or business partners.[12] Such transactions were usually made so that the former owner would not lose his property to creditors. Under these circumstances, profits were not always an important consideration. Nevertheless, money was clearly lost on 41 per cent of the real transactions involving vacant lots for which we have complete price information. This figure would be considerably higher if we were able to take all possible carrying charges into account.[13] It does not seem unreasonable, therefore, to conclude that people who invested in already platted building lots had no better than an even chance to make money in the Victorian city. As the land economist G.B.L. Arner noted

12. About 20 per cent of the properties studied were subject to less-than arm's-length transactions. Shady deals were recorded for a similar proportion of the lots. In all, 38 per cent of the properties contained one of these "deals" in their transaction histories.
13. The list of carrying charges includes advertising costs, legal fees, property taxes, mortgage interest and interest lost because capital was tied up in land. On other avenues of investment during this period, see Robert P. Swierenga, *Pioneers and Profits: Land Speculation on The Iowa Frontier* (Ames: Iowa State University Press, 1968), pp. 205-209.

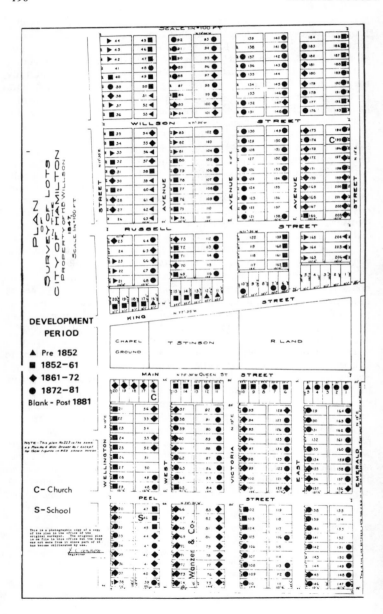

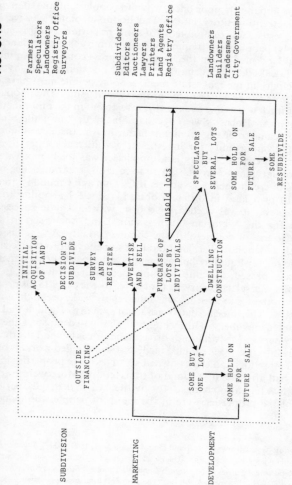

## COMPONENTS

SUBDIVISION

MARKETING

DEVELOPMENT

## PROCESS

INITIAL ACQUISITION OF LAND

DECISION TO SUBDIVIDE

OUTSIDE FINANCING

SURVEY AND REGISTER

ADVERTISE AND SELL

unsold lots

PURCHASE OF LOTS BY INDIVIDUALS

SPECULATORS BUY SEVERAL LOTS

SOME HOLD ON FOR FUTURE SALE

SOME RESUBDIVIDE

SOME BUY ONE LOT

SOME HOLD ON FOR FUTURE SALE

DWELLING CONSTRUCTION

## ACTORS

Farmers
Speculators
Landowners
Registry Office
Surveyors

Subdividers
Editors
Auctioneers
Lawyers
Printers
Land Agents
Registry Office

Landowners
Builders
Tradesmen
City Government

FIGURE 4

LAND DEVELOPMENT IN THE NINETEENTH-CENTURY CITY

sixty years ago, "the holding of vacant land is not a profitable form of investment, except for short periods of time in especially favourable locations."[14]

Given the apparently poor odds, why did people continue to invest in land in nineteenth-century cities? The boom mentality, as we have already suggested, was undoubtedly of some importance in this matter. Quite simply, people were frequently caught up by the feverish activity in the local land market. In this regard, the press could play an important role. Local editors constantly employed boosterism and 'hype' to describe activity in the real estate market. Some people must have been influenced by their enthusiasm for property ownership. Then, too, land was one of the few genuine avenues for investment in the mid-nineteeth-century city. As Robert Swierenga has observed, "until the modern corporation came to be the dominant factor in American economic life, the principal opportunity for investment was in real estate."[15]

Perhaps, however, not every land purchaser was solely motivated by thoughts of speculative profits. There is certainly clear evidence in the documents of credit agencies, such as R.G. Dun and Company, that the elite had to own land in order to obtain credit.[16] The ability to gain capital whenever it was needed, of course, was the key to success and expansion for the Victorian businessman. Land, then, served as an important source of collateral in the commercial city. As one historian has recently noted, however, land did lose some favour in this regard as other more readily liquidated investment possibilities began to emerge.[17] As the years passed, the elite certainly became less involved in land ownership in our selected surveys (Table 6). By the early 1880s property ownership in these subdivisions was much more widespread than it had been initially. The intermediate transactions, many of which had been made between members of the elite, had eliminated most of the non-resident owners but they had not driven land prices beyond the incomes of the working class. Skilled, unskilled, and semi-skilled workers owned four times more lots and clerks owned six times more properties in 1881 than they had after the first round of transactions. Moreover, the distribution of lots had also changed, for by the end of the study period 54 per cent of the properties were in the hands of people who owned no

14. Arner, "Land Values," p. 579.

15. Swierenga, *Pioneers and Profits*, p. 49.

16. See Bertram Wyatt-Brown, "God and Dun & Bradstreet, 1841-1851," *Business History Review*, 40 (Winter 1966), pp. 432-50; James H. Madison, "The Evolution of Commercial Credit Reporting Agencies in Nineteenth-Century America," *Business History Review*, 48 (Summer 1974), pp. 164-86; D.A. Muise, "The Dun and Bradstreet Collection: A Report," *Urban History Review*, No. 3-75 (February 1976), pp.23-26; and Michael B. Katz, "The Entrepreneurial Class in a Canadian City," *Journal of Social History*, 8 (Winter 1975), pp. 1-29.

17. J.A. Bryce, "Patterns of Profit and Power: Business, Community and Industrialization in a Nineteenth Century Canadian City," graduate research paper (Department of History, York University, 1977), pp. 26-28.

more than two building lots (Table 8). The speculative phase had come to an end in these subdivisions.

Who built the Victorian city? Dyos and Warner have both stressed the small-scale character of the nineteenth-century construction industry, though the former has suggested that there was a shift towards larger firms by the end of the century.[18] The Hamilton data basically concur with these findings. Few Hamilton builders erected more than six dwellings in any given year. Yet, towards the end of our study period, there were some new directions in the city's construction industry. In 1874 several prominent Hamiltonians received a charter from the provincial government for the Hamilton Real Estate Association for the multiple purposes of

buying and selling, building upon and leasing Real Estate in the City of Hamilton and County of Wentworth, . . . using portions of said Real Estate as a Public Park and Gardens and as a Hotel and carrying out all business incidental thereto.[19]

Thus, the firm was to be much broader than most building companies of the period. In fact, it was the forerunner of the modern, vertically integrated development corporation, something that was reflected in more than the mere purpose of the company. Directors of the Hamilton Real Estate Association were affiliated with enterprises that could help them in their development goals. They served on the boards of insurance companies, which were already becoming important sources of capital, the Hamilton Street Railway, a utility that could be manipulated to enhance the value of real estate, and two of them served on the city council. Unfortunately, I have not yet uncovered documents that will let me explore these interrelationships in more detail. We should not overemphasize the importance of the Hamilton Real Estate Association, however, for it would be many years before firms of this nature would come to dominate the local construction and development scene.[21]

During the period in question, there were very few formal controls over the development process in Hamilton. Occasionally, the city council passed by-laws to prohibit the construction of frame structures in certain parts of the city, and some subdividers included restrictive covenants against frame dwellings in their deeds.[22] But few attempts were

18. Dyos, *Victorian Suburb*, p. 124, and Warner, *Streetcar Suburbs*, pp. 117 and 184.

19. Records of the Provincial Secretary, Charter Book of Companies, IV, folio 27, MSS, RG-8, Ontario Archives.

20. See *Revision and Consolidation of the By-Laws of the City of Hamilton* (Hamilton: City Council, 1887) and registered instrument G 703, Hamilton Registry Office.

21. On the later importance of large-scale building enterprises in Hamilton see John C. Weaver, "From Land Assembly to Social Maturity. The Suburban Life of Westdale (Hamilton), Ontario, 1911-1951," in this volume and *Hamilton: An Illustrated History* (Toronto: James Lorimer & Co., forthcoming), Chapter 1.

22. See *Revision and Consolidation of the By-Laws of the City of Hamilton* (Hamilton: City Council, 1887) and registered instrument G 703, Hamilton Registry Office.

made to control either land use or the timing of development. The period under examination was clearly an era of *laissez faire* in relation to the land-development process. This lack of regulation, in combination with the small-scale character of the building industry and the increasingly dispersed nature of vacant lot ownership, produced fragmented patterns of development. These could be seen in the neighbourhoods that developed in our selected sibdivisions.

The decision to develop property in the chosen surveys was extremely decentralized, with almost seven hundred different people involved during the period in question (Table 10). Almost 81 per cent of these people were involved in no more than two development decisions in these surveys (Table 11). Evidently, speculative building did not dominate the local construction picture. Houses were being built for specific individuals. Nevertheless, certain types of property owners were much more-likely to be developers than were others. For example, only one-fifth of the non-resident landowners decided to develop a lot, compared to one-third of the Hamilton natives. There were also some interesting differences for the members of the various occupational groupings employed in this analysis. As might be expected, those who were associated in some way with the development industry were the most likely of all the groups to decide to develop their parcels of vacant land, with

Table 10: Proportion of Land Developers by Occupational Category and by Place of Residence

| Occupational Category | Total land buyers | Total developers | Developers as % of buyers |
|---|---|---|---|
| Development industry | 418 | 183 | 43.8% |
| a) Builders | 313 | 144 | 46.0 |
| b) Servicers | 105 | 39 | 37.1 |
| Government officers | 38 | 8 | 21.0 |
| Merchant/industrial bourgeoisie | 386 | 114 | 29.5 |
| White collar | 84 | 33 | 39.3 |
| Skilled workers | 303 | 107 | 35.3 |
| Unskilled and semi-skilled | 154 | 52 | 33.8 |
| Professionals | 56 | 13 | 23.2 |
| Landed gentry | 326 | 76 | 23.3 |
| Women | 297 | 72 | 24.2 |
| Others | 121 | 21 | 17.4 |
| *Place of Residence* | | | |
| Hamilton | 1880 | 623 | 33.1 |
| Elsewhere | 303 | 63 | 20.8 |
| Totals | 2183 | 686 | 31.4 |

SOURCE: compiled from Hamilton Registry Office records, 1847-1881.

Table 11: Total Lots Developed by Occupational Group

| Occupational Category | Number of lots developed | | | | | No. of people | Total Lots | | Avg. no. of lots developed |
|---|---|---|---|---|---|---|---|---|---|
| | 1 | 2 | 3-5 | 6-9 | 10+ | | no. | % | |
| Development industry | 110 | 29 | 29 | 10 | 5 | 183 | 442 | 30.9 | 2.3 |
| a) Builders | 94 | 22 | 23 | 5 | - | 144 | 248 | 18.2 | 1.7 |
| b) Servicers | 16 | 7 | 6 | 5 | 5 | 39 | 174 | 12.7 | 4.5 |
| Government officers | 5 | 2 | 1 | - | - | 8 | 12 | 0.9 | 1.5 |
| Merchant/industrial bourgeoisie | 57 | 20 | 25 | 6 | 6 | 114 | 313 | 22.9 | 2.8 |
| White collar | 23 | 6 | 3 | 1 | - | 33 | 53 | 3.9 | 1.6 |
| Skilled workers | 83 | 18 | 6 | - | - | 107 | 138 | 10.1 | 1.3 |
| Unskilled and semi-skilled | 42 | 6 | 6 | - | - | 52 | 68 | 5.0 | 1.3 |
| Professionals | 8 | 4 | 1 | - | - | 13 | 19 | 1.4 | 1.5 |
| Landed gentry | 43 | 14 | 12 | 6 | 1 | 76 | 171 | 12.5 | 2.3 |
| Women | 53 | 8 | 10 | 1 | - | 72 | 107 | 7.8 | 1.5 |
| Others | 21 | 2 | 3 | 1 | 1 | 28 | 61 | 4.5 | 2.2 |
| Totals | 445 | 109 | 94 | 25 | 25 | 13 | 686 | 1364 | 99.9 |
| Totals | 445 | 109 | 94 | 25 | 13 | 686 | 1364 | 99.9 | 2.0 |

SOURCE: Compiled from Hamilton Registry Office records, 1847-1881.

almost one-half of those belonging to the actual building arm of the industry becoming developers during the study period. This reflects the small-scale nature of most firms in the industry. Such builders purchased a few lots at a time, erected a few houses, and used the profits from these to finance additional projects. Few, apparently, speculated in vacant land. White collar, skilled, unskilled, and semi-skilled workers were also quite likely to develop property. More likely, in fact, than professionals, members of the merchant/industrial bourgeoisie, and the landed gentry, who, as we have already suggested, often purchased real estate for reasons other than development. It seems clear that as we move away from the initial period of subdivision and speculation, the influence of speculators upon the morphology of a small Victorian city such as Hamilton diminishes.

Almost 1,350 houses lined the streets of our surveys by 1881. The resultant streetscapes mirrored the process of land development as it operated in the Victorian city (Figure 5). In a word, heterogeneity characterized the neighbourhoods that emerged in these subdivisions. Most contained churches, schools, and shops, and many also housed industries, noxious and otherwise, and transportation facilities. Nor was there any really overbearing architectural uniformity of the sort emphasized by Warner for Boston's late Victorian suburbs. Most Hamilton blocks took at least twenty years to really develop and many contained vacant lots thirty years after they had been platted. As a result, quite different architectural styles were often juxtaposed along the streets of sample subdivisions. Perhaps uniformity of style came only after what some have called the industrialization of the building industry.[23] Finally, the socioeconomic character of the neighbourhoods, as measured by their occupational make-up, was mixed (Table 12). Rich and poor frequently lived in relatively close proximity even in the newest subdivisions of the period. Later, of course, this would change, especially as development moved beyond the confines of central Hamilton.[24]

Speculators were not without some influence upon the nineteenth-century land-development process. In Hamilton, this was especially true of subdividers, those who were responsible for the first step in the conversion of vacant land into urban use. These individuals were speculating on the future growth of the city, and by subdividing raw land into

23. On the industrialization of the North American housebuilding industry in the late nineteenth century see Gwendolyn Wright, *Moralism and the Model Home: Domestic Architecture and Cultural Conflict in Chicago, 1873-1913* (Chicago: University of Chicago Press, 1980).

24. For a discussion see *Mountainview Survey: Hamilton's New High-Class Residential District* (Hamilton: Home and Investment Realty, Ltd., n.d.), Hamilton Public Library and John C. Weaver and J. Martin Lawlor, "The Making of a Suburb: Westdale, Hamilton, 1910-1950," paper presented to the fall meeting of the Ontario Historical Geographers, Toronto, October, 1976.

Table 12: Occupational Categorization of Assessed Residents by Survey, 1881

| Occupational Group[1] | Survey Number[2] | | | | | | | | | | | | | | All surveys | Entire city |
|---|---|---|---|---|---|---|---|---|---|---|---|---|---|---|---|---|
| | 1 | 2 | 3 | 4 | 5 | 6 | 7 | 8 | 9 | 10 | 11 | 12 | 13 | 14 | | |
| Professional/ Proprietor | 5.5 | 32.0 | 4.0 | - | - | 5.4 | 7.5 | 20.6 | 5.5 | 7.1 | 36.4 | - | 4.6 | 8.8 | 9.7 | 6.9 |
| White collar | 12.3 | 40.1 | 18.5 | - | 11.1 | 17.9 | 17.3 | 53.4 | 7.3 | - | 18.2 | 2.1 | 16.1 | 26.4 | 22.1 | 18.5 |
| Skilled | 51.4 | 12.2 | 50.8 | 38.5 | 40.7 | 44.6 | 53.7 | 16.0 | 50.9 | 42.9 | 27.3 | 72.3 | 41.4 | 39.6 | 42.1 | 40.4 |
| Semi-skilled | 4.1 | 3.4 | 6.5 | 7.7 | 5.6 | 14.3 | 5.6 | 1.5 | 5.5 | 7.1 | - | 12.8 | 13.8 | 6.9 | 6.2 | 7.0 |
| Unskilled | 14.4 | 1.4 | 8.3 | 38.5 | 27.8 | 12.5 | 7.5 | - | 21.8 | 28.6 | 4.5 | 8.5 | 12.6 | 11.3 | 9.8 | 13.4 |
| Unclassifiable | 12.3 | 10.9 | 12.0 | 15.4 | 14.9 | 5.4 | 8.4 | 8.4 | 9.1 | 14.2 | 13.6 | 4.3 | 11.5 | 7.0 | 10.1 | 13.8 |
| Total assessed | 146 | 147 | 325 | 13 | 54 | 56 | 214 | 131 | 55 | 14 | 22 | 47 | 87 | 159 | 1470 | 7477 |

SOURCE: Calculated from the City of Hamilton Assessment Roll, 1881, Mss.

NOTES:

[1]Based on the occupational classification presented in Theodore Hershberg *et al.*, "Occupation and Ethnicity in Five Nineteenth-Century Cities: A Collaborative Inquiry," *Historical Methods Newsletter*, VII (June 1974), pp. 174-216.

[2]For the locations of the various surveys, see Figure 1.

road allowances and building lots they provided the framework within which that future growth could occur. Future generations, in fact, would be able to do little to alter these plans, with the possible exception of the alignment of streets in adjacent subdivisions. And historians like John Reps have been quite correct to criticize their unimaginative and virtually universal adoption of the grid system for their plans of survey.[25] Yet, the subdividers were not the only people who were involved in the Victorian land-development process. They got things under way, but in most cases others carried out the actual development of the lots. And in Hamilton at least, there is litle evidence of the speculative building that characterized the development process in some larger places.[26]

How should we assess the process of land development in the Victorian city? It seems evident that the process was essentially decentralized, uncoordinated, and unregulated. Hundreds of different people were involved in the various stages of urban property development (Figure 6). With the domination by large, integrated development corporations, the same is no longer true today. In the nineteenth century, the lack of supervision at the subdivision stage – and particularly the absence of servicing requirements – made the next two stages in the process of development, marketing and building, both more important and more uncertain than they are at present. Moreover, the dispersed nature of the decisions to develop land in the typical Victorian city meant that streets could take a relatively long time to become built-up. Today we favourably contrast the resultant variety of architectural designs in these older areas with the sterility and sameness in our current batch of subdivisions. Yet, the way that cities developed in the nineteenth century meant that the provision of services such as sewers, water, and sidewalks could not be carried out in the most efficient manner. Quite simply, there were too many gaps in the built-up areas of the city. At a later date, developments in these open spaces would have to be connected to the service infrastructure.[27] Sprawl, with its irregular patterns of development,

25. John W. Reps, *The Making of Urban America: A History of City Planning in the United States* (Princeton: Princeton University Press, 1965), 294-324. On the use of the grid on the Canadian frontier see Ronald Rees, "The Small Towns of Saskatchewan," *Landscape*, 18 (Fall 1969), pp. 29-33.

26. On this point see, for example, John T. Saywell, *Housing Canadians: Essays on the History of Residential Construction in Canada*, Discussion Paper No. 24 (Ottawa: Economic Council of Canada, 1975) pp. 37-41, and H.J. Dyos, "The Speculative Builders and Developers of Victorian London," *Victorian Studies* 11 (Summer, 1968):641-690. For some comments on the overall quality of the houses that were built in Victorian Hamilton see *Hamilton, Canada: Visitor's Handbook* (Hamilton: Assessment Commissioner's Department, 1960), pp. 61-64, and Herbert Lister, *Hamilton, Canada: Its History, Commerce, Industry, and Resources* (Hamilton: City Council, 1913), pp. 47-51.

27. For a discussion of these servicing problems, see Alan F.J. Artibise, "An Urban Environment: The Process of Growth in Winnipeg," Canadian Historical Association *Historical Papers* (1972), p. 122.

probably created far more problems in the Victorian city than did the presence of land speculators. Indeed, a century ago, the worst current consequence of land speculation, continually escalating prices, was largely mitigated by relatively easy entry into the subdivision stage of the development process. Competition kept urban land prices down during most periods of the nineteenth century. To halt investigations into nineteenth-century urban development at the subdivision stage, then, is to miss much of the activity of the Victorian land market.

# Land Subdivision in Toronto, 1851-1883

ISOBEL K. GANTON

The growth of Toronto in the nineteenth century has been discussed by both contemporaries and later writers in terms of the amount and composition of population increase, and of changing political, social and economic institutions and roles. Most references to physical growth have been limited to statements in words or maps about the extent of the built-up area or the extension of the city boundaries through the annexation of suburban municipalities. Only a few have touched on the subdivision stage of the growth process, offering some insight into the factors that influenced the nature and distribution of the resulting street and lot patterns.[1] These comments are true of most studies of urban growth elsewhere as well. With some valuable exceptions, subdivision tends to be treated as little more than a "given" in the development process.[2] And yet it was the creation and marketing of urban lots and their associated street allowances that provided the necessary framework for construction, which in turn provided the framework for human occu-

1. See especially Peter Goheen, *Victorian Toronto, 1850 to 1900: Pattern and Process of Growth* (Chicago: University of Chicago, Department of Geography, Research Paper No. 127, 1970); and Jacob Spelt, *Toronto* (Toronto: Collier Macmillan, 1973).
2. Studies focusing on urban subdivision include the following: Philip H. Cornick, *Premature Subdivision and Its Consequences* (New York: Institute of Public Administration, Columbia University, 1938); Jerome Fellman, "Pre-Building Growth Patterns of Chicago," *Annals of the American Association of Geographers*, 47 (March 1957), pp. 59-82; Homer Hoyt, *One Hundred Years of Land Values in Chicago: The Relationship of the Growth of Chicago to the Rise in its Land Values, 1830-1933* (Chicago: University of

pancy and activity.[3]

The study upon which this summary report is based not only concentrates on this neglected aspect of urban development, but also differs in another important respect from most studies of specific cities. Whereas these tend to restrict analysis to a part of the whole (the suburbs, a sector or a particular district), consideration here is given to concurrent activity throughout the central city, suburbs and urban fringe. The particular topics singled out for summary treatment in this paper are the amount, timing, location and some characteristics of subdivision in the Toronto area between 1851 and 1883, with general observations about the decision-makers in the process.[4]

Most of the subdivisions surveyed in the Toronto area by 1883 can be documented through registered plans (R.P.s) deposited in first the county and later the city Registry Office. The circumstances requiring registration of plans, the time allowed between survey and registration, and the machinery and level of enforcement changed several times between the introduction of the registration concept in 1846 and 1883.[5] The stages of development portrayed in the Sandford Fleming map of 1851, the Boulton *Plan* of 1858 and Goad's *Atlas* of 1884 all show evidence of unregistered subdivision, but also attest to general com-

5. Legislation affecting the registration of urban lot survey plans was contained in the Registry Act of 1846 (9 Vic. c. 34), the Survey Act of 1849 (12 Vic., c. 35), the Registry Act of 1859 (CSUC, c. 89), the Survey Act of 1859 (CSUC, c. 93), the Registry Act of 1865 (29 Vic., c. 24), the Registry Act of 1868 (31 Vic., c. 30), the Survey Act of 1868 (31 Vic., c. 20), and the Survey Act of 1877 (RSO, c. 146).

Chicago Press, 1933); Helen Corbin Monchow, *Seventy Years of Real Estate Subdividing in the Region of Chicago* (Evanston: Northwestern University, 1939); and David Ward, "The Pre-Urban Cadaster and Urban Pattern of Leeds," *Annals of the American Association of Geographers*, 52 (June 1962), pp. 150-66. Warner's study of the urban development process as a whole in a large segment of Boston emphasizes construction and occupancy at the expense of the analysis of subdivision as a phenomenon with distinctive patterns and characteristics. See Sam B. Warner, *Streetcar Suburbs: The Process of Growth in Boston, 1870-1900* (New York: Atheneum, 1972). Several large-scale studies of urban development in England devote more attention to subdivision. See, for example, C.W. Chalklin, *The Provincial Towns of Georgian England: A Study of the Building Process, 1740-1820* (London: Edward Arnold, 1974); and H.J. Dyos, *Victorian Suburb; A Study of the Growth of Camberwell* (Leicester: Leicester University Press, 1961).

3. Charles S. Sargent, Jr., "Toward a Dynamic Model of Urban Morphology," *Economic Geography*, 48 (October 1972), pp. 357-74, demonstrates the significance of this sequence in a study of Buenos Aires.

4. In this paper Toronto and its suburbs, the Toronto area or the study area refers to the area bounded by the Humber River, Eglinton Avenue, Victoria Park Avenue and Lake Ontario, i.e., the area covered in Goad's *Atlas* of 1884.

pliance with the current law – particularly with regard to surveys creating new urban lots and streets.[6]

The registered plans, with some omissions, record the names of landowners (or their agents) and surveyors, the dates of survey and registration, and the dimensions and orientation of lots, lanes and streets.[7] Occasionally they show existing buildings or the names of purchasers of land inside or outside the plan area, give clues to the motives behind the decision to subdivide, or state the perceived attractions or potential uses of the land.

It is important to the understanding of nineteenth-century patterns of urban subdivision to realize that, unlike many modern developers, the subdividers of land into new urban lots seldom built on these lots themselves. Their interest was in marketing land, not houses, and the patterns they created reflected their perception of the utility and marketability of the property in question. The societal interest in certain areas for certain uses at certain times, as revealed in patterns of construction and occupancy, might or might not demonstrate the same timing, directional emphasis, or lot divisions as the underlying subdivision pattern. But through the obvious interest of most subdividers in predicting and providing for the eventual expansion of the built-up area, a two-way relationship may be recognized: subdivisions contained construction and occupancy, and these activities, through the preferences shown in their choice of location and use of land, could be important influences on the decision to subdivide.

Toronto at the end of 1850 was a city of approximately 30,000 people, most of whom lived and worked in the urban area laid out in grid plan

6. These three map sources all show details of lots and construction. Their title in full: *Topographical plan of the city of Toronto in the province of Canada*, from actual survey by J. Stoughton Dennis, Provin'l land surveyor. Drawn, comp. and engraved on stone by Sandford A. Fleming, Provin'l land surveyor, 1851.

   *Plan of Toronto* surveyed and comp. by W.S. and H.C. Boulton, Toronto, c. 1858.

   *Atlas of the City of Toronto and Vicinity.* Toronto: Charles E. Goad, Ltd., 1884.

7. In this study the earliest known date that could be assigned has been used for each plan. In most cases this was the recorded date of survey, but there was evidence on some plans or on the 1851 Fleming map of an even earlier date for some surveys. In approximately fifty cases no survey date was given and the date of registration was used. The aim in every case has been to come as close to the date of the decision to subdivide as possible.

government surveys in the broken front south of Queen Street, or in later privately commissioned surveys on the park lots north of Queen (Map 1).[8] The Fleming map shows that although construction and occupancy were concentrated within a few blocks of Yonge Street, there were already extensive areas of virtually vacant urban lots both in the government surveys and in several park lots to the north and west. This is evidence of a condition that persisted thoughout the study period and has been observed in nineteenth-century urban development elsewhere: the supply of lots was in considerable excess of the demand as expressed in construction.[9] In Toronto, as elsewhere, this surplus was based in large part on an enthusiasm for the speculative potential of unimproved urban land that encouraged subdividers, especially in areas less likely to attract immediate construction, to create and promote lots as commodities in themselves as much as for building purposes.[10] The difficulty of defining the precise details of the amount and location of construction in Toronto for much of the period before 1883 makes it impossible to more than postulate a cyclical pattern in the size of the surplus similar to that demonstrated for other cities.[11]

Plans surveyed by 1850 illustrate the three basic categories of subdivision that prevailed later as well. Though not a universal or necessary preliminary to urban subdivision, some lots of the township survey peripheral to contemporary urban development underwent an intermedite division into large lots of more than one acre. Labelled here "transitional," these were suitable for sale to land speculators, for estates, or for small-scale agricultural activities such as market gardening. A few plans close to the developed city centre have also been categorized as transitional because they established street allowances only, these enclosing blocks of undivided land.

8. The broken front refers to the area between the base line of a survey (in this case Queen Street) and a shoreline (Lake Ontario). With the exception of the southern halves of park lots 1 and 2, the thirty-two one-hundred-acre park lots were originally granted to government officials and leading citizens, and before urban development took place were in most cases used by them or subsequent purchasers as country estates or farms. The remainder of the study area was surveyed in two-hundred-acre farm lots. Both park and farm lots were organized in blocks 1¼ miles square, bounded by allowances for concession (east-west) and side (north-south) roads.

9. See, for example, Cornick, *Premature Subdivision*; and Monchow, *Seventy Years of Real Estate Subdividing*, and Doucet, chapter 7 in this volume.

10. Lillian F. Gates, *Land Policies of Upper Canada*, Canadian Studies in History and Government, No. 9 (Toronto: University of Toronto Press, 1968), p. 44, points out that urban land became popular as a speculative investment in Upper Canada in the late 1830s. Many newspaper advertisements and flyers promoting lot sales made specific reference to speculators as potential purchasers.

11. Building permits were not required in Toronto before 1882, and then only for a limited central area. Unofficial but purportedly accurate lists of new buildings appeared in the *Globe* for most years after 1868, but in only a few instances can precise locations be determined.

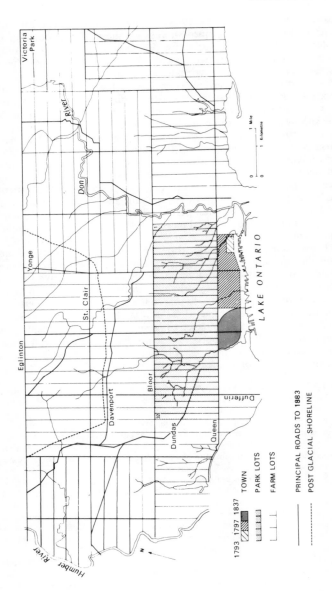

Map 1          Toronto: Framework for Development I.

The second type of subdivision, labelled "initial urban," created urban lots (one acre or less in size) where none had existed before. These not only greatly broadened the range of potential buyers and builders by providing smaller, and therefore cheaper and more functionally viable units of land, but also made the most significant contribution to the creation of the urban street pattern.

Some subdivisions contained both transitional and urban lots. Of these "mixed" plans that were essentially transitional in character, the urban lots usually fronted on major roads between the city and its hinterland. Other mixed plans that were largely urban in content contained lots that, notably due to topography or the persistence of the subdivider's own home and grounds, were larger than one acre in size.

The third basic type of subdivision identified, "resubdivided urban," was principally concerned with creating smaller lots from those of a preceding urban plan. Such plans seldom produced new streets (most exceptions being in the large blocks south of Queen Street), tending instead to be contained within lot and street lines already established by the initial urban survey. Resubdivision had a close relationship with the construction phase of land development, serving as a tool to adapt lots to the realities of the housing market. By 1850 the areas surveyed by the government long before into acre and half-acre lots with frontages of one hundred feet or more were increasingly unsuited to the requirements of a rapidly growing population in a predominantly pedestrian city with employment concentrated in the central business district and near the lake. The move to smaller lots is reflected on the 1851 Fleming map not only by the resubdivision evident south of Queen, but by the narrower lot frontages in most of the private surveys of initial urban lots in park lots near the city centre. By the early 1850s large lots with frontages over one hundred feet had become characteristic of the urban fringe rather than a central city norm. Regarded largely as objects of speculative investment, most were destined to undergo resubdivision in their turn.

Almost all the urban subdivisions surveyed by 1850 lay within the city limits (Maps 2a, 2b and 3).[12] Beyond these, most of the rest of the study area preserved the farm lots and road allowances of the township survey. There were also important roads that ignored the survey grid, and major rivers, minor streams and the post-glacial lake shoreline north of Davenport Road that interrupted the general topographical regularity (Map 1). Within this framework, and limited by little more than the legal requirement to employ provincially licensed surveyors and use the English system of measurement,[13] subdividers in the 1850s began an

---

12. From 1834 to 1883, the limits of Toronto (including before 1859 some Liberties) were formed by Dufferin Street, Bloor Street, the Don River, Kingston Road, the boundary between the second and third lots in the broken front west of Victoria Park Avenue, and Lake Ontario.

13. Re surveyors: 12 Vic., c. 35 (1849); re measurement: 32 Geo. 3, c. 3 (1792).

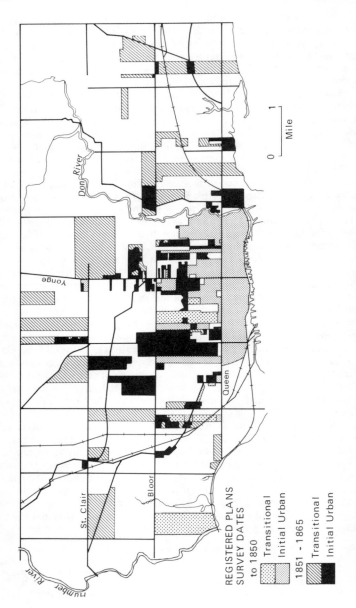

Map 2a  Toronto and Suburbs: Subdivision by Registered Plans to 1865. Transitional and Initial Urban Lots.

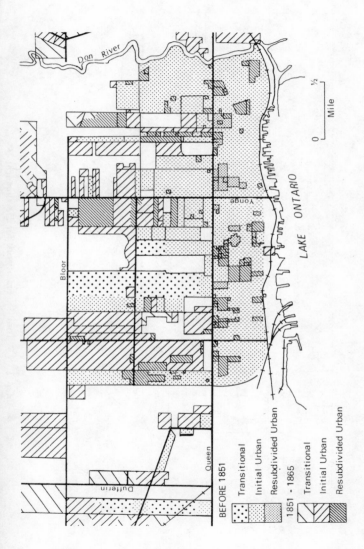

Map 2b Toronto: Subdivision by Registered Plans to 1865. All Registered Plans for Central Area

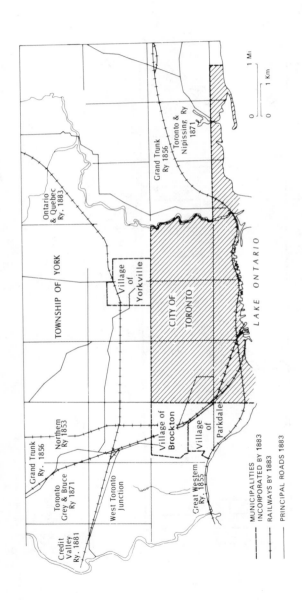

Map 3     Toronto: Framework for Development II

expansion of the urban area unprecedented in its extent but based on already established patterns and practices.

That the types of plans noted (transitional, initial urban, mixed and resubdivided urban) continued throughout the period from 1851 to 1883 is clear from the histogram of registered plans surveyed during those years (Figure 1).[14] It must be emphasized that this shows numbers of plans, i.e., numbers of decisions to subdivide, and on the surface tells nothing about the amount of land added to the urban stock. When the plans for each year were mapped, however, there was a general correspondence between the relative number of plans and the relative amount of land involved. This observation is reinforced by the tabulation of the number of urban lots created each year.[15] Peaks and troughs in amounts platted, urban lots surveyed and numbers of plans came in the same years.

Given this relationship, there was an obvious cycle from trough to trough between 1851 and 1865, with the major amount of subdivision activity occurring in the years 1853 to 1857 inclusive. Far more land was surveyed into transitional and initial urban lots in the five peak years of this cycle than had previously been similarly transformed since the private sector had begun subdivision of the park lots in the early 1830s (Map 2a). As early as 1853, the *Globe* was expressing concern over this expansion and the enuthusiam for investment in new urban lots, claiming there were already enough lots vacant in Toronto to absorb twice the population. The paper reminded its readers of the large amounts of subdivided land abandoned on the fringe of American cities after similar over-speculation contributed to the crash of the late 1830s.[16]

The amount of subdivision in Toronto during the 1850s was based on contemporary evaluation of both the anticipated and actual effects of several growth factors. The population was expected to grow immediately, the result of continued immigration, the return of the Canadian capital to Toronto (1855-1859), and the commerce and industry attracted by the building of canals and, especially, railways. Not only would this new population require housing, but the activities that attracted many of them were expected to bring new capital, new general prosperity, and higher prices for land. As the land boom gathered momentum, the *Globe* noted that "the only subjects of conversation in our streets are the price at which such and such a property was bought, what it is held at

---

14. Plans categorized as "Other" in Figure 1 include those adding or deleting lanes, making minor corrections in earlier R.P.s, or involving factory sites, rail yards, water lots, a cemetery and a small pleasure ground. These are not shown on the maps based on registered plans.

15. Occasional references are made to maps or tabulations that could not be included here, but form part of the analysis for the ongoing doctoral dissertation on which this paper is based.

16. *Globe*, June 7, 1853.

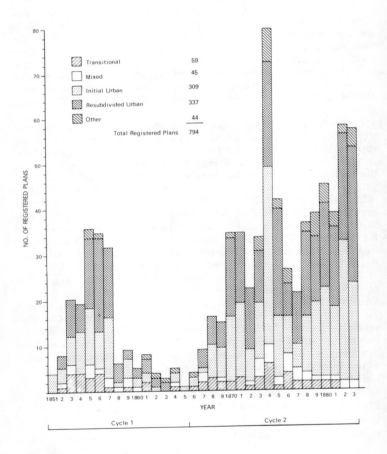

Figure 1     Toronto and Suburbs Registered Plans, 1851-1883
             By Earliest Known Date

now, and what it is expected to bring a year or two hence."[17] During the decade the population did indeed increase by 14,000, and a spate of building activity increased the number of houses from 4,834 to 8,383 (Table 1). Small wonder, then, that the decisions of subdividers reacting to the pressures of a real market demand for both building lots and land for speculative investment, led to an oversupply. Over half the land surveyed into new urban lots during this first cycle had not been built on even by 1883, and most of the transitional lots had not yet been converted into urban lots.[18]

New urban lots filled many of the gaps between pre-1851 surveys during the first cycle (Maps 2a and 2b), though some land within the city limits was held in large blocks until the next cycle of subdivision and, indeed, had not all been absorbed into the registered urban stock by 1883. This reflected the importance of individual decisions to the location and timing of subdivision.[19] Some urban lots were now also laid out in the concession north of the city limits and, in continuation of the trend noted earlier, along main roads out of the city. Although no strong directional preference can be detected for individual years, the map of total new urban subdivision between 1851 and 1865 shows an emphasis on growth to the north and west (Map 2a). Considered in combination with land surveyed in the pre-1851 period, a definite southeast to northwest axis is obvious that, while varying somewhat from the inverted T-shape traditionally associated with the expansion of Toronto's built-up area, conforms to contemporary perception of the main direction of the city's growth.[20]

In contrast to the attraction of major roads, there is no discernible relationship, either positive or negative, between the location of transitional or initial urban subdivision and the location of railway tracks constructed during this period (Maps 2a and 3).[21] Large-scale industry was not the norm in pre-Confederation Toronto, and what new industry did develop could still find adequate space close to the railway terminals and yards in the east and west ends of the city, without having to seek siding access along the lines themselves. Location near the terminals

---

17. *Ibid.* In other issues, e.g. January 26, 1850 and June 1, 1850, the expected increase in urban property values was given as a reason for supporting railway construction.

18. Lots that had been improved were, as might be expected, heavily concentrated in the subdivisions closest to the city core.

19. The persistence of the residential estates of several early park lot owners, notably the Allans, the Boultons and the Denisons, accounted for some of the delay. Other residential estates and institutional lands were also involved, but in some cases it seems to have resulted from a deliberate decision to delay the marketing of land.

20. Spelt, *Toronto*, p. 44, points out this popular characterization of built shape. The *Globe*, September 12, 1855 and August 4, 1856, stresses the westward emphasis of growth during the boom period of the 1850s.

21. The Northern Railway opened in 1853, the Great Western Railway in 1855, and the Grand Trunk Railway in 1856 (Map 3.)

also placed them close to harbour facilities and the existing labour force. New employment with these industries, with the railways themselves, and with the cartage and commercial firms associated with them, accentuated the desirability of land within walking distance for the construction of working-class housing. Resubdivision of the large lots south of Queen Street thus continued at an accelerated pace, with noticeable concentration in the as yet sparsely occupied sections closest to the railway yards (Map 2b).

Resubdivision became important north of Queen Street as well during the 1850s. Growing awareness of a working-class market for smaller lots is exemplified by the changing focus of advertisements for one major subdivision. First offered for sale as large quarter-acre lots for "genteel residences" or speculative investment in 1853,[22] by 1855 some of the lots had been divided by their purchasers and promoted as suitable for the homes of "mechanics and others" – and the time of sale had been moved from midday to the evening so they could attend.[23] By 1857, the original subdividers had drawn up plans to divide each unsold lot in four, and were also aiming at the working-class buyer.[24]

The economic depression that began in 1857 must be seen as the prime factor in the sudden decrease in land subdivision in 1858. The *Globe* noted that house prices in 1859 were half what they had been four years earlier,[25] and attributed many business failures to inability to sell land bought at high prices at the peak of the boom.[26] Even after business conditions had begun to recover in 1859 and money on first-class security was available at 9 or 10 per cent, unimproved land of somewhat speculative value could not attract mortgage money at 12 or 15 per cent.[27] The sheer quantity of land added to the urban stock during the mid-1850s no doubt contributed to the low levels of subdivision that continued throughout the 1860s (Figure 1), but the effects of the oversupply were reinforced by a decline in the rate of population growth (Table 1). The government had moved from Toronto once more, and railway construction had ceased by the end of the 1850s. There were no new railway projects to arouse expectations of growth and profit until late in the 1860s.[28]

The new cycle of subdivision that gathered slow momentum in the late 1860s experienced an intermediate peak of activity in 1873 and 1874

22. *Globe*, April 9, 1853.
23. *Ibid.*, January 27, 1855 and April 27, 1855.
24. *Plan of the Crookshank Estate north of Queen Street in the city of Toronto*. J.O. Browne, civil engineer, P.L. surveyor, Toronto, 25 February 1857.
25. *Globe*, September 12, 1859.
26. *Ibid.*, January 25, 1860.
27. *Ibid.*, July 16, 1859.
28. The Toronto Grey and Bruce Railway and the Toronto and Nipissing Railway opened in 1871 (Map 3).

followed by a drop back to more modest levels during the rest of the 1870s, and was rising toward a cyclical high in 1889 by 1883 (Figure 1).[29] The level did not sink to a low similar to that of the mid-1860s until 1896.[30] Once again there was general agreement between the timing of subdivision activity and recognized periods of boom and depression in both the general economy and construction.

The building boom that began in 1867 probably reflected the generally increasing prosperity as well as the accumulated effects of population growth.[31] Despite the lower rate of growth and the lower absolute increases in the 1860s, there were still over 11,000 more people to be housed in 1871 than in 1861, and the early 1870s brought a further great increase in population. The experience of American cities would suggest that once the supply of available building lots remaining from the previous period of major subdivision activity had been reduced to a point where a real or perceived potential market for more existed, another boom in subdivision would take place.[32] Such a boom began in Toronto in the 1870s, spurred once again by a demand for lots for both building and speculative investment (Figure 1). While the oversupply of transitional and initial urban lots placed on the market through subdivision in the second cycle paralleled that of the first,[33] there is evidence of greater investor caution induced by painful memories of the effects of the real estate excesses of the 1850s.[34]

By 1873 a new economic depression gripped the United States, and some of the many decisions to subdivide in Toronto that year and the next may have been motivated by a desire to put land on the market while conditions remained good. The decline in subdivision that began in 1875 corresponded with the appearance of the depression in Toronto, and with a period of reduced levels of construction.[35] Opinions vary as to the intensity of the effects of this depression on the city, but for subdivision at least they were less drastic than those of the depression leading to the nadir of the mid-1860s. This may be related to less financial distress over a shorter period, or to the pressure of continued large increments to the city's population: between 1871 and 1881, 30,323 peo-

29. 1873, 1882 and 1883 were exceptions to the statement made earlier: in these three years the amount of land subdivided was considerably greater than would be expected from the number of R.P.s surveyed.

30. K.S.H. Buckley, "Urban Building and Real Estate Fluctuations in Canada," *Canadian Journal of Economics and Political Science*, 18 (February 1952), p. 43.

31. D.C. Masters, *The Rise of Toronto, 1850-1890* (Toronto: University of Toronto Press, 1947), pp. 98-99.

32. Fellman, "Pre-Building Growth Patterns of Chicago," and Hoyt, *Land Values in Chicago*.

33. *Globe*, October 31, 1891, quoted in Buckley, "Urban Building and Real Estate," p. 50.

34. *Globe*, May 22, 1866; October 11, 1872. *Monetary Times*, October 23, 1874; November 6, 1874.

35. Masters, *Toronto*, pp. 97, 142.

Table 1: Population and Houses, 1848-1881

City of Toronto

| Year | Population | Increase | | Houses | | | | Increase in total houses | |
|---|---|---|---|---|---|---|---|---|---|
| | | No. | % | Inhabited | Vacant | Being Built | Total | No. | % |
| 1848 | 23,503 | | | 3,795 | 341 | — | 4,136 | | |
| 1851 | 30,775 | 7,272 | 30.9 | 4,621 | 103 | 110 | 4,834 | 698 | 16.9 |
| 1861 | 44,821 | 14,046 | 45.6 | 8,358 | 22 | 3 | 8,383 | 3,549 | 73.4 |
| 1871 | 56,092 | 11,271 | 25.1 | 9,787 | 160 | 333 | 10,280 | 1,897 | 18.5 |
| 1881 | 86,415 | 30,323 | 54.1 | 16,387 | 911 | 536 | 17,834 | 7,554 | 73.5 |

Incorporated Suburban Municipalities

| | Yorkville | | Brockton | | Parkdale | |
|---|---|---|---|---|---|---|
| Year | Population | Total houses | Population | Total houses | Population | Total houses |
| 1861 | | 258 | | | | |
| 1871 | 2,203 | 458 | | | | |
| 1881 | 4,825 | 1,046 | 786 | 157 | 1,170 | 289 |

SOURCE: Census of Canada

ple were added (Table 1). Continued population growth and a partial
return of prosperity in the early 1880s accompanied a distinct increase in
subdivision in those years.[36]

During the early years of the second cycle, the few new urban lots
were added in plans scattered within the city limits (Maps 4a and 4b).
Most filled gaps in the area of earlier urban subdivision, a practice that
continued for the rest of the study period. Accretions to the fringe of the
urban area appeared in the late 1860s, and with the increased subdivision activity of the 1870s, initial urban lots covered large sections
beyond the city limits. Resubdivision occurred chiefly within the city
limits throughout the second cycle, though with the growing scope of
preceding initial urban plans subject to such treatment, the radius and
spatial concentration of resubdivision moved outward over the cycle.

Unlike the infill and resubdivision plans in the city itself, plans in the
outlying areas began to be concentrated in different districts at different
times (Maps 3 and 4a). Between 1870 and 1875, one concentration of
activity extended Yorkville to the north and west of the nucleus established in the 1850s, and another resulted in a small cluster of plans just
east of the Don. A major focus of attention between 1873 and 1879 was
the district south and west of Dundas Street, partly within the city but
chiefly in the new suburb of Parkdale. The last and, in terms of the
rapidity and amount of land subdivided, the most dramatic examples of
concentrated subdivision before 1884 were in the areas northeast of
Dundas and Dufferin and in West Toronto Junction. These had undergone some subdivisions throughout the 1870s, but the main outburst of
activity began in 1879 and reached an unprecedented level in 1882 and
1883. Though separated by an area of sparse urban subdivision, these
last two districts were on the same directional line from the city core,
and their combined effect was to reaffirm and intensify the trend to the
northwest noted in the earlier cycle.

The neutral relationship between the location of subdivision and railway lines noted for the first cycle held true for the second as well,
though tracks now more often bounded than simply passed through
subdivision plans (Map 4a). A negative effect claimed for rail lines on
suburban residential construction in Toronto[37] is in contrast to the continued attraction of rail yards and terminals for land development. The
amount and nature of resubdivision and housing construction in areas
with easy access to the lakeshore rail facilities give evidence of this (Map
4b), and initial urban subdivision appeared in conjunction with anticipated industrial agglomeration and railway employment based on the
new C.P.R. yards in West Toronto Junction.[38]

36. *Ibid.*, p. 175.
37. Goheen, *Toronto*, p. 164.
38. Gordon Garland, "Suburbanization and the Transition to Monopoly Capitalism [West
    Toronto Junction]," Masters Research Paper, Department of Geography, University of
    Toronto, 1978.)

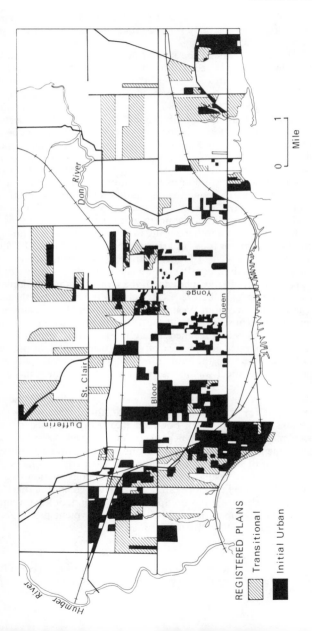

Map 4a   Toronto and Suburbs: Subdivision by Registered Plans, 1866 to 1883

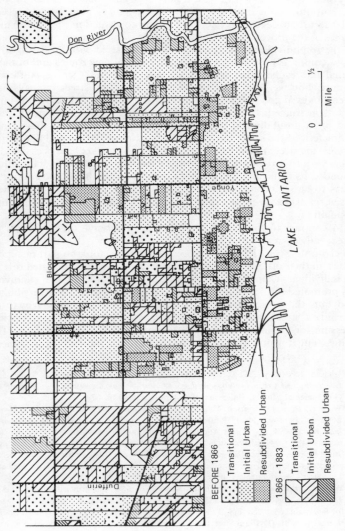

BEFORE 1866

Transitional

Initial Urban

Resubdivided Urban

1866 - 1883

Transitional

Initial Urban

Resubdivided Urban

0    ½
Mile

Map 4b    Toronto: Land Subdivision to 1883

The horse-drawn street railway, introduced in 1861, served only a limited area until the mid-1870s, and even than had little effect on decisions to subdivide (Map 5). It may be argued that the street railway in Toronto, as in other North American cities,[39] tended to precede intense urban development as far as the construction and occupancy stages of the process were concerned. It may also have encouraged some of the resubdivided urban plans closely associated with these stages. A close comparison of the location and timing of new routes and of most initial and resubdivided urban subdivision before 1884 shows a different picture, however, for most of the areas served by street railways had been surveyed and partially, if not intensively, occupied before the lines were constructed. Unlike the rapid and lengthy extensions into vacant and often unsubdivided countryside and to neighbouring villages that characterized the street railways of such cities as Boston and San Francisco,[40] the street railway in Toronto had to be coaxed and occasionally coerced into extending its service, even within the city limits that bounded its franchise.[41] It did not expand beyond those limits until the areas to be served had been annexed to the city, and annexation was based on extensive prior subdivision and burgeoning construction and population.

In contrast to the active participation in subdivision of owners of street railway companies in Boston and San Francisco,[42] the owners of the Toronto Street Railway, before 1884 at least, took no part in subdivision in their official capacity and only one played an unrelated role as an individual. There is some evidence, on the other hand, that subdividers promoted the opening of routes that would enhance the accessibility, and thus the value, of their lots.[43] The relatively cautious growth of Toronto's street railway lines in the face of both public and private pressure for expansion was due in part to financial and franchise difficulties, but it is noteworthy too that in Toronto, unlike Boston,[44] there

---

39. Goheen, *Toronto*, p. 209; Warner, *Streetcar Suburbs*, p. 49; and James E. Vance, Jr., *Geography and Urban Evolution in the San Francisco Bay Area* (Berkeley: Institute of Governmental Studies, University of California, 1964), p. 52.

40. Warner, *Streetcar Suburbs*, p. 23; and Vance, *Geography and Urban Evolution*, pp. 51-54.

41. Louis H. Pursley, "Street Railways of Toronto, 1861-1921," Interurbans Special, *Interurbans*, XVI, no. 2 (Los Angeles: Electric Railways Publications, 1958), and the pages of the *Globe* tell this story. The coercion was based on a term of the Toronto Street Railway franchise that gave the company thirty days to agree to carry out projects proposed by companies seeking new street railway franchises from the city. (See Michael J. Doucet's paper in this volume.)

42. Warner, *Streetcar Suburbs*, p. 60; and Vance, *Geography and Urban Evolution*, p. 52.

43. *Toronto City Council Minutes, 1875*, p. 206, record a petition from R. Denison and others for favourable consideration of the extension of the street railway along Dundas Street to the city limits (Dufferin Street), i.e., across the Denison land shown on Map 6. The line was not built until 1882, however.

44. Warner, *Streetcar Suburbs*, p. 60.

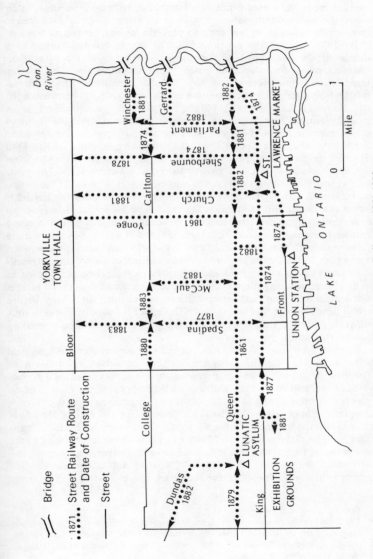

Map 5    Toronto Street Railways, 1861-1883

was no competition to provide an incentive to stake a claim for future market areas. As a profit-motivated company with no immediate interest in the vacant land market, the Toronto Street Railway preferred in most cases to follow rather than to precede the market for its services.

By 1883 most of the city and much of the adjacent suburban area displayed the palimpsest of superimposed surveys that continued to characterize Toronto's development until recent times. The timing, location and characteristics of both individual plans and the accumulated whole were influenced to some extent by the natural features of the site (Map 1). Lake Ontario played a dual role, its attraction as an amenity and its importance for transportation making it the focus of early development, and its very presence precluding much expansion to the south. The wide deep valley and unhealthy marshes of the Don deterred development to the east, though the Humber was too far from the urban nucleus to affect growth to the west before 1883. The post-glacial shoreline ridge to the north was becoming a physical barrier to be reckoned with by road and lot planners by the end of the period,[45] but many of the minor streams that had posed significant barriers to road building in the first half of the century had by then become enclosed sewers that could be largely ignored. During the intermediate years examined here, however, there was an obvious correlation between many gaps in the expanding urban grid and the location of these streams – an observation that would shortly apply to the post-glacial ridge as well. This relationship depended on the persistence of residential estates and institutions that had selected these as desirable sites long before urban development reached them.[46] The effect on subdivision was thus indirect, exercised through the influence of these features on preceding human decisions.

Human decisions, in fact, were of greater importance than natural features in shaping the emerging pattern of urban subdivision in Toronto. Both pre-urban and preceding urban surveys created a framework of lots and roads that, through their presence on the ground or on paper, exerted a strong influence on succeeding subdivision decisions.[47] Despite the approximately 850 R.P.s surveyed by the end of 1883 (794 after 1850), the lot lines of the underlying Crown surveys could, in most cases, still be clearly detected. Major exceptions were to be found in the lots between Parliament Street and the Don River (very early amalgamated into Crown reserve south of Carleton and a private holding between there and Bloor), in the four park lots closest to Yonge Street

---

45. Goheen, *Toronto*, p. 84.
46. The Denison and Givens homes on Map 6 illustrate this point. To give further important examples: Trinity College and Gore Vale on Garrison Creek; Moss Park, Caer Howell, and the University of Toronto on Taddle Creek; and Spadina, Davenport and Oaklands on the brow of the shoreline.
47. cf. Ward, "Urban Pattern of Leeds."

(reorganized by private owners before 1820), and in the section subdivided as West Toronto Junction. In all but a few cases plans did not include land from more than one park or farm lot unless these had come under one ownership well before transitional or urban subdivision took place. With minor exceptions, the original lot lines were perpetuated as streets, plan boundaries or internal lot lines.

Moving forward in the sequence of events, 87 per cent of resubdivided urban R.P.s were entirely within the boundaries of one preceding plan, and in almost every case were bounded by lot lines within it.[48] This nesting tendency no doubt reflected both the relative ease of describing land in deeds and plans in terms of established surveys, and patterns of land ownership and sale: succeeding purchasers would find it easier to deal with one owner only, and contiguous parcels under separate ownership may not have been on the market at the same time. The vast majority of plans subdivided individual parcels of land held by one or more owners, and even the few involving non-contiguous parcels with the same owner, or sponsored by owners of adjoining properties, almost always subdivided land originally marketed with a single preceding subdivision plan.

Much of the explanation for the introduction and extension of a grid plan in Toronto must be found in its many advantages in a speculative land market. It was easy to lay out and extend, easy to describe in deeds and portray on plans, and readily comprehensible in terms of the size, location and utility of lots by buyers subjected to the competitive pressures of auction sales (a common method of marketing real property, especially new lots, in nineteenth-century North American cities). A grid plan provided uniform lots suitable for a compact settlement of rectilinear buildings.[49] And given the model established in Toronto by the government town surveys, and the fact that land parcels north of Queen Street were almost always confined within the boundaries of the rectilinear lots of the township survey, it is little wonder that the grid pattern was perpetuated in subsequent subdivision. Within the long narrow park lots that had never been amalgamated under single ownership, there was little scope for anything else. In a few subdivisions of larger properties in the wider farm lots, some curving streets appeared from an

---

48. Using these observations as a base, maps and occasional references to previous surveys on R.P.s provide evidence of several earlier but unrecorded transitional and urban surveys underlying later plans – on either side of Yonge Street, to give an important example.

49. John W. Reps. *The Making of Urban America: A History of City Planning in the United States* (Princeton: Princeton University Press, 1965), p. 302; and D. Stanislawski, "The Origin and Spread of the Grid Pattern Town," *Geographical Review*, 36 (1946), p. 107.

early date – most of them allowing for topographical irregularities similar to some ignored in division of the park lots.[50]

The clear relationship between the orientation of the longitudinal axes of the park and farm lots and the dominant orientation of the streets platted within them has been noted before.[51] As a result of this correlation, most urban lots surveyed by 1883 faced on north-south streets south of Bloor and between Yonge and Keele north of Bloor, and on east-west streets elsewhere. The inclusiveness of statements claiming a lack of co-ordination between cross streets in adjacent subdivisions and attributing much of this lack to selfish motives is open to question, however.[52] Registered plans and Goad's *Atlas* show that most streets were co-ordinated where possible within the limitations of the size and position of the land being subdivided. If the land in one park or farm lot was sold in blocks differing in shape and size from those of a neighbouring lot, it may have been physically impossible, or impractical in terms of creating viable building lots, for a later subdivider to continue streets laid out in an earlier contiguous plan. Furthermore, Toronto was not platted in a progression of surveys out from the centre, and a subdivider may have had to choose between the varying street patterns established by perhaps widely separated surveys on either side. There are, moreover, many examples of conscious provision for the future joining of streets, even in the layouts of spatially separated plans. Jogs and discontinuities certainly did exist, but were probably of little significance to circulation in the pre-automobile city. They did not affect the oft-repeated contemporary description of Toronto as a city of straight streets and great regularity of plan.[53]

The development of the four park lots shown on Map 6 illustrates well the patterns resulting from the nature and relationships of urban surveys in Toronto before 1884. Park lot 24 was owned and the house shown on Map 6a occupied by members of the Givens family from 1802 to close to the end of the nineteenth century. The three park lots to the west came into the possession of the Denison family between 1815 and 1826, and contained the residences of several members of at least three

---

50. The earliest example was in a large transitional subdivision southeast of Bloor and Jane before 1851. Examples in urban or mixed subdivisions between 1851 and 1883 occurred in Rosedale, on Walmer Road north of Bloor, in the Glen Stewart area in the southeast corner of the study area, and in West Toronto Junction.

51. Spelt, *Toronto*, p. 45.

52. *Ibid.*, p. 44 states that "such east-west streets as were built did not connect with those in adjacent subdivisions" and that cross streets "tended to be resisted if they helped to make holdings of other owners more accessible."

53. Among many may be noted Dr. Charles Mackay, whose 1857 observation is quoted in H. Scadding and J.C. Dent, *Toronto: Past and Present* (Toronto: Hunter Rose and Company, 1884), p. 217; and the 1873 *City Directory*, p. 133.

generations of the family over the years.[54] The first urban lots in the area appeared before 1851 along Queen and Dundas Streets, the main highway between Toronto and the western part of the province (Map 6a). Later subdivisions within the three amalgamated park lots were free to overlap the original lot lines within the block, but like those in the single park lot to the east, did not extend beyond its exterior boundaries (Map 6b). Even internally, the original park lot borders may be traced along streets, plan boundaries or lot lines south of Dundas, and in the 1880 R.P. to the north. All of the resubdivided urban surveys were nested within the urban surveys that preceded them, with lot lines of the later forming their boundaries. None of these resubdivisions affected the street pattern created by the initial urban plans. There is one example of a situation with some parallels elsewhere: the 1879 R.P. on park lot 24 included both initial urban lots (the major portion to the north) and the resubdivision of some lots from an earlier 1857 plan.

Very little construction had taken place north of Dundas Street by 1883, but the more built-up area to the south illustrates two practices typical of other districts as well. There is evidence of some unregistered resubdivision, usually associated with the construction of more than one house on a registered lot, and (in a procedure common from the late 1870s on) evidence that all but two of the resubdivided urban R.P.s were surveyed after construction of houses on the land. It is obvious either from explicit statement or from the portrayal of ground plans accompanied by varying frontages, that the lots were measured from party wall to party wall. It was evidently considered easier to make lots fit attached houses than to make attached houses fit pre-measured lots.

The dominant orientation of the streets (Map 6c) reflects for the most part the north-south orientation of the park lots. There is some departure from this where amalgamation under the Denisons permitted plans with an east-west axis, their location indicating the probable influence of the early construction and importance of Dundas Street on the layout of pre-urban field patterns. With the exception of the lots fronting directly on it, however, the angled portion of Dundas has been treated as an aberration, and the streets and lots of the urban plans based on the orientation of the township survey. The resulting grid has some inconsistencies due to the varied plan axes and patterns of land division and ownership. Others may be traced to the persistence of unsubdivided estates. There are also, however, many examples of both north-south and east-west roads extending across two or more subdivisions, and one plan (because of its linear nature, not shown on Map 6b) specifically

---

54. On land ownership: T.A. Reed, "Memoranda re the Crown Grants of Park Lots in the City of Toronto, Based on the Records of the Registry Office." Mss. notebook compiled 1926. On residence: John Ross Robertson, *Landmarks of Toronto* Vol. I (1894), p. 1 and Vol. II (1896), p. 745; and Toronto *City Directories*.

Map 6            Toronto: A Case Study of Park Lot Development
                        Park Lots 24-27, 1851-1883

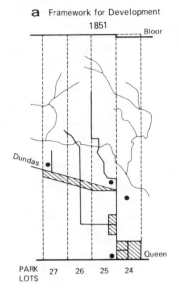

**a** Framework for Development
1851

LEGEND FOR MAPS 6a, 6b, 6c

| | |
|---|---|
| - - - - - - | PARK LOT BOUNDARIES |
| ———— | INITIAL URBAN SURVEYS |
| - · - · - · | RESUBDIVIDED URBAN SURVEYS |
| ▨ | URBAN LOTS SURVEYED BEFORE 1851 |
| ⸬ | NO URBAN LOTS SURVEYED BY 1883 |
| ⌇ | GARRISON CREEK |
| ● | MAJOR DWELLING |
| ———— | ROADS |
| 74 | SURVEY DATE OF REGISTERED PLAN |

SCALE

EACH PARK LOT = 10 CHAINS × 100 CHAINS
660′ × 6600′

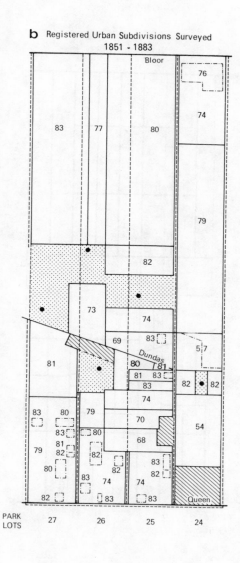

**b** Registered Urban Subdivisions Surveyed
1851 - 1883

PARK LOTS: 27  26  25  24

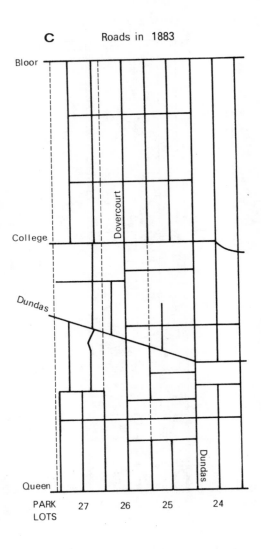

**C**        Roads in 1883

designed to open College Street across the properties of several neighbouring co-signers.

The Denison and Givens families, like some other early owner-occupants of park or farm lots in the study area, were active in the conversion of their land into urban lots – but more typically they also sold large blocks that were subdivided by others. Just as the surveys sponsored by them were typical of subdivision in most of the study area before 1884, so were these subdividers representative of the major decision-making groups involved.

The 794 R.P.s surveyed between 1851 and 1883 were sponsored by institutions, companies and individuals acting either alone (64 per cent), or in concert. In all, thirteen institutions (government, religious, educational, hospital, social and cemetery) signed thirty-eight plans, including two in conjunction with individual subdividers. Over half of these plans consisted of fewer than twenty lots each or made minor alterations to existing plans. Three divided two-hundred-acre glebe lots into transitional lots, and two created large lots suitable only for the mansions of the elite from the fringe of the University property in Queen's Park.

More significant in terms of the creation of streets and building lots were the activities of the eighteen companies (chiefly land, building, banking or loan companies)[55] that were sole sponsors of thirty-one R.P.s and co-signers of seven more. Eighty per cent of these plans were surveyed during the last nine years of the period under analysis, years which also included all eight company plans with over ninety lots apiece. Half of the company plans, like the institutional ones, were under twenty lots of a minor nature, but the growing involvement of companies in subdivision towards the end of the period, especially of companies formed for the express purpose of investing in land development, presaged a change in the development process that would become increasingly significant after 1883.

The subdivider most typical of the years 1851 to 1883, however, was the private individual. Here analysis immediately becomes more complex, for instead of a few corporate entities whose actions can be assumed to be entirely profit motivated and based on a consensus assessment of general economic conditions, there is a multitude of individuals whose economic motivation may have been triggered or directed by a variety of personal aims or circumstances. Many subdividers were active speculators, i.e., they bought land expressly for subdivision and sale.

In most cases the speculation was in vacant land, though small resubdivisions often included houses constructed by or for the owner. Other subdividers were divesting themselves of inherited land or of land originally acquired for personal use, or were acting on behalf of others as

55. A few simple partnerships, e.g. of two carpenters or two founders, were categorized under the names of the individual partners rather than as companies.

Table 2: Occupational Categories of Individuals Associated with Ownership of Registered Plans Surveyed in Toronto and Suburbs, 1851-1883

| Occupational category | Number of owners | % of total |
|---|---|---|
| Government | 50 | 6.7 |
| Law | 79 | 10.6 |
| Other professions[1] | 20 | 2.7 |
| Gentlemen | 20 | 2.7 |
| Business | 185 | 24.7 |
| Commerce | 96 | 12.8 |
| Finance | 48 | 6.4 |
| Manufacturing | 29 | 3.9 |
| Transportation | 12 | 1.6 |
| Land Development | 98 | 13.1 |
| Construction | 52 | 7.0 |
| Real Estate | 30 | 4.0 |
| Building supplies | 16 | 2.1 |
| Skilled Labour | 23 | 3.1 |
| Unskilled labour | 6 | .8 |
| Miscellaneous[2] | 36 | 4.8 |
| Women | 132 | 17.7 |
| Unidentified | 99 | 13.2 |
| Total | 748 | 100.1 |

[1] Army, church, medicine.
[2] Includes farmers (16), hotelkeepers (6), publishers (4), musicians (2), comedian, policeman, veterinarian, riding school owner, vestry clerk, lunatic.

NOTE: Of the 794 R.P.s surveyed between 1851 and 1883, 719 bore names of individuals. Included were those whose names appear only as insane (1) or deceased owners (38), executors, trustees or assignees (71), or who appear to have signed only as spouses of owners (37) or combinations of these (3). The remaining 499 identified subdividers may have been associated with R.P.s in one or more of the above capacities, but signed at least one R.P. on their own behalf.

executors or trustees. This is not to say that this group was any less interested in making money on the land, or that it was not responsive to general market conditions, but individual circumstances such as the need to settle an estate or to raise funds to meet a personal financial crisis were more likely to be factors in the timing of its subdivisions than was the case with the first group.

The relationships that could be sought between categories of subdividers and categories of plans could fill a volume, but attention here must be limited to the components of the group, with reference to a few important variations in their involvement in subdivision. Table 2 shows the primary occupational categories[56] of the individuals who signed R.P.s as owners: 748 people from a wide socioeconomic range. The high proportion of businessmen stands out, a not unexpected relationship at a time when land still represented one of the principal opportunities for the investment of capital. The land development occupations have been separated from their business activities here in order to emphasize their important connection with subdivision, but like the other business interests and unlike most of the other categories, their involvement was considerably greater in the second cycle of subdivision than in the first.[57] The large number of women whose names appeared on R.P.s is noteworthy, though they were on par with those in the government occupations in having the lowest percentage (66 per cent) of their number acting on their own behalf. Most the rest signed as executors or spouses only.

Besides differences in the roles of the signers, there was variation in the number of plans signed by each. Sixty-eight per cent of the names appeared on one plan only, and when the small size of many plan areas is considered, and the fact that many signers were not the sole signers of the plan with which they were associated, it becomes obvious that the contribution to subdivision of many of the individuals in the owners group was minor. This situation was by no means confined to members of the lower socio-economic categories, though most of them fell in this group. In contrast, when the forty-three individuals who signed five or more plans are identified, it becomes evident that some occupational categories were more active than their numbers indicate.[58] This is espe-

---

56. Many of these are known to have also been officers of financial institutions or railways, or to have been involved in politics – especially at the municipal level. A few, who are categorized here according to their Directory identification at the time they signed R.P.s, later changed their occupations, especially to ones in the land development class.

57.

|  | Per Cent of All Subdividers | |
|---|---|---|
|  | Cycle 1 | Cycle 2 |
| Business |  |  |
| Business | 17.6 | 25.6 |
| Land development | 6.8 | 15.0 |

58. Not all of these signed as owners in every case: some plans were signed as executors or trustees.

cially true of lawyers and gentlemen, particularly those who were the sons of prominent early land accumulators.[59] Although eight of the most active subdividers were from the land development group, the members of this category, especially those in the construction trades, were heavily concentrated in association with resubdivision plans of fewer than ten lots each, most showing extensive construction already completed. Concomitant with the low percentage signing more than one plan is the fact that only 52 of the 748 names appear on R.P.s surveyed in both cycles of the study period. One hundred and sixty-nine names appear on plans of Cycle 1 only; 527 are found only on plans of Cycle 2.

Evidence of the great variation in the contributions of subdividers, together with an awareness of the large number of plans that had little or no effect on street creation or lot orientation, indicates that more detailed analysis would produce a greatly reduced list of plans and subdividers of meaningful significance to the evolution of Toronto's urban pattern. This, plus the influence doubtless exerted by the limited number of surveyors employed in creating the plans, must be considered as a contributing factor to the remarkable consistency of the grid that resulted from the large number of decisions to subdivide.

In summary, the timing of subdivision in Toronto between 1851 and 1883, in terms of both numbers of decisions and amount of land, corresponded to periods of economic prosperity and depression and of population growth, and reflected the effects of these on construction. The location and nature of subdivision was less influenced by natural features than by preceding man-made constraints, and some of these bore less relation to subdivision than to construction. In a definite progression toward the north and west, the subdivided area both provided the framework for further stages of development and was encouraged by their selected course and intensity. It was a relationship that depended on a multitude of decisions of varying significance made by a wide socioeconomic range of landowners and builders. The result was the creation of patterns on the land that still dominate much of central Toronto a century later.

---

59. Eleven of the seventy-nine lawyers and seven of the twenty gentlemen are found among the forty-three whose names approved on five or more R.P.s.

# The Maps

The size of the published page has made necessary considerable modification of the nature and detail of the maps of the original paper. This is especially true of Maps 2 and 4, where single complex maps have been separated into an overall representation of transitional and initial urban subdivision over the entire study area, and a detailed portrayal of subdivision for the central area only.

All but Map 6a are on a base derived from a Goad map of 1884 based on his *Atlas* published in the same year. The lake frontage shown thus represents a situation resulting from some infill beyond the natural shoreline. In a few instances where more than one plan affected the same land during the same cycle, only the latest subdivision action is evident on Maps 2 and 4. Omitted from these same maps are plans categorized as "Other" and, on Map 4, plans that subdivided the Island during Cycle 2.

The maps were compiled from the following sources:

*Atlas of the City of Toronto and Vicinity*; City of Toronto and Suburbs. Charles E. Goad, 1884.
   (shoreline, rivers, streams, roads, railways, corporation boundaries, R.P. locations)
*Globe*, 1861-1883.
   (street railways)

Ontario. Ministry of Natural Resources. Township of York. Disposition of Crown Lands, Plan T-2539.
   (crown township survey)

Pursley, L.H., "Street Railways of Toronto, 1861-1921."
   (street railways)

Registered Plans to 1883.
   (R.P. survey dates, boundaries and locations)

*Sketch of the town of York made for William Willcocks, esq., by permission of his honor the President*, 23 July, 1799. Surveyor-General's Office.
   (Crown town surveys)

Spelt, J., *Toronto*, p. 6.
   (post-glacial shoreline)

*Topographical Plan of the city of Toronto . . .* Sandford A. Fleming, 1851.
   (streams, R.P. survey dates, Map 6a)

*Toronto Military Reserve*, from the original document as laid out in Town lots to the East of the Ravine by Capt. Bonnycastle and Resurveyed by William Hawkins, Dy. Surveyor, February 21, 1837.
   (Military Reserve survey)

# Building Halifax, 1841-1871

SUSAN BUGGEY

At its incorporation in 1841, Halifax was a modest colonial town, with a population of about 15,000. With a few prominent exceptions like the fine Government House and the elegant neo-classical Province House, it was a wooden town. Despite a decade of building activity and a rash of local boosterism in the 1840s, at mid-century the editor of the *Acadian Recorder* could still deplore "its long rows of old, dirty, dingy, shaky, wooden houses, all built originally in the tea-chest order of architecture," and its "reputation of being the meanest looking city in the civilized world, in proportion to its wealth and other advantages."[1] By 1871, when Halifax was first measured as a city in the new Dominion of Canada, it had both grown and changed remarkably. Its population had doubled to nearly 30,000, and its physical form had altered substantially.[2] Government building, commercial building, and residential building each made a distinctive contribution to the transformation of the city between 1841 and 1871, while the roles of the architect, the builder and the artisan in the building trades changed in response to new fashions, new laws, new materials, new technologies and new organizational arrangements.

SOURCE: *Acadiensis: Journal of the History of the Atlantic Region*, vol. 10 Autumn 1980), pp. 90-112. Reprinted by permission of the author and the Department of History of the University of New Brunswick.

1. *Acadian Recorder* (Halifax), 30 May 1857.
2. *Evening Express* (Halifax), 17 August 1864. The picture of Halifax by the late 1860s is well drawn in Phyllis R. Blakeley, *Glimpses of Halifax 1867-1900* (Halifax, 1949), ch. 1.

The mid-nineteenth century was, in general, a period of prosperity in Halifax, and building was one of the most visible evidences of this prosperity. A nearly continuous program of government building between 1850 and 1870 reflected not only the initiation of municipal government in the town but also the mid-Victorian social conscience and concern for the quality of urban life. Markets, water towers, drains, sidewalks and street improvements all marked civic efforts to upgrade the quality of the urban fabric. Buildings to maintain the social order were prominent among the new government structures: Rockhead Prison (1857), the Halifax County Court House 1858), and the County Jail (1863) were all built to supplement or replace existing buildings for the control or implementation of due process of law. A number of new government buildings were also directed to meeting social needs. The Lunatic Asylum, contracted in 1856, was followed in 1857 by the City Hospital and in 1867 by a new Poors' Asylum. By locating in the urban fringe where land use was extensive rather than intensive, government building contributed to enlarging the operative areas of the city.[3] The new structures required not only land assembly, but design, purchase of materials and construction. The scope, continuity of demand and security of payment in these areas provided significant employment in the building trades and economic stimulus in the local market.

The evangelical spirit also resulted in building activity, in an industrial school, homes for the aged and for fallen women, a deaf and dumb institute, and a blind asylum. Private philanthropy stimulated both new building, like the cost-shared Blind Asylum in 1868, and the conversion of existing buildings to new uses, like Edward Billings' villa residence purchased for the Deaf and Dumb Institute in 1859.[4] It likewise found expression in church building by all denominations, especially in the 1840s, as well as in substantial church enlargements and new religious-associated buildings such as glebe houses and schools. Another important stimulus to public building was the British army and navy presence in the town. The largest single military project was the fourth Citadel, begun in 1828 and costing, upon completion in the mid-1850s, £252,122.[5] Construction of the Wellington Barracks in the 1850s had a major impact on the Halifax building situation, while other buildings and

3. Contracts for these buildings were noted respectively in *Novascotian* (Halifax), 20 April 1857: *Acadian Recorder*, 9 October 1858; *Evening Express* (Halifax), 16 September 1863; *Novascotian*, 14 April 1856 and *Evening Express*, 4 october 1865; *Acadian Recorder*, 25 July 1857; *Novascotian*, 6 May 1867. A laudatory report of city activity in the 1850s is found in the first *Report of the Several Departments of the City Government of Halifax, Nova Scotia for 1857/58* (Halifax, 1858). On public building in the urban fringe, see J.W.R. Whitehand, "Building cycles and the spatial pattern of urban growth," *Transactions of the Institute of British Geographers*, 56 (1972), pp. 39-55.

4. *Acadian Recorder*, 27 June 1868; *Novascotian*, 8 August 1859.

5. See John Joseph Greenough, "The Halifax Citadel, 1825-60: A Narrative and Structural History," *Canadian Historic Sites*, no. 17 (Ottawa, 1977), p. 9.

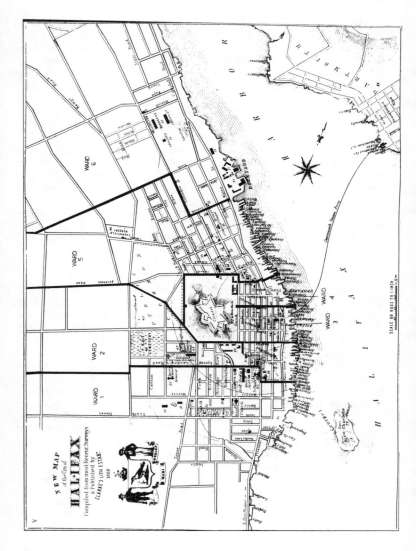

*Figure 1.*

"New Map of the City of Halifax," 1869, with ward boundaries superimposed. (National Map Collection, Public Archives of Canada).

repairs for the Ordnance, Commissariat and Royal Navy services afforded further construction contracts to master craftsmen and builders.

Prosperity also stimulated commercial building and upgrading. By 1841 Granville Street, three blocks west of the harbour, had begun to challenge Water Street as the prime commercial row in the city. Bank building and improvement on Hollis Street and prime commercial construction on Granville Street in the next fifteen years emphasized the trend. At the same time, suburban expansion attracted population away from the crowded core. While the consolidation of the business district can be traced throughout the period from 1841 to 1871, it focused on rebuilding in the districts burnt over in three major fires between 1857 and 1861.[6] Commercial renewal on Hollis Street in 1857 and 1858 was stone and brick in accordance with new building standards, three and four storeys high in reflection of business confidence. Rebuilding the stores in the north block of Granville Street in 1860, to uniform architectural designs in freestone and cast iron with Italianate detailing, gave focus to Halifax building capacity, experience and aspiration. Because whole blocks were rebuilt at nearly the same time, there was opportunity to create coherent streetscapes through similar building heights, aperture patterns and architectural detailing. This momentum was carried over into more modest insurance offices rebuilt in Bedford Row in 1861-62, the new Union Bank on Hollis Street, and the most extravagant building of the early 1860s, the Halifax Club. Other commercial rebuilding for prominent merchants in the area in the early 1860s rounded out the transformation of the business core of Halifax from a mainly wooden district to a predominantly stone and brick quarter of new, large and fashionable buildings,[7] although it would be another quarter-century before similar rebuilding would extend to Barrington Street much of its familiar facade.

While stone buildings were by no means unknown before the mid-1850s, their prevalence after that date, and the widespread use of brick and cast iron, marked a substantial departure from the previous reliance upon wood in the city. The introduction, expansion and enforcement of a 'brick district' between 1857 and 1863 virtually eliminated the construction and enlargement of wooden buildings in the central core.[8] Ironstone, the material of all but the finest stone buildings before 1840,

---

6. The fires occurred on New Year's Eve, 1857, 9 September 1859, and 12 January 1861. *Novascotian*, 6 January 1857, 12 September 1859, 21 January 1861.

7. The extent and character of building can be readily traced in the local newspapers. See, for example, *Halifax Reporter*, 3, 24 November 1860, 28 February 1861, 31 January 1863; *Evening Express*, 8, 10, 17 April, 12 July, 13 November 1861; 28 March, 7 May, 6 June, 25 July, 15 September 1862; 15 January 1863; 25 January, 6 April, 15 June 1864. For the visual record, see *Rogers Photographic Advertising Album* (Halifax, 1871), reprinted as *A Century Ago: Halifax 1871* (Halifax, 1970).

8. Nova Scotia, *Statutes*, 20 Vic. c.35, c.36; 21 Vic. c.77; 22Vic. c.65; 24 Vic. c.45; 25 Vic. c.37; 26 Vic. c.49; 27 Vic. c.81.

*Figure 2.*

Commercial rebuilding in the downtown core in the early 1860s emphasized the consolidation and prosperity of the central business district. The rhythmic articulation and Italianate detailing of the new Granville Street facades (top) were soon reinforced in such prominent structures as the Union Bank (left) and the Halifax Club (right). (top: Public Archives of Nova Scotia; left: Archives of Ontario; right: *Illustrated London News*).

was largely supplanted by freestone and granite. In addition to stone brought in from regional quarries, local builder Robert Davis opened the Chebucto Quarries near Bedford Basin. Brick, which had been imported either from abroad or elsewhere in the province, was made locally at Eastern Passage and for a time at Freshwater. Its local manufacture, using up-to-date, imported machinery, made brick a viable alternative for building in the city, while its low cost in comparison with some made it relatively popular in the legislated brick district. Cast iron, popularized by its use in rebuilding Granville Street, spread rapidly through the commercial district where it was perhaps used even more as a means of updating an existing building than in new construction. Mass production and ease of transport also brought other metals to the city where legislated fireproofing, especially on roofs, promoted their sale. The specification that roofs in the brick district be either flat or crowned gave impetus to the use of American-produced patent felt roofing.[9]

Major growth outside the waterfront areas both north and south of the central core was primarily residential in character. In most cities, the south and west areas have proven to be more fashionable districts than those to north and east. While the Halifax Citadel formed a seemingly impenetrable barrier against concentrated westward settlement before 1871, the south suburbs comprised the most fashionable residential district of the town; in 1836 only one street registered an average real estate valuation lower than £250,[10] and development resembled more closely the popular English ornamental villa than the town house. While the south suburbs attracted mid-century population growth, what most drew residential building to the area was the number of large, accessible estates convenient for subdivision. By the 1850s building activity was occurring in Spring Gardens, named for the fashionable London suburb. In 1862-63, Smith's Fields (between South and Inglis Streets) was sold off in building lots; by the late 1860s subdivision was extending along Tower Road and west to Robie Street, and some construction had already occurred further west near the present site of Dalhousie University.[11] Both quality speculative housing and contracted residences were erected in these areas shortly after subdivision. Further south and west, prime residential development in the 1850s and 1860s along the North West Arm conformed relatively closely to the ornamental villa common

9. See, for example, *British Colonist* (Halifax), 6 March 1858, 25 January 1862, 2, 25 April 1863; *Novascotian*, 21 May, 22 August 1855; *Morning Sun* (Halifax), 24 Arpil 1863; *Halifax Reporter*, 3 November 1860; *Acadian Recorder*, 2 August 1862.

10. Data were derived from Halifax Assessments, RG 35A, vol. 3, no. 12(1836), Public Archivves of Nova Scotia [hereafter PANS].

11. See, for example, *Novascotian*, 13 April 1847, 1 November 1852; *Acadian Recorder*, 8 March 1856, 7 June 1862, 14 March 1868; *Evening Express*, 20 April 1863; *Morning Chronicle* (Halifax), 2, 16 March 1868. See also *City of Halifax, Nova Scotia, 1865* (Reflex Copy of Church Map E-14-15, sheet 2).

in the urban fringes of English country towns in the early nineteenth century.[12]

While residential development in the south suburbs primarily post-dates incorporation, by 1841 nearly 500 proprietors had taken up lots along fourteen streets in the north suburbs, and by 1851 ward 5, comprising most of the north suburbs, contained nearly one-third of the population. Development continued more slowly thereafter, and by 1871 ward 5 had been overtaken in size by ward 1 in the south suburbs.[13] Until 1850 the north peninsula, beyond North Street, was characterized by farms and villa residences representative of the extensive land use of areas more distant from the city core. Subdivision of the area followed construction in the 1850s of the Wellington Barracks just north of the Dockyard. Proximity to the Intercolonial Railway yards, as well as the Barracks, made the district attractive for building throughout the 1860s and 1870s.[14]

Central to the expansion and transformation of mid-century Halifax were the architects, builders and artisans who carried out the building process. Their skills complemented one another in the construction of a building. The architect was the creator of the building insofar as he established its form, style and character in his designs. Usually a man of education and taste, he was trained not only in the practical rudiments of building, but also in its aesthetics. His three functions in a building project were to prepare a design acceptable to his client, to provide specifications and working drawings from which the building would be erected, and to supervise the actual construction to ensure that the design, layout and structural detail were implemented in accordance with the specifications. While architects, like other men of the arts, were attracted to the colonial town from time to time, none appears to have established residence in Halifax before mid-century. Few Halifax buildings in the first half of the nineteenth century were, then, architect-

12. John W. Regan, *Sketches and Traditions of the Northwest Arm*, (2 cd. ed., Halifax, 1909); *Acadian Recorder*, 23 February, 17 May 1856; Heritage Trust of Nova Scotia, *Founded Upon a Rock. Historic Buildings of Halifax and Vicinity (Halifax, 1967), pp. 66, 92, 94, 98;* T.R. Slater, "Family, society and the ornamental villa on the fringes of English country towns," *Journal of Historical ¡Geography,* 4 (1978), pp. 129-44.

13. Figures based on Halifax Assessments, RG 35A, vol. 3, no. 12, PANS: Nova Scotia Assessment and Census Returns, Halifax, 1851, RG 1, vol. 451, PANS (PAC, reel M3137); *Report of the Secretary of the Board of Statistics on the Census of Nova Scotia, 1861* (Halifax, 1862), App. 7; Census of Canada, 1871, Halifax, N.S., Schedule 3, PAC, reel C10550 [hereafter cited as Halifax City census, 1851, 1861, 1871]; *Evening Express,* 22 February 1861; see also J.G. McKenzie, *The City of Halifax* (1841) in PANS and *City of Halifax, Nova Scotia, 1865.* For the character of residential building, see *Founded Upon a Rock*, pp. 70, 90 and *The West House Brunswick Street Halifax* (Halifax, 1973).

14. Halifax Assessments, RG 35A, vol. 3, no. 12, PANS; *Novascotian,* 16 December 1850, 12 October 1857, 14 June 1858; *City of Halifax, Nova Scotia, 1865; City Atlas of Halifax, Nova Scotia* (Halifax, 1878).

designed. Most were created by skilled local builders adapting designs in local materials from the published pattern books which proliferated during the period. The most skilled practitioner of this art appears to have been Henry G. Hill, a Halifax carpenter who, having built the successful Brunswick Street Methodist Church in the mid-1830s, announced that he would thenceforth "offer his services as an Architect, Draughtsman and Builder, and [would] be prepared to furnish accurate working plans, elevations and specifications for buildings of every description." He continued to do so for the next twenty years. Drawing on the neo-classical forms popular in Georgian England and the United States, he provided designs for churches, suburban cottages, villa residences, banks and stores as well as for Halifax's only Temperance Hall and the City Prison.[15] He was probably the most active designer of buildings in Halifax prior to the permanent establishment of professionally trained architects in the late 1850s.

The fate of Hill's design for the Halifax County Court House, in fact, marked the transition from the self-trained architectural designer to the professional architect. As early as 1851, the Nova Scotian government had passed legislation authorizing funds for the construction of a new county court house in Halifax. Hill's 1854 plans for the building provided for a structure "perhaps more enlarged and more costly than was originally contemplated . . . [but] not out of proportion to the scale on which such an edifice should be erected, both with reference to architectural appearance, and the requirements of the capital of the province." But when local artisans refused to tender for the construction, Hill was sent back to the drawing board with an increased outside figure of £6,700, for which he produced "a very pretty design . . . providing all the necessary accommodation." Before Hill's new structure could be erected the Hollis Street fire of 1 January 1857 sent ripples of alarm through Halifax at the thought of housing the country's legal records in a wooden building, while new legislation passed in April 1857 prohibited large wooden buildings within the developed part of Halifax. An amending act, passed on 1 May 1857, exempted a major wooden extension to an existing church, but specifically excluded the new court house from the exemption. Hill was paid the balance of sums due him in connection with the court house designs and effectively relieved of the commission. When an architectural competition was subsequently called to provide new designs for a stone or brick court house, the selection went not to

15. *Novascotian*, 5 February 1836. See also, for example, 2 June, 1 September 1842, 13 May 1844, 28 August 1848, 7 December 1857, 15 February 1858; *Acadian Recorder*, 15 December 1855, 17 May 1856; *Morning Sun*, 30 November 1859; Plan of Rockhead Prison, May 1855, PANS: *Dictionary of Canadian Biography*, vol. XI (Toronto, 1982), pp. 405-406. On pattern books available in Halifax, see *Novascotian*, 18 April, 16 May 1859.

*Figure 3.*

Significant government building in the 1850s created both design and construction opportunities in Halifax which attracted architects and builders to the city.
The tradition of local design in construction, such as Rockhead Prison (top), was
challenged by the work of outsiders who soon outstripped local artisans in prime
building competitions such as the Halifax County Court House (bottom). (top
and bottom: Public Archives of Nova Scotia).

Hill, whose principal working material was wood, but to the prestigious Toronto architectural firm of William Thomas & Sons.[16]

William Thomas was an Englishman who, after building experience at home, particularly in the Midland towns and spas, had emigrated to Toronto about 1840. In the course of the next fifteen years he designed numerous Upper Canadian buildings including churches, commercial structures, and houses as well as such important public buildings as St. Lawrence Hall, Victoria Hall in Cobourg, Guelph City Hall and the Court House in Niagara-on-the-Lake.[17] It was St. Matthews' Church which brought a representative of the Thomas firm to Halifax. The committee responsible for rebuilding the church after destruction of its century-old building in the 1857 fire resolved in May to invite Thomas' son, W.T. Thomas, to the city. Shortly thereafter they settled that the church should "be built after the design of St. Andrews Church, Hamilton," one of William Thomas' most admired works.[18] With two Thomas commissions in Halifax already underway, the Granville Street fire, which wiped out the north block of the street in September 1859, induced the Thomas firm to open a Halifax office. It was first advertised in October 1859 while Thomas was busy preparing unified designs for seven proprietors rebuilding the centre of the city's prosperous dry goods trade. Thomas' individual building designs, unified by massing, height, detailing and materials, created a distinctive streetscape which reflected contemporary mercantile confidence. The cast iron store fronts which Thomas introduced along most of the Granville Street block had not only the well-recognized advantages inherent in the material but also the prestige of manufacture by the leading New York producer of architectural cast iron, Daniel Badger. The buildings popularized the Italianate detailing and forms which would soon dominate prime Halifax building of every type. The Thomas office remained open until 1863 when, with Granville Street rebuilt and a subsequent design for the rebuilding of the Union Bank completed, C.P. Thomas, the resident Thomas partner, left Halifax for Montreal.[19]

16. Nova Scotia, *Statutes*, 14 Vic. c.3, 16 Vic. c.20, 17 Vic.c.48, 19 Vic. c.65, 20 Vic. c.35, c.36, 21 Vic. c.13, 23 Vic. c.46; Halifax Court House, MG 4, Ms. file, no. 3, PANS: Nova Scotia, *Journals of the House of Assembly, 1854-55*, App. 15; 1857, p. 371; 1860, pp. 681-2; *Novascotian*, 1 September 1856; *Acadian Recorder*, 28 August 1858.

17. T. Ritchie, "The Architecture of William Thomas," *Architecture Canada*, 44 (May 1967), pp. 41-45; Neil Einarson, "William Thomas," *Dictionary of Canadian Biography*, vol. VIII (in press); *Halifax Reporter*, 3 November 1860.

18. St. Matthew's Church, MG 4, vol. 61, 1857-1860, PANS.

19. *Novascotian*, 19 September 1859; *Illustrations of Iron Architecture, made by the Architectural Iron Works of the City of New York* (New york, 1865; DeCapo Press, 1970), plates 68, 74, 75, 76, 78, 79; Heritage Trust of Nova Scotia, *A Sense of Place. Granville Street. Halifax, Nova Scotia* (Halifax, 1970); Harold D. Kalman, "NSCAD Moves to the Waterfront," *Rehoused in History. Nova Scotia College of Art and Design* (Halifax, 1979); *Halifax Reporter, 3, 24 November 1860; Evening Express*, 13 November 1861; *The Halifax, Nova Scotia Business Directory for 1863*, compiled by Luke Hutchinson (Halifax, 1863).

While the Thomas firm was the most prestigious architectural firm in Halifax at mid-century, Scottish-born architect David Stirling's work was both popular and widespread. Arriving in 1850 from St. John's, Newfoundland, where he had designed the handsome Bank of British North America, he soon prepared designs for the Halifax branch of the same bank and for the long-discussed city market. Although he subsequently departed to Upper Canada, he returned to Halifax in 1862 to design a large number of well-known buildings, including the Halifax Club, churches, stores, residences such as Alexander Keith's, and the Provincial Building. A number of his buildings combined to lend distinctive architectural quality to downtown Halifax. Well versed in the newly popular Italianate forms which he handled with considerable skill, Stirling has been described as "probably the best architect" in Nova Scotia by the late 1860s.[20]

By this time the extent and quality of building in Halifax had attracted other architects to the city. Of those identified, only the firm of Henry Elliott & Henry Busch appears to have possessed the formal experience, the wide range of commissions and the continuity to have achieved an impact comparable to Stirling or Thomas. Elliott & Busch's documented work includes alterations and designs for several churches, the county jail, several prominent stores, and two distinctive residences, "Oaklands" for William Cunard in 1862 and "the Octagon House" in Dartmouth in 1871.[21] Although most buildings erected in Halifax were still not architect-designed, the substantial number of large, prominent and stylish buildings which had been erected in the city since the late 1850s had made architect-designed buildings an integral part of the Halifax townscape.

While an architect designed buildings, in the normal course of events he did not construct them. This was the role of the builder. In 1838 building craftsmen comprised 8.96 per cent of the heads of families in the city; in 1861, with increased labour specialization, 12.7 per cent of the labour force were employed in the construction industry.[22] The building industry was composed of a number of highly skilled and closely controlled crafts. Members commenced their careers as apprentices to a master craftsman, graduated to become journeymen, and might,

20. Notebook of David Stirling, architect, Queen's University Archives, Kingston (xerox copy, PANS); *Novascotian*, 29 April, 11, 25 November 1850; *Evening Express*, 12 May 1862; 30 October, 9 November, 28 December 1863; *Acadian Recorder*, 15 July 1865, 16 March 1867; Robert C. Tuck, *Gothic Dreams. The Life and Times of a Canadian Architect. William Critchlow Harris 1854-1913* (Toronto, 1978), pp. 18-20.

21. See, for example, *Novascotian*, 9 January 1860; *Halifax Reporter*, 28 February 1861, 31 January 1863; *Evening Express*, 31 January 1862, 26 August 1863, 10 February, 6 April, 25 May 1864; *Acadian Recorder*, 8 May 1865, 20 April 1867; *Founded Upon a Rock*, pp. 92, 108.

22. The 1838 figure was calculated from data in Halifax City census, 1838, RG 1, vol. 448, PANS. The 1861 figure appears in L.D. McCann, "Staples and the New Industrialism in the Growth of Post-Confederation Halifax," chapter 4 in this volume, table 2.

*Figure 4.*

Stirling's "Keith Hall" (top), built for Alexander Keith on south Hollis Street, and Elliott's "Oaklands" (bottom), built for William Cunard on the North West Arm, exemplify the transfer in the early 1860s of the new design and construction skills in Halifax from government and commercial to residential building. The Italianate forms and detailing lent a more opulent character and more complex visual interest to the architecture of the city. (top: author; bottom: Public Archives of Nova Scotia).

if skilled and ambitious, become master craftsmen themselves in time. The system had its roots in the medieval guilds and much of it remained firmly intact in pre-industrial areas in the nineteenth century.[23] Apprenticeship remained the principal manner of training in the skills and mysteries of the trades, and fathers passed their knowledge to their sons and nephews, although with increasing specialization the sons of labourers, probably themselves serving in the construction industry, also became building artisans.

Differences in the economic and social status of journeymen and master craftsmen make their differentiation of some significance in analysing the building industry. Some masters, like Henry G. Hill, mason Thomas Saunders and carpenter Peter Artz, can be identified clearly from organizational records, wage agreements, and court cases. Those identified to date, however, represent a very small percentage of the building artisans working in Halifax. In early nineteenth-century Philadelphia Thomas Smith has found, by linking census data with entries in city directories and with total tax assessment and property tax, that 19.1 per cent of artisans were employed in the construction industry.[24] Very rough calculations based on 1838 and 1861 Halifax census data yielded 26.79 per cent and 27.58 per cent respectively of artisans employed in the building trades. Smith also calculated the ratio of masters to journeymen at 1.0:0.88, which is not dissimilar to the oversimplified 1.0:1.04 ratio derived from Halifax directory and census entries for 1870 and 1871.

Acceptance in the construction trades was marked not only by apprenticeship but also by cooperation with others in one's trade. The largest of the construction trades in Halifax – the carpenters – had organized as early as 1978 and incorporated in 1850. Their rules and regulations as well as their quarterly meetings were intended to reinforce a strong sense of workmanship and artisanal pride.[25] Wage agreements, such as the increase to 7/6s. per journeyman's 10-hour day negotiated in 1854, were a well-established practice in Halifax by mid-century, when legislation against the combination of workmen specifically exempted from its restrictions those who voluntarily met together to establish or set wages, hours and other regulations regarding their work.[26] By then the number of carpenters working in the city had grown enormously,

23. There is scant evidence on which to evaluate its practice in mid-nineteenth century Halifax, but see Jean-Pierre Hardy and David-Thiery Ruddel's valuable study of apprenticeship in Quebec, *Les Apprentis Artisans à Québec 1660-1815* (Montreal, 1977), 2nd. part; Ian McKay, "The Working Class of Metropolitan Halifax 1850-1889" (honours thesis, Dalhousie University, 1975), ch. 2.

24. Thomas Smith, "Reconstructing Occupational Structures: The Case of the Ambiguous Artisans," *Historical Methods Newsletter*, 8 (1975), pp. 134-46.

25. *Rules and Regulations of the Brothers Carpenters' Society of Halifax, Nova Scotia* (Halifax, 1833); *Novascotian*, 10 November 1845; Nova Scotia, *Statutes*, 13 Vic. c.43.

26. *Morning Journal*, 19 June 1854, 30 March 1857; Nova Scotia, *Statutes*, 27 Vic. c.11.

from 11 in 1838 to 408 in 1861 and to 518 in 1871. All work beyond sawing – such as trimming, moulding and carving – was carried out by carpenters, and the extent of labour involved in this work as well as in the final installation of these products in the structure made carpenters the largest skilled trade group in Halifax in 1838. Nonetheless, while their numbers grew substantially, their proportion of the building trades fell from 63.1 per cent in 1838 to 51.5 per cent in 1871,[27] as some of their labour-intensive work was subsumed by the four planing and moulding factories, primarily steam-powered, then operating in the city. In the early 1860s the increase in the number of carpenters, the building boom with its opportunities in a larger market for specialization, and local labour tension all contributed to formal subdivision among carpenters. Incorporation of the House Joiners' Union Society of Halifax and of the Shipwrights' and Caulkers' Association of Halifax and Dartmouth in 1864 marked a trade consciousness which was demonstrated in a carpenters' strike for higher wages later that year.[28]

While the number of carpenters increased rapidly in mid-century Halifax, the numbers of masons and stonecutters grew even more quickly. From 1838 to 1861, they increased from 50 to 223.[29] The three successive fires in the downtown core, which led to strict building codes and extensive rebuilding in stone, may have made the numbers artificially high, but thirty years of work on the massive stone Citadel and large-scale government and institutional building in stone in the 1850s attracted and held masons in the community. Following a strike for higher wages in 1864, the Stone Cutters' and Masons' Association of Nova Scotia was incorporated in 1865 with a constitution and by-laws closely based on those of the Journeymen Stone Cutters' Association of the more mature Boston market.[30] While trades like painters, plasterers, and plumbers, whose crafts were discrete, played only a limited role in the construction of any building, these trades expanded significantly in response to the extensive local building activity of the 1860s.[31]

E.W. Cooney has identified four types of building firms in the early

27. Figures based on Halifax City census, 1838, 1861, 1871.
28. Nova Scotia, *Statutes*, 27 Vic. c.24, c.33; *Evening Express*, 3 June 1864; *British Colonist*, 4 June 1864.
29. Figures based on Halifax City census, 1838, 1861.

30. *Evening Express*, 29 June 1864; Nova Scotia, *Statutes*, 28 Vic. c.75; The Journeymen Stone Cutters' Association of Boston, Roxbury, Charleston and their Vicinities, *Constitution and By-Laws* (Boston, 1863); *Constitution and Bye-Laws of the Stone Cutters' and Masons' Association, of Halifax and Nova Scotia* (Halifax, 1865); *British Colonist*, 30 May 1868.
31. See, for example, Halifax city directories for 1858/59, 1863 and 1869/70; Halifax City census, 1861 and 1871.

nineteenth-century construction industry in Britain.[32] During the 1840s
most Halifax building firms roughly corresponded to Cooney's first
type. They were operations run by a master craftsman – usually a carpenter or a mason – undertaking work only in his own trade and usually
employing only a small number of journeymen and apprentices. The
expanded construction activity in the late 1840s and early 1850s appears
to have stimulated development of a second type of building firm similar to that distinguished by Cooney, as a master craftsman, usually a
carpenter or a mason, undertook responsibility for the construction of a
whole building, carrying out himself those parts related to his own trade
and sub-contracting the other parts to appropriate other master craftsmen. This was presumably the nature of the operation usually ascribed
to the term "builder" in the city directories and the census. The twenty
builders listed in the first Halifax business directory in 1858 and the fifty
enumerated in the 1871 census may reasonably be accepted as representing minimum numbers of this type.[33]

Until the 1850s all building firms in Halifax apparently operated as
one of these two firm types. Moreover, in the 1840s, when the demand
for building designs and the fee for supervising their construction (normally 6 per cent of the construction cost) was not sufficient to provide
an income, Henry G. Hill fulfilled the roles of both architect and builder
for several structures including villa residences and the Temperance
Hall. These operations appear to have been carried out as two branches
of a single firm; Hill himself did the design and other architectural work,
while master carpenter John Mumford, in Hill's employ, directed the
actual building activity. Whether the artisans in Hill's employ constituted a permanent work force or whether they were independent journeymen carpenters hired for each undertaking is not entirely clear. The
extent of Hill's work, particularly his involvement in speculative house
building, suggests that there was probably a small regular workforce
which was supplemented as occasion required by independent journeymen carpenters.[34]

32. E.W. Cooney, "The Origins of the Victorian Master Builders," *Economic History Review*,
    VIII (1955), pp. 167-8; "The Speculative Builders and Developers of Victorian London:
    A Comment," *Victorian Studies*, XIII (1969/70), pp. 355-7. Halifax does not appear to
    conform to the distinctions between building enterprises outlined by John Burnett, *A
    Social History of Housing 1815-1970* (Newton Abbot, 1978), p. 20. The local market was
    not large enough to operate firms primarily distinguished by public building or commer-
    cial construction, although many smaller firms may have confined themselves to the
    third category, housing.
33. Halifax city directories for 1858/9, 1863 and 1869/70; Halifax City census, 1861 and 1871.
    Of the 25 builders reported in street entries in the Halifax city directory for 1869/70, 12
    were carpenters and 6 were masons; the trade affiliations of the other 7 remain unidentified.
34. See, for example, *Novascotian*, 1 September 1842, 13 May 1844, 28 August 1848, 1 Nov-
    ember 1852, 5 September 1859, 26 November 1860; *Acadian Recorder*, 17 May 1856, 3 Octo-
    ber 1857; *Morning Journal* (Halifax), 19 June 1854.

The role of the builder in mid-century Halifax was a precarious one. If a building was constructed by contract, tendering was highly competitive. The builder had to estimate accurately his competition, his markets for both supplies and labour, and any technical problems. Usually the contractor was left to finance his own operations, to find materials and labour, and to complete construction against an established deadline, which often incorporated a penalty clause for failure to meet the date. George Lang's contract for a Granville Street store, for example, had a penalty of £90 per week for non-completion; payment was to be made 85 per cent on completion and the balance three months later. When the proprietor proved unable to pay the building costs, he gave Lang a mortgage on his property which Lang, in due time, assigned to his own creditor.[35]

In addition to construction by contract, a substantial quantity of residential building was erected on speculation. While prosperity and rising population in Halifax strongly encouraged such construction, it was also the outlet for a builder's skills and labour when contracted work was not available. Acquiring suitable land was the first stage in the building process. Its availability, its profitability as speculative property, its cost in relation to building costs all influenced the location and timing of building. Lots, primarily in the north and south suburbs, were pruchased on speculation and usually mortgaged immediately for most of the purchase price. Subsequently they were either resold, sometimes to other builders, or built upon as the market and the builder's resources dictated.[36] Once building was completed, the structure might be sold or rented. As a large portion of the nineteenth-century population was never able to achieve home ownership, substantial quantities of rental properties existed in the city at any time. The majority were no doubt owned by investors for rental income, some of whom at least had commissioned their construction for that purpose. Whether by choice or by circumstance, a number of Halifax builders retained ownership of some houses which they had built and which they subsequently let.[37] In the early 1860s, the unusual level of sustained commercial construction created a shortfall in residential construction and stimulated a noticeably large number of independent builders to respond to market demand with speculative building for rapid sale in mid-decade. Effective timing

35. See, for example, *Acadian Recorder*, 9 October 1858; *Evening Express*, 10 May 1865; Halifax Provincial Building (New), vert. mss. file, PANS; Notebook of David Stirling, Queen's University Archives; #1314, Thomas vs. Romans and #1653, Murdoch vs Romans and Lang, RG39C, 1862, PANS.
36. Halifax County, Index to Deeds, 1836-66, PANS, See, for example, entries for Henry G. Hill, Robert Davis, George Lang, George Blaiklock.
37. See, for example, John Mumford's ads in *British Colonist*, 22 February 1862; *Acadian Recorder*, 26 October 1862; 21 March 1863; also, *Novascotian*, 1 November 1852; *Evening Express*, 26 February 1862.

and locating of speculative building were a major key in determining a builder's success or failure in the speculative housing market.

Repairs and enlargements, such as those carried out on the Brunswick Street Methodist Church in 1857-58 at a cost of £3,000,[38] constituted the final principal area of the builder-craftsman's operation. Traditionally, this type of work constituted a large portion of a builder's and a craftsman's labour and income, and the large number of wooden buildings, the numerous enlargements to existing buildings, and the demand for fashionable renovation suggest that Halifax was no exception. Although Halifax builder-craftsmen undertook not only complete buildings but also contract work in their own trades, it is not clear what portion of the work of such firms was actually comprised of the more elementary, direct trade type.

Because of the precarious nature of building and the traditionally high bankruptcy rate in the building industry, financing was often a builder's most difficult problem. Like the English speculative builders,[39] the Halifax builder-craftsman must have found financing through a network of local sources including his sureties on contracts, mortgages on his property, and after 1850 the local building society. Some evidence of the important role played by the Nova Scotia Benefit Building Society can be seen in its transactions in mortgages in the decade 1852 to 1861. In these years it financed 452 mortgages and received on assignment a further 20; against this it released only 122 and assigned 4. Although a second building society was formed in 1863, the business of the earlier society appears to have increased rather than decreased through the mid-1860s.[40]

The third type of builder described by Cooney – not a craftsman but perhaps a merchant, erecting complete buildings on the basis of contracts with master craftsmen in the various trades – appears to have been rare in Halifax, except where a merchant was building for his own use either directly in a personal capacity as in residences or stores or in a personal investment capacity for rental purposes. Estate agents, such as Samuel Gray and William Allen, may have undertaken some such activity to feed their busy real estate agencies, but the one clearly identifiable builder of this type in Halifax was auctioneer J.D. Nash. More detailed examination of Nash's activities through property transactions, taxation and court cases is required to determine the extent and success of his operations but one example of his undertakings was his huge Victoria Buildings erected on Hollis Street shortly after the 1857 fire. While its

---

38. *Novascotian*, 15 February 1858; *Acadian Recorder*, 20 February 1858.

39. Burnett, *A Social History of Housing*, p. 24.

40. For discussion of building societies and their advantages, see *Novascotian*, 5 April, 10, 17, 24 May, 14, 28 June 1847; 30 August 1852; for Halifax societies, see *British Colonist*, 9 February 1850; *Halifax Reporter*, 9 August 1860; Index to Deeds, 1852-1866, PANS; *Acadian Recorder*, 28 February, 9 May 1863.

designer and builder are both unknown, a notice in the *Novascotian* in June 1857 indicated that Nash was joining Doull & Miller and Maurice McIlreith in the erection of "fine brick buildings with cast iron fronts on the premises recently purchased." In January 1861 Nash's building was offered as a good opportunity for investment from which a return of 9 to 10 per cent might be expected. As the notice went on to explain, "the proprietors [were] only anxious to sell that with the proceeds they may build up more Brick Buildings in the business parts of the city." How many, if any, of the buildings which Nash auctioned also represented a partial investment on his part is, at this stage, unknown. In May 1862, with the residential market more opportune for investment than the commercial one, Nash purchased the field between the Horticultural Gardens and Camp Hill Cemetery on which he intended to erect fourteen two-storey brick dwelling houses, "tenements for the masses." In noticing the purchase, the *British Colonist* lauded the much-needed "enterprise and public spirit' with which Nash had previously erected "buildings at once useful and ornamental."[41]

It is more difficult to assess when the fourth type of builder described by Cooney – the master builder, erecting complete buildings and employing more or less permanently a relatively large body of labourers and workmen in all the principal building crafts – first appeared in Halifax. By the mid-1850s the two circumstances that Cooney associates most strongly with the rise of the master builder both prevailed in Halifax.[42] Large-scale government building coincided with commercial prosperity and its attendant new construction and fashionable improvement of existing structures. The extent and sustained level of this building made possible the development of substantial workshops and relatively large, permanent work forces. Simultaneously, the emergence of the practice of assigning single, fixed-price contracts for large buildings, rather than solely for their individual parts as in the past, favoured the general builder who could capitalize upon economies of scale and the assurance of materials and an established work force. By the 1860s there were half a dozen builders working in Halifax whose operations exemplified characteristics of the master builder. All were regular bidders on major construction contracts in the city in the 1860s.

Robert Davis, a mason who had arrived in Halifax by 1840, may have been the first to establish an operation which began to move beyond the limits of his own trade. By the mid-1850s he was employed on successive government contracts while also carrying out private commercial building. The prominent new City Market (1854) was followed by the Lunatic Asylum (1856), the City Prison (1857) and the City Hospital (1857). The large scale of government construction, its regular system of payments,

---

41. See, for example, *Novascotian*, 15 June 1857; *Evening Express*, 30 January 1861; 23 May 1862; *British Colonist*, 27 May 1862; *Evening Express*, 14 July, 8 September 1865.
42. E.W. Cooney, "Origins of the Victorian Master Builders," pp. 170-6.

and the sequence of projects which he undertook provided Davis with the opportunity to establish a large regular work force. When it was not occupied on government projects, it was engaged in prime commercial construction which included the Bank of Nova Scotia, the adjoining new store for Messrs Boggs & Ross, and alterations to the Union Bank two blocks distant. In September 1856 the *Novascotian* reported Davis' weekly wage payment at £315; if that was his average weekly wage figure for the building season, his business was paying out at least £10,000 a year in wages alone. If Davis was, however, operating beyond the limits of his own trade by the late 1850s, it is surprising that he was not more successful in competing in the expanding construction market, where he was already well established, against ambitious newcomers to the Halifax scene. In the prime building period from 1860 to 1863 Davis continued, in partnership with Andrew Barton, to take some substantial contracts, but by 1865 he seems to have largely withdrawn from major building and to have limited his principal role to the less capital-intensive supply of building materials.[43]

George Lang, whose Halifax career lasted only from 1858 to 1865, epitomizes the builder-craftsman working in Halifax on a scale larger than his own craft, in this case masonry. The designer and sculptor of the Welsford-Parker Monument in 1860, Lang established himself in Halifax as the contractor for Thomas' county court house and the three Thomas-designed stores on the west side of Granville Street. His commercial and institutional building in the early 1860s, especially structures designed by Stirling, culminated in his successful competition for the $100,000 Provincial Building in 1864. By this time Lang had his own work force, sustained by both the scale and quantity of his commissions, a well-established creditor in James Forman, cashier of the Bank of Nova Scotia, and a masons' workshop, a carpentry and building practice, a slate quarry and a steam engine. In addition to contracts for the erection of buildings of all sizes, he was prepared to provide plans and specifications for buildings and to do all kinds of jobbing. Lang's building establishment, however, was shattered in 1865 when, through a combination of circumstances, including his own overextension, he became bankrupt in the course of erecting the Provincial Building and retired to brick-making at Shubenacadie.[44]

Two more successful builders who arrived in Halifax in the 1850s were George Blaiklock and Henry Peters. Both Blaiklock and Peters came to the city in 1852 as partners in the firm Peters, Blaiklock & Peters, Quebec principals contracted by the British army to erect the

---

43. *Acadian Recorder*, 17 July 1841, 8 April 1865; *Morning Journal*, 19 April 1854; *Novascotian*, 15 September 1856, 20 April 1857, 2, 9 January 1860; *Evening Express*, 8 April 1861; *Morning Chronicle*, 11 March 1868.

44. Susan Buggey, "Building in Mid-19th-Century Halifax: the Case of George Lang," *Urban History Review*, vol. 9 (October, 1980), pp. 5-20.

Wellington Barracks. This contract introduced a new scale of building activity in the city. In its first year Peters, Blaiklock & Peters employed three hundred artisans and labourers in clearing the site and commencing operations.[45] The following year, unable to obtain locally a sufficient quantity and quality of bricks for the required work, they established a brickyard at Eastern Passage, equipped with American machinery, which by 1855 mechanically produced the first numerous, quality supply of bricks available to the city; in 1861 it reported over a million bricks of various types for sale. By the early 1860s, other brick manufacturers following Peters, Blaiklock & Peters' example made construction of brick buildings a feasible option to stone or wood.[46] To provide the interior and exterior finishing materials for the barracks, Peters, Blaiklock & Peters also established the first steam-operated door, sash and blind manufactory in Halifax, which was in operation by late 1855. Again, they brought to Halifax a new class of machinery rapidly coming into commercial production in the United States. Sold to long-established local carpenter Alexander Bain in 1860, the mill offered "Doors, Blinds, Sashes and Mouldings, of every description constantly on hand, and executed at shortest notice [as well as] all kinds of Carpentry and Joiner's work made to order." It too was followed by similar manufactories which, "with the help of a few attendants, could duplicate the work of dozens of joiners working with hand planes, and could quickly turn out large volumes of mouldings of extraordinary low cost and exact profile."[47]

The Wellington Barracks established Peters, Blaiklock & Peters firmly in the Halifax building world. Not only did they become prime suppliers of manufactured building materials in the local market, but in the course of extending their contract in 1855, they obtained reputable financial backers in William Hare and M.B. Almon.[48] Upon completion of the barracks in 1858, they moved from military construction to the civil sphere by taking the £9000 contract for erecting the new St. Matthews Church, "universally spoken of as, beyond comparison, the most

45. *Novascotian*, 6 September, 11 October 1852.
46. *Novascotian*, 22 August 1855, 21 April 1856, 14 June 1858; Royal Engineers' Papers, vol. 48, passim, p. 171 ff, esp. pp. 258, 264, MG 12, PANS; *Evening Express*, 31 July 1861, 19 March 1862; *Morning Sun*, 24 April 1863.
47. *Acadian Recorder*, 1 March 1856, 17 November 1866, 12 January 1867; *Evening Express*, 4 January 1858, 19 May 1865; *Novascotian*, 26 December 1859, 5 March 1860; Halifax County deeds, bk. 126, f. 522, PANS; John H. Englund, "An Outline of the Development of Wood Moulding Machinery," *Bulletin* of the Association for Preservation Technology, X-4 (1978), p. 20.
48. Detailed correspondence relating to the progress and problems of the Wellington Barracks contract is found in Royal Engineers' Papers, vols. 48 and 49, MG 12, PANS.

*Figure 5.*

The Wellington Barracks (top), built in the North Suburbs to replace the North Barracks on Citadel Hill burned in 1850, and the new St. Matthews Church (bottom), moved from the town centre to the edge of the South Suburbs, marked the mid-century decentralization of Halifax. These two major construction contracts in the 1850s established Quebec builders Peters, Blaiklock & Peters in the Halifax building industry where George Blaiklock and Henry Peters firmly situated themselves as leading local builders throughout the 1860s. (top: National Army Museum, London; bottom: author).

elegant specimen of ecclesiastical architecture in the city."[49] Dissolution of the firm in 1860, and the departure of its senior partner, Simon Peters, left George Blaiklock and Henry Peters to re-establish themselves as individual builders in Halifax.

Indicating his willingness "to enter into contracts for the erection of every description of buildings," Blaiklock moved immediately to a number of prime commercial contracts for rebuilding in the burnt districts on Granville, Hollis and Prince Streets and was soon recognized as a "contractor well and favourably known as the builder of some of our finest buildings." Blaiklock's 1865 contract for Trinity Church involved him in the classic problems of the Halifax building world; despite the building committee's dedicated efforts to create a building fund equal to the costs, a year after completion Blaiklock had still not received $2,500 of his contract price,[50] although his scale of operation was sufficient to absorb this delay in payment. As early as 1862, Blaiklock moved into the purchase of building lots and quality speculative house building in the rapidly opening south suburbs. In 1864, for example, he offered for sale "Two commodious and well-built Brick Houses on Queen street, now in course of completion, with Brick Stables and Coach houses in rear," featuring not only a "beautiful situation" and "modern style of construction" but "Bathrooms fitted with pipes for hot and cold water; Water closets, &c." Later that year he commenced a terrace of six pressed-brick and freestone dwellings on Pleasant Street, which he had completed and disposed of by late 1865. By building and selling as rapidly as he could and holding his houses for rental only when he could not sell them, Blaiklock was apparently able to operate a revolving capital fund which continuously financed at least a portion of his building activity. Other funds were obtained by mortgaging properties to investors or to one of the local building societies. At the same time he was building substantial residences by contract, such as the Mary, Queen of Scots house on Queen Street. By 1871 Blaiklock's building establishment was reported to have a capital investment of $30,000, thirty-three employees earning aggregate yearly wages of $7,000 and products defined as "Houses & Warehouses, Repairing, &c." valued as $30,000 per annum.[51]

In the 1860s Henry Peters moved even more firmly than his former partner, Blaiklock, towards the status of a master builder. Peters' activities were more diversified, more secure, and more extensive. Early in the decade, when commercial building was most active, Peters undertook a

---

49. *Novascotian*, 25 July 1859.

50. *Halifax Reporter*, 12 January 1863; *Novascotian*, 26 December 1859; *Evening Express*, 10 April 1861, 25 January 1864; *Morning Chronicle*, 5 March 1868.

51. *Evening Express*, 15 January 1864; Halifax County deeds, index, 1861-71, PANS; see also *Evening Express*, 26 February 1862, 20 April 1864, 24, 27 November 1865; *British Colonist*, 26 January 1864; Halifax City census, 1871, Ward 2, Schedule 6.

series of prominent brick and stone stores as well as the prestigious Union Bank. In 1862 he successfully built the new skating rink in the Horticultural Gardens, firmly establishing his reputation as a reliable, economic builder; in 1863 he took contracts for the naval hospital and the new jail; in 1865 for the addition to the Lunatic Asylum; in 1867 for the Poors' Asylum; in 1868 for the Blind Asylum. At the same time he built Benjamin Wier's handsome Italianate residence on south Hollis Street. From at least 1862, he was buying building lots on speculation in south Hollis, Lower Water, Pleasant and Kent Streets, mainly areas of the south suburbs prime for the quality speculative housing which he was to erect on them as his other building contracts allowed. In 1865 he moved even further towards the role of developer by purchasing with two others the "Bremner property," which they proceeded to lay out in building lots according to a plan prepared by Peters. By 1871 his "Carpenter & Building Establishment," with an invested capital estimated at $24,000, used lumber, bricks, building stone, plaster, lime, and portland cement valued at $7,750; it employed 32 men at aggregate yearly wages of $11,085 and produced "Houses, &c." valued at $28,000.[52]

The undisputed Halifax master builder in 1871 was Samuel Brookfield who had succeeded his father at the latter's death the previous year. John Brookfield, primarily a railway contractor, had come to Halifax in the early 1860s and subsequently constructed an engine house, several wharves and some short railways in the civil sphere. His largest works were a series of contracts for the British army totalling over $500,000 and including a magazine, a military hospital, two barracks, four batteries, and work on the defences at George's Island. In 1866 he also took over Lang's incomplete Provincial Building and finished it satisfactorily at a contract price of over $80,000. In 1871 the firm's building establishment, with a capital investment of $16,000, employed sixty-two workers for aggregate yearly wages of $35,000. The work made use of "Building Materials of every Kind" in the production of "Building of Every Description" valued at $70,000 a year. While no evidence has been found to show that he provided his own plumbing and painting services rather than purchasing these services by contract from one of the numerous, well-established local firms, the scale, credit, capacity and labour force

52. *Evening Express*, 12 July 1861, 15 December 1862, 2, 16 September, 23 November 1863; 4, 25 January, 15 June 1864; 27 March, 19 April, 2 June, 4 October 1865; *Halifax Reporter*, 28 February 1861, 31 January 1863; *Morning Chronicle*, 2 March 1868; *Acadian Recorder*, 27 June 1868; Halifax County deeds, index, 1861-71, PANS; Halifax City census, 1871, Ward 2, Schedule 6.

of the Brookfield establishment were in other respects those of a "master builder."[53]

Changes in the Halifax building world were an integral part of the mid-nineteenth-century transformation of the city. Large scale and sustained demand for building services in the government, commercial and residential sectors stimulated the formation of architectural practices whose principals replaced self-trained designers as the architects of the city's most prominent new structures, as had already occurred in the largest Canadian cities. Continuous and substantial construction also facilitated the emergence of the building firm characterized by a relatively large, permanent work force, a reliable supply of materials, steady capitalization, and the capacity to undertake building of whatever type and scale required. The Brookfield firm, still in operation a century later, epitomized this direction. Increasing preference by proprietors for assigning single fixed-price contracts for whole structures to masons or carpenters who would orchestrate the total construction further altered traditional practice in the industry and the craftsman's role in it. The building boom in the late 1850s and early 1860s, which attracted increased numbers of building artisans to the city, encouraged specialization within their trades, which was reflected both in the extent of services available and in stronger trade organization. At the same time extensive demand for building supplies spurred mechanization in local production while commercial manufacture of new products elsewhere introduced a greater diversity of materials in the Halifax market. Although depression in the 1870s severely affected the building trades, the impact of these mid-century changes was seen in the evolution of the industry in the 1880s and 1890s as a more industrialized component of the social order and in the physical fabric of the late nineteenth-century city. More immediately, the influence of changes in the mid-century building world was evident in the transformation of a predominantly wooden town into a more substantial, more sophisticated, and more coherent cityscape, important segments of which remain in place today.

---

53. "Credentials for Private Circulation" (copy in possession of Parks Canada, Halifax Defence Complex); *Halifax and Its Business* (Halifax, 1876), pp. 150-1; A.B. Johnston, "Defending Halifax: Ordnance 1825-1906" (Parks Canada, Manuscript Report Series #234), p. 31; *Evening Express*, 28 April, 15 November 1865, 5 March 1866; *Acadian Recorder*, 16 February 1867, 27 June 1868, 26, 27 October 1870; Halifax City census, 1871, Ward 1, Schedule 6.

# Reshaping the Urban Landscape?
# Town Planning Efforts in Kitchener-Waterloo, 1912-1925

ELIZABETH BLOOMFIELD

> Canadian town planners will take courage in the midst of much
> cause for discouragement from the fact that at last a Canadian city of
> the size and importance of Kitchener has passed into law a compre-
> hensive town plan for the regulation of its future growth and develop-
> ment on the scientific lines advocated by town planning science.[1]

"We have been making Town Planning History."[2]

In the mid-1920s, Kitchener enjoyed the reputation of a pioneer in the
Canadian town planning movement, as the first Ontario muncipality to
adopt a modern town plan and enact an associated zoning by-law. The
plans by Thomas Adams and Horace Seymour for Kitchener and
Waterloo were on view at the British Empire Exhibition (Wembley) in
1924; and Adams in his speech there described the two municipalities as
"the most advanced in Canada in regard to town planning."[3] Kitchener

SOURCE: This is a revised version of an article from the *Urban History Review*,
Vol. IX (June 1980), pp. 3-48, entitled "Economy, Necessity, Political Reality:
Town Planning Efforts in Kitchener-Waterloo, 1912-1925." Reprinted by permis-
sion of the author and the editors of the *Urban History Review*.

1. "Kitchener Plan Becomes Law," Town Planning Institute of Canada, *Journal* IV, 1
   (1925), p. 1.
2. H.L. Seymour, "Report on the Plan of the Town of Waterloo," 30 April 1924, p. 16
   (Seymour Papers, Public Archives of Canada).
3. Town Planning Institute of Canada, *Journal* III, 3 (1924); Seymour, "Report, Plan
   . . . of Waterloo," foreword.

was often cited as a model by Seymour and others in the *Journal* of the Town Planning Institute of Canada, for its adoption of a comprehensive plan and zoning by-law and for its initiative in convening meetings of Ontario municipal leaders to press the provincial legislature for wider zoning powers.[4]

At the moment of triumph, Adams and Seymour glossed over the long struggles, the frustrations and compromises they and other planning advocates had endured to gain acceptance of the plan. For Adams, though he had already taken up his work for the Regional Plan of New York, the Kitchener plan was a practical vindication of his seven years as Town Planning Adviser to the Canadian Commission of Conservation. For Seymour, and the planners associated with him in the Town Planning Institute of Canada, the Kitchener-Waterloo plan was a test case to prove that a professional town plan could be accepted by the citizens of two smallish, ordinary municipalities, and also that a comprehensive zoning by-law could be enacted in Ontario. The sequence of procedures followed in the Kitchener and Waterloo plans illustrated the general pronouncements of Adams, especially in the 1920s. They also provided a basis for elaboration in other Canadian town plans, notably that of Vancouver for which Seymour was resident engineer in the later 1920s.[5]

Kitchener and Waterloo must have seemed unlikely places to be acclaimed the most advanced in Canada for their attitudes to town planning. They were small in population (with a total of only 30,000 people in 1924, four-fifths of whom lived in Kitchener), highly industrialized, unpretentious and practical in municipal policies.[6] Both were "Main Street" towns, in which

---

4. Brief references to aspects of the Kitchener case have also been made by recent writers on planning and urban reform. See, for example: Walter Van Nus, "The Fate of City Beautiful Thought in Canada, 1893-1930," in Gilbert A. Stelter and Alan F.J. Artibise, eds., *The Canadian City: Essays in Urban History* (Toronto: McClelland and Stewart, 1977), p. 179; John Weaver, " 'Tomorrow's Metropolis' Revisited: A Critical Assessment of Urban Reform in Canada, 1890-1920," in Stelter and Artibise, *The Canadian City*, p. 405; Thomas I. Gunton, "The Ideas and Policies of the Canadian Planning Profession, 1909-1931," in Alan F.J. Artibise and Gilbert A. Stelter, eds., *The Usable Urban Past: Planning and Politics in the Modern Canadian City* (Toronto: Macmillan, 1979), p. 182; Walter Van Nus, "Towards the City Efficient: The Theory and Practice of Zoning, 1919-1939," in Artibise and Stelter, *The Canadian City*, p. 239.

5. For a recent survey of the role of Commission of Conservation, see: Alan F.J. Artibise and Gilbert A. Stelter, "Conservation Planning and Urban Planning: The Canadian Commission of Conservation in Historical Perspective," in Roger Kain, ed., *Planning for Conservation: An International Perspective* (London: Mansell, 1981), pp. 17-36.

6. Processes of urban development in Kitchener and Waterloo have been examined in Elizabeth Bloomfield, "City-Building Processes in Berlin/Kitchener and Waterloo, 1870-1930," Ph.D. thesis, University of Guelph (1981).

It is exceedingly difficult to persuade men who are obsessed by a small private interest that the freedom of the whole community to live its own life in clean air and sunlight and to live and play in a pleasant environment is more important than the freedom of any individual to do as he likes with a piece of land which he considers his own . . . and that the efficiency and economy of industry and commerce may be involved in proposals for a more orderly and scientific development of towns and cities.[7]

This case study of efforts to plan the physical development of Kitchener (Berlin until 1916) and Waterloo from 1912 to about 1925 emphasizes local perceptions of the urban environment and of the need for planning, planning as an issue in local politics, and the interaction between outside planners and the local community. Themes include the significance of key individuals and the press in leading public opinion, the reluctance of municipal councils to antagonize the voters, and recurrent suspicions of the motives of those who advocated planning. Attitudes to planning are considered in relation to each town's commitment to economic growth, and its sense of rivalry with other Ontario municipalities. The study has wider interest too, in illustrating the diffusion of ideas about planning as of other aspects of urban reform from larger metropolitan centres to small cities and the transition in these concepts from "City Beautiful" ideals to a more economically acceptable and politically realistic type of plan.

The history of the planning movement is considered in several parts: its origins particularly in the general awareness of civic problems around the time Berlin was seeking city status in 1911-12 and the formation of the Berlin Civic Association; the Leavitt plan, 1913-14; the revival of concern for planning in the Kitchener City Planning Commission from 1917; negotiations for a new comprehensive plan carried out by Adams and Seymour 1921-24; and an assessment of the significance of these planning efforts.

## ORIGINS OF THE PLANNING MOVEMENT

Before 1911-12, concern for the aesthetic and orderly arrangement of the urban environment was subordinated to the prevailing growth ethos. The creation and maintenance of public parks was one of the few ways such concern was expressed. Waterloo preceded Berlin in adopting the provincial Public Parks Act in 1890 which gave the power to appoint a Board of Park Commissioners and acquire land for development for park purposes. Waterloo's success convinced Berlin that it must follow

7. "Kitchener Plan Becomes Law" . . ., p. 1.

suit, in 1894-96, not without considerable disagreement and some changes in mind. Each town had one main park before 1910.[8]

General statements on the scope of city planning were rare, but in one example, a *Daily Telegraph* editorial on "Laying out Towns" in 1906 lamented that "when Berlin was laid out . . . there was no general idea of its future needs . . . . This was an oversight that can never be remedied." It proposed "the system adopted in many German towns" of separating "factory, business and dwelling districts" each with its "peculiar requirements."

> It is perfectly plain that a dwelling or residential part of a town should not be confused with a factory or manufacturing part, that these should be distinct and separate as much as possible, the manufacturing institutions placed by themselves to the east of the residential section so that smoke and soot may blow away without injury to the residential part. . . .[9]

A small measure of protection for the amenity of at least some residential districts had been achieved by Berlin's adopting the "Building Line Act" in 1904. When petitioned by a majority of the property-owners on a particular street or section of street, the municipal council could by a vote of two-thirds or more of its members, declare that to be a "residential street." A building line would be specified, beyond which houses could not be built any closer to the street. This provision was first used in 1904, then more frequently from 1909, with about seventy such building line by-laws passed by 1920. This zoning by local petition did protect some of the better residential neighbourhoods from unwanted incursions of industry, but was useless in helping the poorer house-holders.[10] Suggestions of separate "factory districts" were also made from time to time, more from consideration of efficient provision of services than to protect the residential areas.[11]

It was only when Berlin was about to attain city status that interest in

---

8. Waterloo Town Council, Minutes 4 August 1890; Town of Waterloo By-law 110; Town of Berlin, By-law 561, 1 October 1896. Berlin had previously owned the "old Town Park" (also called "Woodside") of about 28 acres since 1873, which was administered (or neglected) directly by the council and had no funds for its upkeep. It was rather remote from the urbanized area at the time. A brief historical sketch was given in *Daily Telegraph*, 6 March 1905. Berlin also passed By-law 419 (6 May 1889) "to provide for the establishment of boulevards in Berlin and plant sod and shade and ornamental trees."

9. *Daily Telegraph*, 16 July 1906.

10. Municipal Amendment Act 1904, *Statutes of Ontario*, 4 Edw. VII, ch. 22, sect. 19. Waterloo seems not to have used this provision, until a resolution in 1912 that "people who propose building on residential streets be and they are hereby restrained from placing their new building nearer to the street line than other buildings on the same street." Waterloo Town Council, Minutes, 1 April 1912.

11. For example, by Dr. W.L. Hilliard to the Waterloo Board of Trade (Minutes, 25 March 1907); *Daily Telegraph*, 15 January 1910 (for Berlin).

city planning became more widespread and sustained. Several factors were involved in this new awareness of planning. One was pride in the city's remarkable growth during the previous decade. The 1911 census revealed that Berlin's population had grown by 56 per cent since 1901, and that the value of industrial production had increased by 408 per cent between 1890 and 1910 — the eighth fastest rate of growth among Canadian municipalities with over 10,000 population in 1911, and second only to Sault Ste Marie in Ontario. Berlin had one hundred factories and 60 per cent of the labour force were engaged in manufacturing and mechanical occupations.[12] So much had been achieved despite the town's lack of natural advantages. Success was to be explained by the one great asset of a frugal, industrious workforce and by the unremitting effort of town council and Board of Trade to attract and hold enterprising manufacturers.

Allied to pride in its successful growth was a dedication to continued municipal self-improvement, necessary for Berlin to compete with other Ontario towns and cities in the pursuit of economic growth. Berlin had been early to adopt the principle of municipal ownership and operation of the urban utilities. City planning, with housing and public health a prominent issue in the larger cities of the day, now appeared an appropriate challenge as Berlin joined the league of larger municipalities. This mood is caught in the following declaration from the volume commemorating cityhood published in mid-1912.

> In this City Progressive, there are indications pointing to a widening of public spirit and a deepening of interest in the general welfare. A leaven is at work which has for it object the making of Berlin a more beautiful city, a better city and a model city.[13]

A third element, implied in some of the terms used in the passage just quoted, was the growing knowledge of city planning ideas and organizations in larger cities of the day. The daily newspapers carried news of such developments and reprinted articles and speeches on the subject. Both the grandiose City Beautiful and public health aspects of contemporary city planning were presented to local readers.

During 1912, Berliners were also made aware of some of the social consequences of rapid urban-industrial growth in their own city, especially those associated with overcrowded housing. Thus they glimpsed the much graver problems of larger cities which must be expected with continued growth. All these factors reinforced the perennial practical frustrations of congestion on King Street, dangerous railway crossings, the awkward street layout and housing shortages to bring planning to the forefront as a topic of public discussion.

For the first year of intensive concern with city planning, activity was

---

12. Census of Canada, 1911, vol. II, p. xiv; Census of Canada, 1911, vol. IV, table VI.
13. *Berlin: Celebration of Cityhood* (Berlin: German Printing and Publishing Co., 1912) 72.

limited to Berlin, and was not co-ordinated by any formal organization. Two of the strong local advocates of city planning took the lead from an early stage, and were supported by extensive press coverage and comment on the subject. While city planning was discussed in terms derived from elsewhere, notably European and American cities, most of those involved at this stage seem to have been more interested in specific local problems which they thought might be painlessly rectified by general city planning. Until the spring of 1913, there was remarkable agreement in support of general city planning, perhaps because its implications were not yet understood.

A summary of some of the events of 1912 illustrates the variety of strands involved in the interest of planning. The year began with a *Daily Telegraph* editorial on "Town Planning," reporting C.E. Hodgetts in Montreal on the subject of the "effects of the haphazard laying out of cities."[14] The editor declared that now Berlin was about to become a city, the subject of planning should be considered by bodies such as the town council, Board of Trade and Horticultural Society. In mid-January the Berlin Horticultural Society sponsored a special meeting on "Civic and Suburban Beautification." Mrs. Dunnington-Grubb, associate of landscape-planner Thomas Mawson in his Toronto office, gave an address illustrated by stereopticon views on the theme: "It pays to beautify a city."[15] At the February 1912 meeting of the Berlin Board of Trade, President H.L. Janzen raised the matter of city planning with reference to the need for a new post office and customs building, the widening of King Street and the wasteful lack of co-ordination among public utilities and the Engineer's Department, evident when a newly laid pavement was torn up to lay sewer pipes or water mains.[16]

In July, at the time of the cityhood celebrations, Mayor W.H. Schmalz raised the subject again in relation to the need for a new city hall to match Berlin's new civic dignity.[17] The Rev. Mr. Crews of the Trinity Methodist Church preached on "The City Problem," with its aspects of government, public health, slums and moral conditions (notably "booze" and illegitimacy).[18] The need for comprehensive city planning was elaborated by D.B. Detweiler, who also condemned the lack of control over the buildings of dwellings on lanes and alleys, and sug-

---

14. *Daily Telegraph*, 4 January 1912.
15. *Daily Telegraph*, 17 and 23 January 1912. S.J. Williams and W.H. Schmalz of the Horticultural Society, which had been formed in 1910, had hoped to have Thomas Mawson himself visit Berlin and advise on its beautification. Mawson was giving a special course of six lectures at the University of Toronto, and his lectures were printed verbatim in the paper. *Daily Telegraph*, 10 and 11 November 1911.
16. Berlin Board of Trade, Minutes, 8 February 1912.
17. *Daily Telegraph*, 15 July 1912.
18. *News Record*, 15 July 1912.

gested that members of Toronto's Civic Guild should be invited to Berlin to talk about city planning.[19]

More specific proposals were made to the city council in November 1912 by W.H. Breithaupt, local engineer and entrepreneur with American experience and connections. He recommended that Blucher Street, a narrow and discontinuous alignment of road in the North Ward, be improved into a wide, boulevarded thoroughfare, as the beginning of a system of boulevards around the outskirts of the city. He defined city planning, blending its aesthetic and functional aspects, and observing that "any city . . . was beautiful that had its business, manufacturing and residential sections and also provided for pleasure grounds and driveways." But most of Breithaupt's address and the subsequent discussion dealt with specific problems of Berlin's street system, notably the discordance between the section parallel to the business zone on King Street and the block of streets parallel to and generally north of the Grand Trunk Railway tracks, the lack of alternatives to congested King Street for traffic between Berlin and Waterloo, and the danger of level crossings. The *Daily Telegraph* reported that while no resolution was passed in connection with the proposition it was generally agreed that steps should be taken to adopt a general system of city planning.[20]

With an editorial "A City Plan" a few days later, the *Daily Telegraph* began a long series of editorials, special reports and reprinted articles defining and advocating planning. It commended "the principle of making permanent improvements according to a definite and comprehensive plan," as approved by the city council, as "a most encouraging sign of

---

19. *Ibid.*, 9 August 1912. Daniel Bechtel Detweiler (1860-1919) son and grandson of Mennonite ministers; travelling salesman for J.Y. Shantz Button Co. 13 years; from 1901, vice-president of G.V. Oberholtzer Shoe Co. (later renamed Hyrdo City Shoe Co.); from 1908 developed Algoma Power Co. from large waterpower resources at Michipicoten on north shore of Lake Superior. Advocate and publicist for Niagara Power movement from 1902, convening representatives of Western Ontario municipalities in key meetings; used similar approach in promoting civic planning and improvement (1912— ) and the Great Waterways Union of Canada (1914— ) Active on the Board of Trade, and the Light Commission and in the promotion of the beet sugar industry. A collection of Detweiler's papers in the Kitchener Public Library contains some useful items on the early city planning movement.

20. Berlin City Council, Minutes, 3 November 1912; *Daily Telegraph*, 5 November 1912. William Henry Breithaupt (1857-1944) second son of Louis Breithaupt founder of Breithaupt Leather Company of Berlin; qualified engineer at Rensselaer Polytechnic, Troy, N.Y., bridge engineer for Pennsylvania Railroad, C.P.R., Chicago, Santa Fé and California till 1890; consultant bridge and structural engineer New York City till 1899; 1900 returned to Berlin to manage Breithaupt family interests in Berlin Gas Works, Berlin and Waterloo Street Railway and extensive real estate. Built Berlin and Bridgeport Street Railway 1902 and developed riverside park at Bridgeport. Chief mover in city planning movement 1912-22; very active on Berlin/Kitchener Library Board, founder of Waterloo Historical Society and of Grand River conservation scheme.

the growing recognition of the futility and unwisdom of longer continuing the haphazard and wasteful methods of municipal improvement."[21] The editorial also advised engaging an expert without delay, and warned that the city had insufficient powers under the Municipal Act to do much about planning. The *News Record* published reports by the city assessors on the "well nigh scandalous . . . living conditions of the masses" with the comment that the situation called for a general city plan and better building regulations as well as an effort by "the manufacturers and monied men of the city" to provide more housing.[22]

"An important meeting of prominent citizens" was convened by D.B. Detweiler on 18 November 1912 in the city hall, "to inaugurate a town planning policy for Berlin." As this was Berlin's first public meeting on town planning, it is interesting to consider what kinds of people attended. Of the thirty-two persons present, at least eighteen were members of the Board of Trade, including its current president. Seven were city aldermen in 1912, and seven others had been at some time or would shortly afterwards serve on the Berlin Council. Ten were manufacturers, as either owners or managers of small to middle-sized enterprises, one was a successful merchant, while thirteen were engaged in some professional occupation (three lawyers, four in education, two engineers, two doctors, one accountant and one clergyman). There was one banker, one florist and major property-owners, one newspaper publisher, the G.T.R. freight agent, one bank teller and one industrial worker, and two were retired "gentlemen."[23] Fourteen of those present could be classified as community leaders on the basis of their above-average participation in the economic, political or associational life of the city. Only four had above-average wealth or business success, but these included representatives of the largest property-owners in Berlin, the Breithaupts and the Janzens. In the absence of the leading industrialists and workingmen and in the preponderance of members of the middle class, and especially of professionals, the meeting was typical of the group most active in the urban reform movements of the time.[24]

Approving references were made to the drastic replanning of Euro-

---

21. *Daily Telegraph*, 9 November 1912. These included a whole-page article on "Town Planning in Germany," by Frederic C. Howe, reprinted from the *New York Outlook*.

22. *News Record*, 13 and 14 November 1912. Two days later a family of Roumanians were reported to have been squatters in Woodside Park for some months, *News Record*, 16 November 1912.

23. Detweiler Diary (Detweiler Papers, Kitchener Public Library) 18 and 19 November 1912; *Daily Telegraph*, 20 November 1912; *News Record*, 20 November 1912. The one industrial worker was perhaps hardly typical of that group. Mr. C.M. Bezzo [sic] had, on the same day as the meeting had a short essay published in the *Daily Telegraph* on planning, appealing for it to be "in the interest of the great mass of toilers." As C. Mortimer Bezeau, he was later to be a mayor of Kitchener (1930-31).

24. Paul Rutherford, "Tomorrow's Metropolis: The Urban Reform Movement in Canada, 1880-1920," in Stelter and Artibise, eds., *The Canadian City*, p. 282.

pean cities and to progressive attitudes to planning in the cities of
western Canada by those who had travelled to Europe and the West.
But most were concerned with particular problems, especially urging the
purchase of land for a federal-civic square with a new post office and
city hall, and the need to extend sewer and street railway systems. It was
generally agreed that "a definite and adequate town planning policy was
needed . . . if Berlin was to expand along the right lines," and there was
unanimous support for the motion: "That this meeting ask the City
Council to stop the issue of building permits for dwelling houses on
lanes [as these are] the slum districts of the future."[25] It was also
resolved that a more general meeting be arranged for 11 December with
the double aim of a launching a city-planning movement in Berlin and a
larger provincial City Planning Association. The committee elected to
organize this meeting consisted of the mayor, president of the Board of
Trade, the most prominent figures in business, manufacturing, real estate and the professions, and was chaired by a newspaper publisher.[26]

Detweiler's powers as publicist and organizer and the contacts he had
first made ten years earlier in the Niagara Power movement were exercised again. He wrote to presidents of boards of trade and mayors of
thirty-five towns and cities in western Ontario, inviting them to a meeting in Berlin on "Town Planning and Beautifying." Detweiler's invitation was followed up by a supporting circular from the secretary of the
Berlin Board of Trade, W.M.O. Lochead, whose sentiments on city
planning were phrased more grandiloquently than was typical of the
Berlin planning movement: "Come and bring others with you so that
the beauty of civic art and the glory of unrestrained and symmetrical
civic expansion may be exemplified and assured for ourselves and for
future generations."

By December, Detweiler had received favourable replies from Preston, Brantford, Waterloo, Galt, Brampton, St. Thomas, Woodstock,
Hamilton, Toronto and Listowel. An illustrated address was to be given
by C.H. Mitchell, the engineer who had assisted Detweiler in the early
organization of the Niagara Power campaign by speaking at the first
meeting at Berlin in 1902, and was by 1912 vice-president of the Toronto
Civic Guild. Mitchell would also drive around Berlin during his visit, to
make specific suggestions for improvements.[27]

Readers of Berlin's newspapers at this time could hardly miss the
major articles on the scope and implications of planning. The *Berlin*

25. *Daily Telegraph*, 20 November 1912.
26. *Ibid.*, 4 December 1912. The committee comprised W.V. Uttley (chairman), W.H.
    Schmalz (Mayor), H.L. Janzen (President of Board of Trade), A.L. Breithaupt, D.B.
    Detweiler, W.D. Euler, J.F. Honsberger, C.H. Mills, George Rumpel, J.A. Scellen,
    H.J. Sims, Robert Smyth, S.J. Williams.
27. Letter quoted in Waterloo Board of Trade, Minutes, 25 November 1912; copy of
    circular in Detweiler Papers, Kitchener Public Library; Detweiler Diary, 13 June
    1912; 20, 21, 25 and 30 November 1912; 10 and 11 December 1912.

*Daily Telegraph* especially featured the "live subject." W.H. Breithaupt elaborated his suggestions for Berlin's street system, urging "a comprehensive general system of streets for the as yet partly occupied territory within and adjoining the city boundaries, and intelligent improvement of existing streets," including the widening of busy, narrow streets and the cutting of additional streets to give easier access from the outlying areas to the centre of business.[28] Berlin's city engineer, Herbert Johnston, also discussed the problems of city planning, mentioning progress in major Canadian cities, the laying out of garden suburbs and company towns and the work of the Public Health Committee of the Commission of Conservation.

> The primary objectives should be . . . proper housing accommodation for all citizens with sufficient light, air and space; [to] improve and widen the streets to handle traffic rapidly and safely; . . . to do everything that will make the lives of the people of this city more comfortable, healthy and happy.

To achieve these objects, "the one essential" was the designation of factory, business and residential districts (the last "graded as to the class of homes") — what came to be called zoning.[29] The work of the Toronto Civic Guild was described, notably its influence in having the City and Suburbs Plans Act passed by the Ontario Legislature.[30]

The City Planning and Civic Improvement Congress held in Berlin in 11 December 1912 linked the local advocates of town planning with the wider Canadian movement. The 150 persons present passed a resolution that a representative committee of fifteen work towards the formation of a provincial organization. Three members of the committee were Berliners, with other members from Galt, St. Catharines, Toronto, Waterloo, Sarnia, Welland, Preston, Brantford, Ingersoll and Ottawa.[31] Dr. C.E. Hodgetts of the Commission of Conservation stressed the housing problem in civic improvement and outlined the best methods to prevent slums becoming established. C.H. Mitchell used examples from English and American cities to illustrate the possibilities of civic improvement. He adapted his message to his audience by urging municipalities to plan while still fairly small in size "if they are to escape the mistakes of older cities," and also commending co-operation among towns and cities "to impress upon the Dominion and Provincial authorities the great desirability of legislation to encourage and assist various kinds of civic improvements." Mitchell deprecated

> the common idea in American cities until quite recently that city planning has been almost exclusively identified with city beautifying. This view is not fair to

28. *Daily Telegraph*, 7 December 1912; *News Record*, 7 December 1912.
29. *Daily Telegraph*, 9 December 1912.
30. *Ibid.* This act applied only to cities of over 50,000 population and thus could not help Berlin.
31. *News Record*, 12 December 1912.

the whole subject, because it loses sight of the practical sides of the question, which are very many and complex. . . . City planning should mean the acquiring of a city convenient, useful, economical and healthful as well as a city beautiful.[32]

Mitchell's particular suggestions for Berlin, based on his drive around the city, concerned the site of a civic centre and the extensions of several central streets; he also urged the restriction of parts of the city to industrial or residential uses, and praised proposals to buy additional parklands.

In early 1913, both the president of Board of Trade and the new mayor took up the issue of city planning in important speeches. President Janzen, who owned substantial real estate, especially in Berlin's business district, declared that city planning had three aspects: street planning, the "most vital," as Berlin's "streets were laid out to accommodate the traffic of a village or, at most, a small town"; the design and layout of adequate public buildings, federal and civic, to serve a city of 50,000 population; and an alternative to level railway crossings. "Berlin is now too large a place to permit the Grand Trunk Railway to use part of the principal street as a car-shunting yard."[33] Mayor Euler, in his inaugural address to the city council, exhorted citizens to

lay broad and deep the foundations of a greater Berlin . . . in our civic work we should plan for a future population of 40,000 to 50,000. . . . The City Planning Movement in its various phases of street widening and straightening, civic centre, location and architecture of public buildings, parks and playground development, and general beautification, deserves special encouragement.

He suggested a joint meeting of the city council, Board of Trade, Trades and Labor Council, Horticultural Society and Women's Canadian Club with the "object of forming a local city planning organization."[34] The mayor's call for a public meeting was endorsed by the council at its next meeting, and $50 voted towards the organizing expenses.[35]

"Bigger, better, beautiful, busy Berlin is the aim of new organization" was the headline of one local press report of the representative meeting held on 6 February 1913.[36] Chaired by the mayor, and attended by more than fifty citizens, the meeting included discussion of the general scope of city planning as well as "problems of practical necessity." Particular

32. *Daily Telegraph*, 12 December 1912.
33. *Ibid.*, 10 January 1913.
34. *Ibid.*, 13 January 1913. Just after Mitchell's visit the *Berlin News Record* published a special bulletin of the Civic Guild (12 December 1912), and there were several suggestions for a local organization modelled on Toronto's. The engineer H.J. Bowman proposed a "joint civic guild for the twin city" to include Waterloo (*News Record*, 16 December 1912).
35. Berlin City Council, Minutes, 19 January 1913.
36. *News Record*, 7 February 1913.

problems discussed were those raised at previous meetings, but the question of possible union with Waterloo, or at least co-operation in planning, was asked for the first time. Finally, following the example of the Toronto Civic Guild, it was decided to elect a "preliminary, working organization committee" of twelve members "to suggest methods . . . for securing a permanent body representative of all the various classes and interests of the city, to take up the work in an energetic manner."[37]

## BERLIN CIVIC ASSOCIATION AND THE LEAVITT PLAN

The preliminary committee quickly attracted support from the Board of Trade, which urged the city council to help the new association.[38] There was a suggestion from the Trades and Labor Council that junk dealers should be segregated as "wherever these junk dealers establish their yards it depreciates the value of the property, and it was generally the property of the workingmen that suffered."[39] During February and early March, the committee worked out the constitution of the new Berlin Civic Association, so that if would have open membership but a governing council including representatives of the city council, Board of Trade and other interested and influential organizations. By mid-March, the finance committee of the city council had recommended a grant of $200 to the Berlin Civic Association.[40]

In early April, the Berlin Civic Association announced its objective of one thousand paid members, and was investigating the best site for a new city hall and civic square. The *Berlin Daily Telegraph* kept up the propaganda for planning with a glowing account, "City Planning in Calgary," of the engagement of Thomas Mawson "to prepare a comprehensive scheme of reconstruction along modern and scientific lines."[41] Two days later, the Civic Association announced arrangements for a visit to Berlin by a "well-known city-planning expert," C.W. Leavitt, Jr. of New York, to address a public meeting and make recommendations for improvements. In addition to the organizations involved already in the formation of the Civic Association, invitations were sent to the Light, Water and Park Commissioners of Berlin, the Board of Educa-

---

37. *Daily Telegraph*, 7 February 1913, *News Record*, 7 February 1913. Those elected were ex-mayor W.H. Schmalz; ex-President of Board of Trade H.L. Janzen; the Rev. F.E. Oberlander (Lutheran), the Rev. Theo Spetz (Catholic); Dr. A.E. Rudell, Mrs. J.A. Hilliard; G.M. DeBus representing business; August Lang and F.S. Hodgins representing industry; and three engineers, H.J. Bowman, W.H. Breithaupt and Herbert Johnston.
38. Berlin Board of Trade, Minutes, 8 February 1913.
39. *Daily Telegraph*, 8 February 1913.
40. Berlin City Council; Finance Committee, Minutes, 12 March 1913.
41. *Daily Telegraph*, 3 and 16 April 1913.

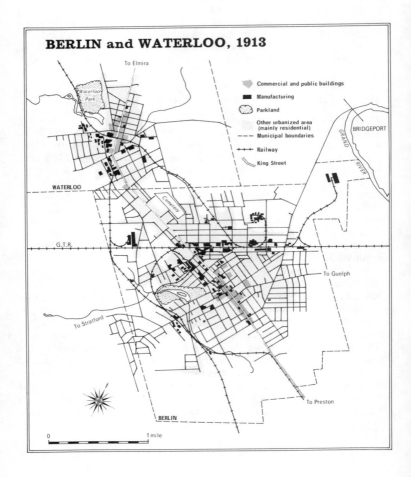

MAP 1          Actual streets and land uses of Berlin and Waterloo in 1913.
               Compiled from "Map of Busy Berlin" and "Plan of . . Water-
               loo . .", 1913, and from directories and fire insurance plans.

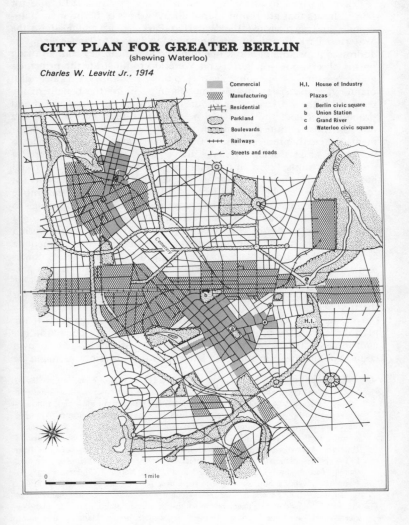

MAP 2    Leavitt's plan for Berlin and Waterloo, 1914.

tion, Daughters of the Empire, and the municipal councils of Galt, Waterloo, Preston and Hespeler.[42]

Leavitt arrived by the 11:30 a.m. train on 18 April 1913, and for several hours that afternoon toured all five wards of Berlin in an automobile, and also visited the Waterloo Park, so he could advise the Waterloo Park Board on improvements in layout.[43] That evening he addressed a meeting of about one hundred in the Library Hall, drawing on his observations of Berlin and Waterloo to illustrate the principles and procedures of planning. He played down grandiose, City Beautiful planning, emphasizing instead that planning is for the benefit of the people. "If the work to bring about [this purpose] is well planned and properly executed, the town will be made beautiful, even if there are none of those features . . . which popularly go by the name of ornaments." Leavitt predicted that "Berlin and Waterloo are destined to become a great residential city," and considered that they offered a great opportunity "to plan comprehensively on the lines of order and harmony with existing conditions." He made practical suggestions for sewage disposal, relief of traffic congestion, abatement of the smoke nuisance, building with reinforced concrete and underground wiring - to improve the appearance, efficiency and public health.[44]

Leavitt also presented the executive of the Civic Association with a written report of his recommendations, particularly that immediate action be taken to secure the services of a professional planner competent "to give you a practical as well as a beautiful plan." It was perceived by the executive that the main stumbling block would be money, as it was believed that the cost of drawing up a plan would be $5,000 for the first year and $1,500 a year thereafter for advising on the working out of the plan. D.B. Detweiler and Mayor Euler thought that if the plan dealt with the problem of level crossings, the city council might contribute $1,000. W.H. Breithaupt hoped that the Waterloo town council might make a grant, and that private subscriptions might be

42. *Ibid.*, 10 April 1913. Charles Wellford Leavitt, Jr. (1871-1928) of New York, civil and landscape engineer, with an impressive record of laying out company towns, resorts, private estates for the wealthy (in New York State, Florida and California), and racetracks. His other Canadian commissions included racetracks at Toronto, Winnipeg and Montreal, the town of Grand Marie, Manitoba, and schemes for the beautifying of suburbs such as Government House property in Toronto and Pelham Park in Hamilton. Leavitt was later elected president of the American Institute of Consulting Engineers and attended most of the National Conferences on City Planning, including that at Toronto in 1914. When he died in 1928 the National Conference on City Planning published a notice of regret in the journal *City Planning* (immediately after a similar notice for Ebenezer Howard), in which it was said that Leavitt's "activities in the realm of City Planning have done much toward the acceptance of the city planning idea through successful practice." (Frank S. So of the American Planning Association helped in tracing Leavitt's biographical details).

43. Waterloo Board of Park Management, Minutes extraordinary meeting, 18 April 1913.

44. *News Record*, 19 April 1913; *Daily Telegraph*, 19 April 1913.

persuaded from those interested in laying out real estate blocks. It was evidently assumed that Leavitt should be invited to draw up the plan. One member asked whether there was not someone in Canada to do the work, but it was believed that "no man in Canada claimed to be an expert."[45]

Parallel with its urging of a comprehensive plan, the Berlin Civic Association encouraged local efforts to overcome the housing shortage. In late April and May, a housing company modelled on the Toronto Housing Company was organized "to buy land and build houses for sale to mechanics and workingmen." Under Ontario legislation, the municipality might guarantee the bonds of such a housing company up to 85 per cent, and dividends were limited to 6 per cent. W.E. Gallagher, a foreman who was also a city alderman, became chairman of the provisional board of directors. Other directors included the mayor, eight manufacturers, three merchants, two builders and a physician. On the whole, the more substantial businessmen were more active, as was the Board of Trade, in this cause than on the Berlin Civic Association itself. Some opposition was expressed by workingmen that such a housing company would tend to keep down their wages, but generally the Trades and Labor Council was in favour.[46]

From this point, the city planning movement ran into difficulties. Instead of being discussed at a council meeting in early May, the Leavitt report was merely referred to a later finance committee.[47] Changed attitudes were clear at the Board of Trade on 8 May 1913. A resolution from the Toronto Board of Trade was endorsed:

> that a Government Commmission be appointed to make a thorough investigation of city planning, subdivision of property, laying out of streets and all things of interest to cities and towns.

But President Janzen then launched into a sweeping condemnation of the local planning association. City planning had, he declared,

> drifted into the hands of a promiscuous and caluminous lot of people of not very large vision who seemed to think that cty planning is spending money from the city treasury.[48]

Among the reasons for the change in attitudes to city planning were personality clashes, distrust of the motives of those enthusiastic about planning by others less involved and a new awareness of the high costs of a corrective approach to planning which would require the expensive acquisition and clearance of built-up blocks in order to extend streets or

45. *Daily Telegraph*, 30 April 1913; *New Record*, 3 May 1913.
46. *Daily Telegraph*, 29 April, 23 May, 13 June 1913; Berlin Board of Trade, Minutes, 5 June 1913.
47. Berlin City Council, Minutes, 5 May 1913; *Daily Telegraph*, 6 May 1913.
48. Berlin Board of Trade, Minutes, 8 May 1913.

open squares. In vain did D.B. Detweiler defend the Civic Association at the Board of Trade meeting, and W.H. Breithaupt argue in a letter to the *Daily Telegraph* that Berlin had a unique opportunity to reshape the central business section, and to guide new peripheral development. He appealed to businessmen in these terms:

> What is $1,000 or $5,000 in comparison to the gain in values which will at one accrue if a proper comprehensive plan is adopted? The advertising value of such a plan will be worth many times the cost.[49]

By the end of May the general mood was unpromising. The finance committee recommended that city council grant $1,000 for a comprehensive plan "on condition that the Waterloo council guarantee a proportionate amount, and that the proposed plan of the two municipalities is acceptable to both."[50] But a council amendment to strike out this recommendation of the finance committee was narrowly passed by eight votes (including the mayor's) to seven.[51] Five of those who voted against any grant to the Berlin Civic Association had been prominent in the city planning movement within the previous seven months. An attempt by the Civic Association and the *Daily Telegraph* in late May to depict how the proposed federal square and civic centre scheme might look on the King-Frederick-Scott block, may not have had the desired effect of stirring civic pride, but instead may have impressed readers more with the inevitably high costs of such a project.[52]

Rebuffed by a majority on city council and Board of Trade, the Berlin Civic Association vowed to recruit more paying members, and by early July 1913, W.H. Breithaupt and D.B. Detweiler reported they had raised $500 from canvassing friends in order to pay for a definite plan.[53] On the strength of private subscriptions received and hoped for, the Civic Association issued a firm invitation to C.W. Leavitt to carry out a thorough survey and prepare a comprehensive plan. Leavitt arrived in late September for a brief visit. He declared that Waterloo would have to be included in the plan, and asked that small committees of citizens survey (and report by October 18) on specific aspects of city development.[54]

49. *Daily Telegraph*, 10 May 1913.
50. Berlin City Council: Finance Committee, Minutes, 29 May 1913.
51. Berlin City Council, Minutes, 2 June 1913; *Daily Telegraph*, 3 June 1913.
52. *Daily Telegraph*, 27 May 1913. Asked by Detweiler, Miss Watson, sister of artist Homer Watson, made a perspective drawing of the proposed civic centre, which was put on display and published in both daily newspapers. Detweiler Diary, 13 and 21 May 1913.
53. *Daily Telegraph*, 4 June, 7 July 1913.
54. *Ibid.*, 25 September 1913. Chairmen of the citizen committees were: transport, W.H. Breithaupt; streets and driveways, D.B. Detweiler; housing, Ald. W.E. Gallagher; industry, August Lang; playgrounds, G.M. DeBus; public buildings and civic squares, H.J. Bowman; water and drainage, the City Engineer; education, Dr. Honsberger; sanitation and health, Dr. Oberlander.

Leavitt made two interim suggestions at this time which showed how little he understood Berlin's municipal powers. Leavitt proposed that the new civic square be sited off King Street, in the block between Queen and Scott, Duke and Roy Streets, and urged that the city start assembling land. Those of the B.C.A. in favour spoke of the square's advantages such as a firebreak, a place of rest for women shoppers and a "good advertisement for Berlin." But, asked George Rumpel, "where is the money to come from?" He calculated that to open Duke Street alone would cost $100,000 and to acquire only one block from Frederick to Scott would cost $235,000. In predicting great resistance to such spending, he remarked "Berlin is a big city, but small in some matters."[55] That Berlin's electors would be certain to oppose such expensive remodelling of the city centre was proved when they decisively defeated a by-law proposal to enlarge city property at King and Frederick Streets by buying the Bowman House property for only $35,000.[56] Leavitt's tentative plan also emphasized that the implementation of zones would involve expense in "making the first radical changes in the part of the city already built upon wrong lines." Leavitt urged that "the city should immediately acquire all lands to be used for any improvements" such as streets, parks, squares and public buildings. This land would be

> sold at a price measured by the increased value created by the improvement, and may thus be made to defray all the expenses incurred. Only actual experience can show how much land will need to be so used; for the increase in its value will depend largely upon the future of the community, its rate of growth and other local conditions.[57]

Leavitt's assumption that Berlin would be able to acquire such an indefinitely large area at potentially enormous cost reveals his ignorance of the ways in which the city's power was limited by Ontario provincial legislation and by the need to submit all money by-laws to the local electors.

With the waning of general enthusiasm for city planning, the tone of newspaper editorials altered, disavowing the aesthetic and stressing the practical and economic benefits of planning. City planning was not

---

55. *Daily Telegraph*, 2 October 1913.
56. Berlin City Council, Minutes, 6 October 1913; *Daily Telegraph*, 1 November 1913. The property was eventually bought by the city in January 1922 for $80,000 (By-law 1649, 23 January 1922). The mood revealed in this vote influenced the 1914 council at its first meeting to rescind a by-law passed 2 June 1913 for widening Queen Street North between King and Weber streets by six feet which would have had the effect of straightening out a "jog" between Queen Street North and South. (*Daily Telegraph*, 20 January 1914).
57. *Ibid.*, 11 November 1913. Leavitt's assistant, Arthur D. Fisken, visited Berlin in the second week of November, bringing the tentative plan as a basis for criticism.

tearing down a space in the middle of the town and building it over to resemble a section of Dresden or Paris. Such schemes are almost always architectural in their inception. They are expensive and the result is they are not carried out . . . We now realize that real city planning is intelligent direction of the growth of a city, with only such changes in the older sections as are necessary for health and economy.[58]

Alterations of the existing urban fabric to eradicate slums were preceived to be necessary, however, and some former supporters of planning now turned more directly to the housing problem. A paper given at the International Town Planning Conference, Ghent, was reprinted with its message that

Town Planning . . . cannot solve the housing problem because it fails to touch the two root difficulties – the high price of land and the poverty of the working classes.

Reform of the tax system, to a single tax according to value, was necessary to end speculation and the stimulation of high land prices and restrictions on house-building and factory investment.[59] Discussions of the link between housing problems and foreign immigration at the Ottawa meeting of the Conservation Commission were reported, with the comment

As is well known, the foreign quarter of the Canadian town is the section under present conditions where squalor and unsanitary surroundings prevail to the greatest degree.[60]

Berlin City Council acted twice in February 1914 in relation to local housing problems. It passed a by-law forbidding the granting of permits for erecting or remodelling residences on lanes or alleys (defined as streets less than 40 feet wide). It was argued for the by-law that the erection of dwellings on lanes led to slum districts which resulted in the depreciation of neighbouring property, and that such buildings were often hastily erected before the grade of the street was established, which led to inefficient servicing with water mains and sewer pipes. Against the by-law it was said that workingmen could not afford proper lots on street frontage.[61] The council also expressed sympathy for the proposals of the Berlin Housing Company (composed of leading industrialists and merchants) to erect workingmen's housing under the Housing Companies' Act, by which the company would provide 15 per cent of the cost and the city would guarantee bonds for the balance. The main motive was to stimulate industrial growth "generally admitted to

---

58. *Daily Telegraph*, 8 September 1913, "Sane City Planning."
59. J. Fels, "Taxation, housing and town planning," *Daily Telegraph*, 22 November 1913.
60. *Ibid.*, 8 January, 22 January 1914.
61. *Ibid.*, 5 February 1914; City of Berlin By-law 1309, 5 March 1914.

be somewhat retarded owing to lack of houses."[62] From mid-1913, both the Berlin city council and Board of Trade regarded the housing question as of higher priority than a general city plan. The city council did not in fact guarantee the bonds of the Housing Company and the whole scheme seems to have built no new houses for Berlin.

Until early 1914, the Waterloo town council chose to remain aloof from the city planning proposals, and Waterloo did not form an equivalent of the Berlin Civic Association. In mid-1913, W.H. Breithaupt had won a motion of support from the Waterloo Board of Trade for the inclusion of Waterloo in the city planning scheme, though one critic observed that the plan would help Berlin rather than Waterloo and lead to union of the two municipalities.[63] The Board of Trade established a Civic Planning Committee which organized a special meeting to discuss planning. Most of the "large attendance" of twenty-five members agreed that planning was generally beneficial, but were uncertain whether to combine with the Berlin Civic Association in a joint plan, pay $500 or $600 for a separate contract with Leavitt, or wait and see what the plan for Berlin looked like. P.V. Wilson, president of the Waterloo Board of Trade, supported the idea of a proper plan, for

> if the Council had a plan they would have an easier time in directing people who wish to purchase factory sites . . . a plan would also provide residential districts, so that people building homes would know that in the future they would not be under obnoxious smoke.[64]

His arguments for a comprehensive plan, in association with Berlin, carried the day. A representative meeting of Waterloo citizens, held in late January and addressed by D.B. Detweiler and W.H. Breithaupt of Berlin, supported the same policy more enthusiastically and almost unanimously urged the Waterloo town council to grant $400 toward the Leavitt plan. The deciding factor was the desirability of co-ordinating street layouts along the boundary between the two municipalities.[65] A few days later, the town council voted for the grant, though Town Engineer Charles Moogk was sceptical of what could be bought for $400 ("it would remedy the evils on paper but not in fact") and claimed that 90 per cent of the ratepayers would vote against civic planning.[66] Leavitt's assistant Fisken paid a hasty visit to Waterloo so it could be included in the joint plan.[67]

Leavitt's "City Plan for Greater Berlin (shewing Waterloo)" was received on 10 March 1914, and set off another spate of interest in

62. *Ibid.*, 20 February 1914.
63. Waterloo Board of Trade, Minutes, 9 June 1913; *Daily Telegraph*, 11 June 1913.
64. Waterloo Board of Trade, Minutes, 27 October 1913; *Daily Telegraph*, 28 October 1913.
65. *Daily Telegraph*, 24 January 1914.
66. Waterloo Town Council, Minutes, 26 January 1914; *Daily Telegraph*, 28 January 1914.
67. *Daily Telegraph*, 29 January 1914.

planning. The fact that the Berlin Civic Association could afford to pay only $1,500 (including the $400 from Waterloo) for the services and expenses of Leavitt and Fisken perhaps explains why the plan was not personally delivered to Berlin, but had to be collected by W.H. Breithaupt from Leavitt's assistant, Fisken, who came only as far as Toronto.[68] Measuring five by six feet, the plan was displayed at the offices of the Berlin and Northern Railway, the street railway enterprise owned by W.H. Breithaupt; it was also accompanied by a report of fifty foolscap pages. The *Berlin Daily Telegraph* described the plan as "an interesting study" and a "magnificent piece of art" (a phrase often repeated less flatteringly without the adjective), and remarked that "contrary to general expectations there are very few radical changes proposed in the central portions of either Berlin or Waterloo."[69]

Leavitt's plan was a compromise between the "incongruous," "unwieldly and badly balanced" physical layout of the Twin-City, especially of Berlin, and his conceptions of the ideal city:

> Should we have clear field on which to construct the ideal city, we might consider the centre to contain the public, semi-public and busiess office buildings lying as close together as practicable. On the links of transportation, either rail or water, would be the manufactories and other industries. Surrounding these and radiating into the country should be the parks and parkways, separating this busy life from the residential sections and, at the same time, forming natural connections which afford pleasant walks between the business and the home.[70]

Leavitt himself considered his plan to be "an ideal rather than a practical plan for Berlin," "something . . . impressive" which would inspire the citizens. But he also suggested that a city plan, once adopted, should remain a fixed and unchanging framework for all future development.[71]

Visually, the most distinctive feature of Leavitt's plan was the system of boulevards encircling the existing built-up areas of Berlin and Waterloo in a "figure-of-eight" or "dumb-bell" shape, and connecting the existing parks with one another as well as with others which might be acquired and developed. Existing land uses were generalized into broader zones of busines and manufacturing, the latter being greatly increased in total area. In suggesting the layout of future residential expansion, Leavitt gave expression to his preference for curvilinear street patterns, with radiating diagonal arteries connnecting the new suburbs with existing main streets and the system of boulevards. The Berlin civic square was to be off King Street, occupying the area between Queen, Weber, Scott and Duke Street extended, with the new

68. *Ibid.*, 12 March, 16 June 1914.

69. *Ibid.*, 12 March 1914.

70. *News Record* (Forward Edition), 25 April 1914, p. 17.

71. *Ibid.*, also letter from Leavitt to Detweiler, 24 April 1913 (Detweiler Papers, Kitchener Public Library).

city hall on the southeast side of and facing Frederick Street. A Union Station Plaza could be created by removing the G.T.R. freight sheds and clearing buildings from several blocks surrounding the station, and would be connected with the civic square by two avenues, in "quiet dignity." An alternative to King Street was proposed, to run from Berlin's easterly limits through to the new Blucher boulevard in the North Ward. In Waterloo, the area surrounding the existing town hall could be developed into a civic square. To deal with the problem of level crossings, Leavitt suggested that no subways be built until the Grand Trunk Railway had elevated the track by seven or eight feet for most of its length through the urban area; then it would be possible to construct subways on six streets.[72]

The report was the main subject at the first annual meeting of the Berlin Civic Association held on 17 March 1914. President W.H. Schmalz remarked that, in sponsoring Leavitt's plan, the citizens have been given an opportunity to assist

> a branch of civic work, which while it belongs mainly to the municipal council, is usually neglected and forced into the background. . . . Civic planning must eventually prove of great value. . . .; the practical and artistic needs are considered alike.

Rather optimistically, he declared that

> The proper foundation when systematically laid, will enable the city to add from year to year such improvements that are not only necessary but also ornamental, without any additional cost or inconvenience to anyone.

He described the plan as "a work of art" which contained many excellent features, rather than a blueprint to be followed in all of its details. The report he called "an instructive work, historically, geographically, statistically and artistically." It was decided to appeal to the B.C.A. members and the city council for $1,000 so that two hundred copies of the map could be printed and the report published in pamphlet form – "a great advertising medium."[73]

The lack of radical changes proposed in Leavitt's report did not dispose a majority on Berlin City Council to regard it any more favourably than it had in mid-1913. In response to the B.C.A. request for

72. *Daily Telegraph*, 12 March 1914; *News Record* (Forward Edition) 25 April 1914.
73. *Daily Telegraph*, 18 March 1914. Leavitt's plan was printed in a smaller format, and copies survive in the Planning and Development Department of Kitchener City Hall, the Archives of Ontario and the Public Archives of Canada (among the Seymour Papers). The report was not printed, but an anonymous, untitled and undated typescript of 24 foolscap paages among the Detweiler Papers in the Kitchener Public Library would seem to incorporate considerable passages from Leavitt's report and to have been the draft of the proposed pamphlet. The large original map was officially accepted by City Council from the B.C.A. in mid-June and appreciation expressed for the "splendid plan," Minutes, 15 June 1914.

$1,000, the finance committee of council recommended $300.[74] This was vigorously opposed by five aldermen at the next council meeting, especially by ex-Mayor Hahn and Dr. Hett who were both generally spokesmen for the labour interest, if not of the Trades and Labor Council. Alderman Hahn declared that he

> could not see where Berlin would derive any benefit from the plan . . . it looked like some kind of embroidery outline. There is not the least possibility of carrying out the plan.

If the matter were submitted to a vote of the people, he was satisfied it would be voted down. Alderman Hett complained:

> It was perfectly plain that the plan was absolutely impracticable. It was patterned after the city of Paris. . . . It is nothing more than an extraordinary extravagance. . . . Berlin seems to be like a boy with a swelled head. It makes mistakes. This city should not be carried away with these extravagant notions, but should be in the line of retrenchment. . . .; The rate should be low so that manufacturers can be attracted to the city. It is in the interest of the workingmen that the tax-rate should not be high.

Alderman A.L. Breithaupt's remark that Hett was willing to spend $1,000 for cemetery gates detracted from Hett's call for retrenchment; and Breithaupt's claim that the plan though "by no means perfect, gives us something to work towards" helped to get the $300 grant passed by ten votes to five.[75]

In Waterloo, interest in Leavitt's plan had spurred the formation of an advisory committee on town planning, comprising five citizens prominent in commerce, manufacturing or the insurance business, the town engineer, Charles Moogk and Dr. W.L. Hilliard who became the leading advocate of planning in Waterloo.[76] Approval of the "Westmount" subdivision (proposed by A.W. Merner for the Westmount

74. Berlin City Council: Finance Committee, Minutes, 12 March 1914.
75. Berlin City Council, Minutes, 15 March 1914; *Daily Telegraph*, 17 March 1914. Dr. Hett was among the 32 who attended the first planning meeting in November 1912; he shortly afterwards explained his view of Berlin's priorities in a letter to the *News Record*, 5 December 1912. He was convinced that the first objective should be to "make life more comfortable for the men and women who are the workers in our factories and who bring wealth and prosperity to Berlin."
76. Waterloo Town Council, Minutes, 2 March 1914. Other members – P.V. Wilson of the Waterloo Trust and Savings, J.H. Roos of Mutual Life, E.F. Seagram of the Seagram distillery and C. Fischer and S.B. Bricker, merchants – were all active in the Board of Trade. William Louis Hilliard (1868-1966) son of Thomas Hilliard, Waterloo newspaper publisher and founder of Dominion Life Assurance Company (1889—    ) and Waterloo Trust and Savings Company (1912—    ). Practised medicine in Waterloo from 1897; director and medical referee of three insurance companies. Served on Sewer and Water and Light Commissions; chief mover in town planning movement from 1914, chairman of Town Commission from its formation in 1921. Acclaimed mayor of Waterloo 1916 and 1917.

Improvement Company) was deferred by the town council till it could be known that it was compatible with Leavitt's plan for Waterloo.[77] The advisory committee commended Leavitt's proposals, and urged the town council to accept the plan and to carry out the report "as far as possible," especially asking "that no building permits be granted for vacant properties lying in the way of proposed street extensions as shown on the plan."[78] At first, council demurred because of some errors and because Leavitt himself had not done the work; but it did accept the plan later and appointed five more members to the advisory committee.[79] Typical of the practical attitude to planning was the first recommendation of the enlarged advisory committee: that there be a system of garbage disposal for the town, because "before a Town can be beautiful, it must be made clean."[80]

The Sixth National Conference on City Planning at Toronto in late May 1914 attracted two representatives each from Berlin city council, Berlin Board of Trade and Berlin Civic Association.[81] Local newspapers gave considerable space to the Conservation Commission's Canadian town planning act, by which "the whole question of Town Planning and the Housing Problem" would be the responsibility of a federal cabinet minister, to whom chairmen of zones in each province would report. Expert engineering advice would be freely available to each municipality. There was also provision for compensating private property owners adversely affected and for the local authority to recover half the unearned increment if property values greatly increased. [82] Addresses by Frederic Olmstead and Thomas Adams ("the greatest authority in his own line throughout the English-speaking world") were announced.[83] Leavitt's plan for Greater Berlin was on show in Convocation Hall, and was praised by the *Globe* as "perhaps the finest of all" the exhibits of almost three hundred Canadian, American and European centres, for "getting away from the old idea of a gridiron" to "handsome curves arranged in every sort of ingenious fashion to conform with the lay of the land," so making the city "pleasing to the eye, as healthy and park-like as possible while in no way interfering with commercial and

---

77. Waterloo Town Council, Minutes, 2 March, 31 March 1914.
78. *News Record*, 23 April 1914.
79. Waterloo Town Council, Minutes, 4 May, 8 May 1914; *Daily Telegraph*, 5 May, 9 May 1914; *News Record*, 9 May 1914. The new members were F.G. Hughes and C.W. Wells (dentists), E.M. Devitt and W.J. Beattie (merchants) and B.N.W. Grigg (Mutal Life insurance).
80. *Daily Telegraph*, 11 June 1914.
81. *Ibid.*, 16 May 1914. Mayor Euler and Alderman Rudell represented the council, G.M. DeBus and W.H. Leeson the Board of Trade, and S.J. Williams and W.H. Breithaupt the B.C.A.
82. *Daily Telegraph*, 11 and 13 May 1914; *New Record* 13 May 1914.
83. *Daily Telegraph*, 16 and 23 May 1914.

other necessities."[84] Alderman Dr. Rudell commented on his return to Berlin that city planning as discussed at the conference seemed to apply to bigger cities than Berlin, and he assumed that the property-owners and real estate agents would continue to lay out the subdivisions without the municipality having much influence.[85]

What did Berlin-Waterloo's first attempt at city-planning achieve? The Leavitt plan had almost no direct effect, except for the modification of a few subdivision plans in Waterloo. The only way for residents to protect their neighbourhood continued to be for a majority on a specific street to get up a local petition to the city council for their street to be declared "residential" with a by-law setting the "building line." The aesthetic ideals of the early civic planning movement inspired some property developers and the new Waterloo Lutheran Seminary to use landscape architects or their imitators in laying out grounds attractively.[86] The new interest in parkland led to the donation of some land by private owners and the purchase of more by the Berlin Council, and to a movement to conserve woodland, especially along the banks of the Grand River.[87] Financing and presenting the Leavitt plan gave local enthusiasts experience of the substantial local resistance to be expected to radical replanning if it would restrict the freedom of small property-owners and run the city into debt to correct the mistakes of the past.

## KITCHENER CITY PLANNING COMMISSION

Local concern for planning and housing dwindled after the outbreak of war in mid-1914. Some interest continued in planning matters in other cities and in the Conservation Commission. Thomas Adams, Town Planning Adviser to the commission from late 1914, visited Berlin-Waterloo several times at the invitation of the Berlin Civic Association. During the debate on changing Berlin's name, Adams addressed a lengthy letter to the Berlin and Waterloo municipal councils and Boards of Trade, urging that the opportunity be taken to unite the two towns. He argued for economy and efficiency of administration by reducing overlapping and duplicate services, and declared:

> I can conceive how fine the combined city would be as a town planning proposition and perhaps that is one of my chief interests in supporting the union.[88]

84. *Globe*, 27 May 1914; *Daily Telegraph*, 29 May 1914.

85. *News Record*, 28 May 1914.

86. E.g. *Daily Telegraph*, 25 January 1913; *ibid.*, 1913. The most notable new subdivision was "Queen's Park," laid out by Messrs. Grauel and Riener on land bought from Henry Schneider on the southwestern edge of Berlin. Its curvilinear street plan was in marked contrast to the grid plans elsewhere.

87. Berlin City Council, Minutes, 2 December 1912; *Daily Telegraph*, 20 May 1914.

88. *Daily Telegraph*, 1 April 1916 has full text; Waterloo Town Council, Minutes, 6 March 1916.

Berlin was also represented by D.B. Detweiler on the provisional committee which in early 1916 tried to set up a permanent Civic Improvement League for Canada, with others such as Sir John Willison, G. Frank Beer, Noulan Cauchon and Thoma Adams.[89] But the service of the Conservation Commission most appreciated locally was its concern with urban fire hazards – "Canada's Fire Loss" as it was called, a cause pioneered by Detweiler on behalf of the Board of Trade.[90]

City planning groups survived in both Waterloo and Berlin/Kitchener. Waterloo's advisory committee on town planning advised the town council on street extensions and the approval of subdivisions and notably urged the town to acquire the Snider mill-dam property to prevent its being drained and subdivided into building lots.[91] In January 1916, Waterloo Board of Trade came out strongly in petitioning for an Ontario town planning act

to enable cities, towns and municipalities to prepare town planning schemes with the object of securing improved sanitary and hygienic conditions, and convenience and amenity in connection with the layout of building land.[92]

W.H. Breithaupt for the Berlin Civic Association asked Berlin city council in June 1914 to form an advisory council like Waterloo's, but a similar purpose was served by appointing some aldermen each year to the executive of the B.C.A.[93] The city council also, reluctantly, made small grants to the B.C.A. in 1915 and 1916.[94] The B.C.A. advised city council, unsuccessfully on such issues as the widening of King Street, more successfully on the opening of new streets on the edge of the city and ending duplication of street names.[95]

With the passing of the Ontario Planning and Development Act in April 1917, cities as small as Kitchener could at last take advantage of town planning legislation. The act empowered any city, town or village to have made for its approval a general plan of the municipality and of the surrounding "urban zone" (up to a radius of five miles in the case of a city, three miles for a town or village). The main elements of such a

89. *Daily Telegraph*, 6 March 1916.

90. Berlin Board of Trade Minutes, 6 May 1915; Kitchener City Council, Minutes, 4 December 1916; Waterloo Town Council, Minutes, 15 December 1916.

91. *Daily Telegraph*, 9 May 1914; this objective was achieved with Town of Waterloo, By-law 598, 8 January 1917.

92. Waterloo Board of Trade, Minutes, 24 January 1916.

93. Berlin City Council, Minutes, 15 June 1916. The number of aldermen on the Civic Association's Executive increased from two in 1914 to four in 1917, Berlin City Council: Finance Committee, Minutes, 16 April 1914; Kitchener City Council, Minutes, 5 February 1917.

94. $100 in 1915, Berlin City Council, Minutes, 5 April 1915; $200 in 1916, Minutes, 17 February 1916.

95. For example, the opening of Edna Street along the eastern boundary of Kitchener by By-law 1363, 2 November 1914.

plan were described as highways, parkways and boulevards, parks and playgrounds and the emphasis was on regulating future development rather than correcting past mistakes. Any municipal council might also appoint a planning commission to consist of the mayor or reeve and six other persons, required only to be "ratepayers" who would hold office for three years, two retiring each year but eligible for reappointment. Municipal officers such as clerk and engineer were to perform duties at the request of the planning commission and the commission's operating expenses were to be paid. Appeals against plans could be made to the Ontario Railway and Municipal Board.[96]

The act had been welcomed in advance by a unanimous motion of Kitchener city council:

> That recognizing the importance of legislation to facilitate a general scheme for the future growth of the City, including such purposes as widening streets, opening new streets, establishing factory, residence and business districts, and regulating in general the best development of City, we strongly endorse the proposed enactment of legislation to this end by the Ontario legislature at the present session.[97]

One factor in the city council's alacrity was probably a wish to draw attention away from the dissension caused in the community by an attempt to change the city's name back to Berlin at the time of the 1917 municipal elections. There was less agreement on the composition of the new planning commission which Kitchener moved quickly to establish. Finally W.H. Breithaupt and W.H. Schmalz were appointed for three years, Walter Harttung and A.R. Lang for two years and Samuel Brubacher (the only alderman) and Martin Huehnergard (city assessor) for one year.[98]

The Kitchener City Planning Commission (C.P.C.) consisting of much the same members till 1922, with A.R. Kaufman a significant addition from 1920, held monthly meetings at which desirable improvements were proposed. W.H. Breithaupt was chairman till 1921, and its leading spokesman. He adhered to the corrective view of planning and continued trying to carry out aspects of the Leavitt plan. In Breithaupt's view

> Town planning consists largely in rectifying past mistakes and omissions. . . . Most of our cities . . . suffer inconvenience from inadequate and indirect street systems, from lack of provision of space required for public buildings and

96. *Statutes of Ontario*, 7 Geo. V, ch. 44 ("The Planning and Development Act" 1917).
97. Kitchener City Council Minutes, 19 February 1917.
98. Kitchener City Council, Minutes, 5 June 1917. The earlier suggestions of six ratepayers by the Kitchener Civic Association were criticized in the council for including too many large real-estate owners. *Daily Telegraph*, 22 May 1917; Kitchener City Council, Minutes, 21 May 1917; Kitchener City Council: Finance Committee, Minutes, 31 May 1917.

parks; industries, business and residence sections are not localized. Each private owner of land large or small has been allowed to subdivide with regard only to getting the largest number of saleable parcels out of his holding, without consideration of fitting into a general street plan – often blocking what might have developed into an important thoroughfare – without restriction of proportion of parcel or lot, or how much of it may be built on. Factories are located in the midst of residence sections due to change of ownership of the ground, and business locations may be similarly determined.[99]

Progress had been "obstructed" by "prejudice and private interest," and provincial legislation was inadequate to carry out any corrective planning. Breithaupt held the engineer's ideal of the city efficient and, with a touch of the autocrat, had little patience with the need to persuade those less sure about the benefits of planning.

Planning proposals made by the C.P.C. to Kitchener city council between 1917 and 1921 fared poorly. A motion that a new road be cut through the cemetery from King Street east was defeated by council.[100] It was easier to plan for new urban development, as in approving the plan by the Canadian Consolidated Rubber Company for a subdivision of semi-detached houses (financed under the Ontario Housing Act) close to the Dominion Tire factory in northwest Kitchener.[101] The City Planning Commission also took issue with the city council's cemetery committee on the site of the proposed new cemetery.[102] While the city engineer was helpful, the C.P.C. received little support from city council. Criticized for "taking a little too much upon themselves in asking the City Engineer to busy himself with some of their plans for the opening of streets," the C.P.C. was required by a council motion "when entering upon any schemes of street improvement to consult with the Council before proceeding to the point where labour and expense are involved."[103] When a long list of proposed street extensions was received by city council from the City Planning Commission, the mayor remarked sourly that it was easy to see that the C.P.C. had no responsibility for financing their proposals, as only one of them would cost $250,000.[104]

In early 1919, the C.P.C. presented a set of changes as a "proposed

99. W.H. Breithaupt, "Some features of town planning with application to the City of Kitchener," Town Planning Institute of Canada, *Journal*, I, 6 (1921) pp. 5-8.
100. Kitchener City Council, Minutes, 4 February 1918.
101. *Daily Telegraph*, 20 November 1919. In response to the acute housing shortage of 1919, both Kitchener and Waterloo adopted the Municipal Housing Act and appointed Housing Commissions. The immediate result was the construction of 200 houses in Kitchener and 100 in Waterloo for the employees of the Dominion Rubber System.
102. *Daily Telegraph*, 16 January 1919; Kitchener City Counciil: Finance Committee, Minutes, 13 February 1919; *Daily Telegraph*, 23 April 1919.
103. Kitchener City Council, Minutes, 5 May 1919.
104. *Daily Telegraph*, 20 July 1920.

new city plan" under the Planning and Development Act. The plan was
far from being a comprehensive one as it was concerned only with street
layout; about half the proposed changes were for opening up or extending streets in built-up central Kitchener; and about half were suggested
alignments of new streets to provide a framework for suburban development on the northern and eastern edges of the city.[105] Both kinds of
proposals were objected to, by individual landowners and by the Waterloo town council, which was naturally affected by proposals for new
arterial streets on the northern edge of Kitchener and also legally
involved as Waterloo came within Kitchener's "urban zone" under the
Planning and Development Act.[106] A hearing by the Ontario Railway
and Municipal Board, in Kitchener on 7 May 1919, upheld most objections and witheld its approval. The board considered that some of the
proposed peripheral roads cut across too many previous surveys and
infringed the freedom of property-owners to lay out subdivisions; the
cost and injury to private investment of cutting through central areas
was not justified by the uncertain benefits of doing so.[107] The O.R.M.B.
eventually did accept the Kitchener plan as amended, but any specific
proposals also required city by-laws for implementation, and city council showed little disposition to co-operate.

By early 1921, the C.P.C. was realizing that practical progress in
planning would depend on a new joint plan, preferably by an outside
planning expert.[108] Thomas Adams was invited to visit Kitchener, and
among his recommendations was the suggestion that Waterloo should
form its own town planning commission under the Planning and Development Act.[109] Waterloo council acted promptly in appointing six "men
who are practical and representative of all classes" to join the mayor,
and confirmed this resolution with a by-law.[110] By this move, Waterloo
overcame the handicap it had suffered for four years, of having to
submit its own development and subdivisions plans to the Kitchener
C.P.C. and of being poorly organized to react to C.P.C. proposals
which affected Waterloo.[111]

105. *Daily Telegraph,* 20 November 1918; a retrospective view of the plan is given in W.H.
     Breithaupt, "Some features of town planning" (1921).
106. Waterloo Town Council, Minutes, 5 May 1919; *Daily Telegraph,* 6 May 1919.
107. *Daily Telegraph,* 17 May 1919.
108. Kitchener City Council Minutes, 18 April 1921, 6 June 1921.
109. Waterloo Town Council, Minutes, 6 July 1921; *Daily Telegraph,* 7 July 1921.
110. Town of Waterloo By-law 703, 2 August 1921.
111. Persons appointed to the Waterloo Town Planning Commission were Dr. W.L. Hilliard
     (physician) as chairman, and Charles W. O'Donnell (a baker) for three years; Allen
     Bechtel (apparently retired) and John H. Ziegler (a grocer) for two years; and William
     D. Brill (proprietor of Waterloo Shirt Co.) and E.H. Schlosser (customs clerk) for
     one year.

## THE ADAMS-SEYMOUR PLAN

When the City Planning Commission asked Kitchener city council to consent to a new general survey and comprehensive plan, the city began a process of making up its mind which was almost farcically prolonged. However it did lead to a successful contract with Adams and Seymour, and to a thorough and successful plan. The new phase saw the emergence of A.R. Kaufman as C.P.C. chairman and leading proponent of planning. It was also marked by a shift to the preventive emphasis in planning, and a more conciliatory and compromising approach to specific planning objectives.

N.C. Helmuth, secretary of the C.P.C., first wrote to city council in June 1921 urging a new general plan; and in August a letter was received from A.R. Kaufman apparently offering to pay part of the cost.[112] His offer was referred to the finance committee, which resolved:

> That although we appreciate the interest of Mr. A.R. Kaufman and his offer re the city plan, the great amount of important engineering work makes it impossible to undertake a general city plan this year.[113]

In the next few months, Kitchener was given an object lesson in the need for city planning. The grant of a permit to the Dominion Shirt Company to build a factory on Church Street, an entirely residential street in the South Ward, parallel to King Street, aroused the opposition of local residents to the point of threatening legal action. The city quickly revoked the building permit. Kaufman took the opportunity to write to the *Daily Telegraph* pointing out that the Church Street residents could have protected themselves long ago by declaring their street residential and asking council to pass a by-law fixing the building line, but stressing that the C.P.C. was urging council to engage a city planning expert who would divide the city into zones and so protect residential areas from invasion by factories.[114]

Next day, the newspaper reported that city council had decided to bring "town planning expert" Thomas Adams to Kitchener to make a

112. Kitchener City Council, Minutes, 6 June 1921, 2 August 1921. Alvin Ratz ("A.R.") Kaufman, 1885-1979; son of Jacob Kaufman, industrialist, who developed empire of lumber and planning mill products and promoted Berlin's rubber industry, helping to establish three successful rubber companies 1900-1907. The third, the Kaufman Rubber Company, established 1907 by father and son, continued in the family, and made a large proportion of Canada's rubber footwear. "A.R." was a paternalistic employer, resisting attempts to unionize his workers and provoking a general strike in 1960. Philanthropist on a substantial scale, notably to hospitals, YMCA and YWCA; pioneered and financed Canadian family planning movement from 1930. Dominant figure in town planning in 1920s, chairman C.P.C. 1922-1959; served on Kitchener Board of Park Management 1924-1964.
113. Kitchener City Council: Finance Committee, Minutes, 10 August 1921.
114. *Daily Telegraph*, 8 March 1922.

detailed survey of the city and that $3,000 for his fee would be budgetted over the next three years.[115] This was contradicted later by a joint meeting of the city planning committee of council and the C.P.C., when it was decided to use the money budgetted for Adams' visit to pay for an engineer to make a contour map of the city and a landscape architect to make a detailed survey, both under the direct supervision of the city engineer.[116] Two weeks later, at the next council meeting, supporters of an Adams plan helped pass a motion that a contract be made with Thomas Adams for a total cost of $3,000, $300 in 1922, and $1,350 in each of 1923 and 1924 with the stipulation that "no zoning be done until such time as a new and correct survey plan is made" to satisfy the finance committee of council.[117] But it was only narrowly passed over many objections after supporters had argued that $300 was nothing to prevent future controversies like the Church Street factory building permit.[118] The council reversed itself two weeks later, however, in accepting a finance committee recommendation that any contract with Adams be laid over for further consideration and that J.T. McGarry, the assistant town engineer, prepare a re-survey of the city.[119]

In late August another season of vacillation began. The city clerk was instructed to write to Thomas Adams to ask if he would be willing to sign a contract on the same basis as he had with London, Ontario.[120] But when Adams' favourable reply was received, it was referred to the finance committee and again treated in a cavalier and delaying fashion.[121] The city engineer was asked to contact other planners, including Charles Leavitt of New York, and Noulan Cauchon of Ottawa to see if their rates might be competitive.[122] When no replies had been received by late November, and after some uncharacteristic urging by the Board of Trade that the city engage a competent town planner, the finance committee finally recommended that a contract be signed with Adams.[123]

During the year of delay and indecision, the C.P.C. had continued to

115. *Ibid.*, 9 March 1922. Adams had by this time left the Conservation Commission and was in private practice as a town-planning consultant in England. Biographical details of Thomas Adams are not provided here, as they have been published several times elsewhere, most recently in J.D. Hulchanski, *Thomas Adams: a biographical and bibliographic guide* (Dept. of Urban and Regional Planning, University of Toronto: Papers on Planning and Design, No. 15, 1978).

116. Kitchener City Council, Minutes, 20 March 1922; *Daily Telegraph*, 21 March 1922.

117. Kitchener City Council, Minutes, 3 April 1922.

118. *Daily Telegraph*, 4 April 1922.

119. Kitchener City Council, Minutes: Finance Committee Minutes, 13 April 1922; Kitchener City Council, Minutes, 18 April 1922.

120. Kitchener City Council, Minutes, 30 August 1922.

121. *Daily Record*, 17 October 1922.

122. Kitchener City Council: Finance Committee Minutes, 2 November 1922.

123. *Daily Record*, 21 November 1922; Kitchener City Council: Finance Committee, Minutes, 30 November 1922.

make specific recommendations on street openings, widenings and extensions, on building lines and the approval of new subdivisions, as well as to urge the engagement of an outside planner. That the city council was no more reconciled to the potential statutory powers of the City Planning Commission was shown in October 1922 when it reacted strongly to proposed amendments to the Town Planning Act. According to the explanation given to city council by Mayor Greb, city or town planning commissions would be given power until then vested only in the municipal councils, and would be able to undertake works which would place the municipality at great expense as well as to receive a levy of one and a half mills on the tax rate. The mayor and Alderman Ahrens were deputed to attend the convention of the Ontario Town Planning Association in Toronto on 17-18 October, and to register a strong protest.[124] Actually Mayor Greb misrepresented the scope of the proposed amendments, grossly exaggerating the tax levy which would, in fact, have been only one-tenth of a mill. One wonders how much of the widespread opposition to the proposed legislation was similarly based on misunderstanding.[125]

Meanwhile Waterloo's Town Planning Commission (T.P.C.) had little success with recommendations to the town council. Dr. Hilliard, chairman, addressed council in March 1922 on the lack of any complete plan for the town and recommended the establishment of building lines.[126] By October 1922, members of the T.P.C. were so discouraged that they thought of resigning. It took a well-attended meeting of the Waterloo Board of Trade to impel Waterloo's council into serious consideration of a joint plan with Kitchener.[127] At its November meeting, the council agreed in principle but delayed action as no funds had been budgeted for 1922, and the town engineer was fully occupied with other work.[128] During the winter Horace Seymour, now engaged as Adams' associate on the Kitchener plan, and A.R. Kaufman of the Kitchener C.P.C., as well as the Waterloo T.P.C., led by Dr. Hilliard, continued to

124. Kitchener City Council, Minutes, 16 October 1922; *Daily Record*, 17 October 1922.
125. The full text of the proposed legislation was given in Town Planning Institute of Canada, *Journal* I, 12 (1922), pp. 10-17.
126. Waterloo Town Council, Minutes, 13 March 1922.
127. Waterloo Board of Trade, 30 October 1922; *Daily Record*, 31 October 1922.
128. Waterloo Town Council, Minutes, 7 November 1922.

put pressure on the Waterloo council.[129] At the council's March meeting, the T.P.C. asked that Waterloo sign a contract with Adams for a comprehensive plan in conjunction with Kitchener's; Seymour, who was present, submitted a draft contract for the council's consideration.[130] By April, this had been amended to a contract with Seymour himself as consultant, for a plan to cost $2,000.[131] Finally in May, Waterloo town council approved a grant to the T.P.C. of $500 a year "with access to all books, records and plans of the Town as long as they are not removed from Waterloo."[132]

Seymour supervised the systematic survey and planning of Kitchener (from January 1923) and Waterloo (from May 1923), and spent far more time than Adams on the spot.[133] But the stages of work were an object lesson in Adams' methods – a reconnaissance survey of the city and region with maps of transportation, street services, street traffic, assessment values and the existing land use pattern; tentative skeleton plan of the region based on the survey; city survey; and finally a complete working plan of the city adapted to the law of the province.[134] What was more significant in contributing to the acceptance of the Kitchener and Waterloo plans was the interpolation of phases of public presentation and discussion of survey findings and tentative recommen-

129. Seymour himself in his "Report on the Plan of Town of Waterloo, May 1923 to April 1924" (Seymour Papers, PAC) credited the "active insistence" of Dr. Hilliard and the "unofficial, but extremely helpful assistance" of A.R. Kaufman. Horace Llewellyn Seymour, 1882-1940; graduated S.P.S. University of Toronto 1902; surveying 1903-14; with Commission of Conservation 1914-19; 1919- Federal-Provincial Housing Loan; 1922-24 associate to Adams for Kitchener and consultant for Waterloo; 1926-29 Resident Engineer, Harland Bartholomew Ltd., for Vancouver Plan; 1929-32 Director of Planning, Alberta; 1932— housing and town planning consultant, Ottawa and consultant Alberta; with some consulting for Saint John, N.B. and Nova Scotia during the 1930s. During the 1920s, very active in the Town Planning Institute of Canada, holding office and helping produce the *Journal*. "Unofficial biography" in the collection of Seymour Papers, Pubic Archives of Canada; Shirley Spragge, "Biographical Note: Horace Seymoure", *Plan Canada*, 15, 1 (March 1975) 45-6.

130. Waterloo Town Council, Minutes, 5 March 1923.

131. *Ibid.*, 2 April 1923.

132. *Ibid.*, 7 May 1923.

133. During the year that Adams was consultant for the Kitchener plan, he was also engaged in preparing a regional plan for West Middlesex, part of Greater London, and had several contracts in Canada left over from his service for the Conservation Commission. Late in 1923, it was announced that Adams had been appointed Director of Plans for the Regional Plan of New York and was cancelling all his Canadian work except for the Kitchener plan and the new Armstrong Whitworth company town (Corner Brook) in Newfoundland. Town Planning Institute of Canada, *Journal*, II, 6 (Nov. 1923); *Daily Record*, 24 November 1923.

134. Thomas Adams, "Modern city planning: its meaning and methods," *National Municipal Review*, XI, 6 (June 1922), pp. 157-76. A Canadian version was published in Town Planning Institute of Canada, *Journal*, I, 11 (August 1922), pp. 11-15.

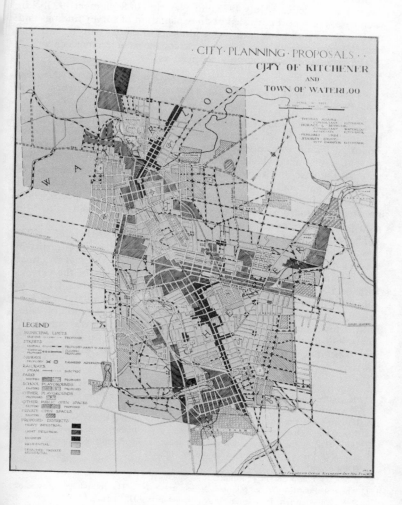

MAP 3    Proposals of Adams and Seymour for street improvements and zones of Kitchener and Waterloo, 1924. Facsimile of map published in *Journal* of Town Planning Institute of Canada, January 1925.

dations. In the first of two such phases, preliminary reports were produced for Kitchener in July 1923 and Waterloo in September 1923.[135] In each case, the report consisted of an explanation of survey findings and draft recommendations, and was addressed to the City or Town Planning Commission; with press releases sketching the main recommendations.[136]

The reports were favourably received as

little expense will be involved . . . the engineers have eliminated everything that does not seem to be attainable within a reasonable number of years at comparatively small cost.[137]

In both Kitchener and Waterloo, the draft proposals were first publicly discussed in special meetings of the Boards of Trade. The Kitchener Board of Trade was addressed by A.R. Kaufman on the benefits of preventive planning which would guard against repetition of the mistakes of the past.[138] The Waterloo meeting, under the joint auspices of the Town Planning Commission and the Board of Trade, was addressed by Seymour in a "very comprehensive and enlightening" speech which inspired the Board of Trade to set up an advisory committee to help gain public and council support for the plan.[139]

The second phase of public discussion in October and November 1923 was more broadly based. The reports were printed in instalments in successive Saturday editions of the *Daily Record*.[140] Special meetings were arranged by the Kitchener C.P.C. with members of the various Kitchener commissions and boards, and by the Waterloo T.P.C. to announce plans for a civic square park.[141] In mid-November, the C.P.C. and T.P.C. together announced the culmination of this phase of "submitting town-planning schemes to the scrutiny of the general public."[142] The publication of the draft plan and zoning by-law on 24 November was to begin a week of ward meetings at which anyone might question

135. Thomas Adams and Horace Seymour, "Report on the City Planning Survey of Kitchener, Ontario," July 1923; Horace Seymour "Report on the Town Planning Survey of Waterloo, Ontario," September 1923 (Seymour Papers, Public Archives of Canada).
136. For Kitchener the press releases featured alternative arterial streets parallel to King Street (*Daily Record*, 27 July 1923), the rationale of the zoning system (*Daily Record*, 28 July 1923) and the proposed annexation of about one square mile of Waterloo Township of Kitchener (*Daily Record*, 18 August 1923).
137. *Daily Record*, 28 July 1923.
138. Kitchener Board of Trade, Minutes, 14 September 1923.
139. Waterloo Board of Trade, Minutes, 24 September 1923. The committee consisted of C.W. Wells (current President), the senior members J.H. Roos and P.V. Wilson and C.D. Collins.
140. The Kitchener report was excerpted in *Daily Record*, 20 October, 27 October and 3 November 1923; the Waterloo report in 3 November, 10 November, 17 November and 24 November 1923.
141. *Daily Record*, 24 October 1923.
142. *Ibid.*,14 November, 17 November 1923.

aspects of the plan or make suggestions. Five public meetings were held: for the North and West Wards of Kitchener 26 November; for the North and West Wards of Waterloo 27 November; East and Centre Wards of Ktichener 28 November; South and East Wards of Waterloo 29 November; with a final general meeting also to serve Kitchener's South Ward on 30 November.[143]

Seymour addressed all meetings, and Adams also spoke at the final meeting which was by far the best attended, the only meeting to attract considerably more than about twenty. Planning was justified in practical, commonsense terms ("we plan our home and factories – why not our cities and towns?"), and there was usually an appeal to the self-interest of householders concerned to protect their investment from the unwanted intrusion of factories or business. But businessmen were also reassured that the zoning regulations would not be retroactive, and non-conforming uses could remain. Adams made a nicely calculated appeal to local pride and sensitivities when he claimed that "the smaller city which was properly planned would be an attraction to industries in the big cities" interested in relocating, and that "the mere fact that a building was a factory should not give occasion to objection because many factories were as attractive as residences."[144]

Generally the plan was well received. It was clear that no attempt would be made drastically to reshape the existing urban landscape, and that the purpose was rather to provide an orderly framework for future urban development. Significant objections were made at one Waterloo meeting and, more strongly, at one Kitchener meeting. Waterloo wanted a larger industrial zone, with most of it away from Kitchener boundary and in the northwest of the town, where land was lower, cheaper and more accessible to both railway lines. It was feared that if the industrial area were located near Kitchener "many workmen would build in Kitchener, and their money would be spent in Kitchener."[145] Residents in part of Kitchener's South Ward objected to the zoning of their area for industrial use (they maintained that the "logical" place for industry was "the territory around the tracks in the North ward"). Three ladies claimed that it was unfair to zone land on Benton and Church Streets for residential purposes only, as when "they wanted to sell their land in future they would be losers" if the land could not be used for commercial purposes.[146] The zoning map was changed to suit all these objections and also to benefit from some useful suggestions for new arterial streets, made at the meeting for Kitchener's East and Centre Wards.

Two Kitchener incidents marred the mood of sweet reasonableness

143. Each meeting was well reported in the *Daily Record* of the following day.
144. *Ibid.*, 1 December 1923.
145. *Ibid.*, 30 November 1923.
146. *Ibid.*, 1 December 1923.

achieved during the phase of public education. The first was a squalid row in the Kitchener city council which showed that resentment and suspicion remained about the motives of those prominent in planning. A.R. Kaufman presented a bill for $600 he had already paid, which had been incurred when he engaged a draftsman from Toronto to prepare the basic survey maps for the new city plan. Kaufman had done this without consulting either city council or the C.P.C. of which he was chairman, because he believed there was "need for urgent haste" if the city plan was not to be delayed a whole year. After an acrimonious discussion in which resentment of Kaufman's wealth and "high-handed" methods was plain, the council voted eleven to three against reimbursing Kaufman.[147] The other incident was of another order, but showed that Kitchener C.P.C. was not of one mind on the proper approach to planning. W.H. Breithaupt wrote a letter to the *Daily Record*, restating his corrective view of planning and outlining desirable street improvements in the already developed parts of the city. Seymour replied in a memorandum to the C.P.C. explaining the practical reasons for stressing preventive planning:

> We have placed emphasis on preventive rather than corrective measures in the interest of obtaining both economy and the most permanent results . . . where we have been unable to adopt suggestions that have the effect of correcting existing defects, it is because we do not think the value of the improvement is worth the expense involved.[148]

Both municipal councils promptly approved the draft plan in general outline and, assisted by Seymour's advice, both sought to amend section 399a of the Municipal Act to give the municipality, with the approval of the Ontario Railway and Municipal Board, power to pass by-laws regulating the use of land or buildings for any purpose, not just to reserve districts of detached private residences, as in the act.[149] Amendment was sought, by private bills for each place, as well as by general legislative amendment. The fact that W.G. Weichel, mayor of Waterloo, also represented North Waterloo in the Ontario Legislature, was helpful in both courses of action. With more general amendment in mind, Kitchener used a time-honoured tactic, inviting representatives of municipal councils, boards of trade and planning commissions and others interested in town planning in southwestern Ontario, to a meeting in mid-February 1924.[150] Nine municipalities in addition to Kitchener and

147. *Ibid.*,2 November, 6 November 1923.

148. Quoted in Thomas Adams and Horace L. Seymour, "Report on the Plan of the City of Kitchener, Ontario, February 1923 to March 1924," 4 April 1924 (Seymour Papers, Public Archives of Canada, pp. 2, 9-11.

149. Kitchener City Council, Minutes, 3 December 1923; Waterloo Town Council, Minutes, 3 December, 15 December, 1923; *Daily Record*, 4 December 1923.

150. Kitchener City Council, Minutes, 5 February 1924.

Waterloo attended, to discuss the need for legislation on town planning – London, Hamilton, Brantford, Galt, Guelph, Elmira, Preston, Bridgeburg and Oakville.[151] A resolution of warm support for the Kitchener-Waterloo amendment was passed, which perhaps helped the amendment to pass the Legislature in April 1924.[152]

In the end, only Kitchener enacted the zoning by-law. The city council first passed the zoning by-law in October 1924, but it had to be amended on appeal to the O.R.M.B., and was finally passed on 26 December 1924.[153] Its passage coincided with the completion and opening of the new city hall, the need for which had been so long felt and associated with the earlier phase of the planning movement. The most disheartening change from the Adams-Seymour draft was the deletion of restrictions on billboards, the "billboard industry" having insisted on maintaining its hold in residential districts and being powerful enough to get its way with O.R.M.B.[154] Subsequent amendments to the Kitchener zoning by-law could only be made with O.R.M.B. permission. From 1925 to 1930, there were twelve such amendments, ten of them permitting small commercial development, such as corner stores, in residential areas.[155] Reviewing the operation of the zoning by-laws after three years, A.R. Kaufman reported emphatic support from the entire city council, and general approval from the public.[156] He cited several specific examples where zoning and the Adams-Seymour plan had resulted in savings and greater efficiency of city services. "The local enthusiasm . . . is based on a realization that the plan is not visionary and impossible of attainment."

Waterloo's town council continued to "lay the matter over for further consideration" and was still doing so in 1930, when the Waterloo Board of Trade and Town Planning Commission held a joint meeting at which the city engineer and building inspector of Kitchener spoke about the way in which the zoning by-law was "working harmoniously" there.[157] The Board of Trade expressed full sympathy and support for the town planning movement, but that was not enough to overcome municipal inertia and lack of a sense of urgent need. The indefatigable Seymour signed an agreement with Waterloo township in 1924 for a township plan but there is no evidence such a plan was implemented.[158]

151. *Daily Record,* 16 February 1924.
152. Horace L. Seymour, "Report on the Plan of Town of Waterloo, Ontario, May 1923 to April 1924," 30 April 1924 (Seymour Papers, Public Archives of Canada).
153. City of Kitchener By-law 1823, amended by-laws 1834 and 1835.
154. Town Planning Institute of Canada, *Journal* IV, 1 (1925), p. 2.
155. City of Kitchener By-laws; O.R.M.B. Files.
156. Alvin Kaufman, "Town Planning in Kitchener after three years trial," Town Planning Institute of Canada *Journal* VII, 6 (1928), pp. 134-37; A.R. Kaufman, "Kitchener Town Plan," *Saturday Night,* 7 August 1937.
157. Waterloo Board of Trade, Minutes, 9 October 1930.
158. Waterloo Township Council, Minutes, 31 May 1924; Town Planning Institute of Canada *Journal* IV, 1 (1925), p. 2.

## SIGNIFICANCE OF THE KITCHENER-WATERLOO CASE

In concluding this essay, three questions are considered. What were the factors in Kitchener's eventual success in passing a zoning by-law, in however weakened a form? In what ways did experience in Kitchener-Waterloo influence planners in their generalizations about priorities? How well does this case study illustrate general trends in Canadian planning?

During these early stages in the process of introducing planning, the role of several agents can be seen, as well as the significance of timing and of changing ideas about the purpose and procedures of planning. In some of these, Berlin/Kitchener and Waterloo were characteristic of other cities; in other respects, the small size and sub-metropolitan functions made the towns distinctive. Local boards of trade and the press were essential in keeping the idea of planning before the local community, especially in the 1912-14 phase, as they were also in maintaining the growth ethos and civic boosterism generally. The Waterloo Board of Trade continued to promote planning in the 1920s, when the Kitchener Board had lost its active enthusiasm for the cause. The daily press was much stronger in its advocacy of planning in 1912-14 than later.

The role of determined individuals was crucial throughout, especially in exerting pressure on the municipal councils and in contacting outside planners and other towns and cities. The personalities of these leading advocates played a part in gaining general public acceptance. W.H. Breithaupt, D.B. Detweiler and A.R. Kaufman were clearly the most important figures, with Dr. W.L. Hilliard active in the smaller Waterloo setting. Breithaupt, Detweiler and Kaufman were all "businessmen"; none ever served on the municipal council (though Detweiler was elected to the Berlin Light Commission), but all were "public-spirited citizens" concerned for Berlin/Kitchener's progressive image. Breithaupt worked longest for planning in this period, his interest as an engineer being especially in the technical and efficiency aspects. Detweiler was more concerned with planning as one progressive cause among several; his contacts with influential people in government and the planning movement kept Berlin in touch with provincial and national trends. Breithaupt was responsible for the continuity between the first and second phases of intensive interest in planning and the 1919 plan was his work; his influence was smaller during the Adams-Seymour phase when his corrective view of planning was replaced by the more practicable preventive approach. Kaufman was the vital lack between Adams and Seymour and the local community, and most active in keeping up pressure on both municipal councils.

The collective attitudes of the municipal councils played a part. The Berlin council in the first phase was especially apprehensive about committing itself to a grandiose scheme of civic improvement which would raise the tax-rate and frighten away prospective manufacturers and good workmen; but it was also susceptible to the idea that a progressive plan would have advertisement value. Berlin was caught between its self-image as a progressive, rapidly growing, industrial city and its reputation for low property taxes. It was also very conscious of its image among other towns and cities and very competitive with them. Berlin-/Kitchener's initiative in calling Ontario planning conferences in December 1912 and February 1924 echoed the role it prided itself on taking with the beginnings of the Niagara Power movement. Waterloo was much less active in the planning movement, usually responding to suggestions from Berlin/Kitchener, rather than taking the initiative itself. But Waterloo's town council and Board of Trade usually reached a consensus more readily and were more consistent in planning matters – except in the final step of adopting the 1924 plan and zoning by-law.

It has been suggested that property developers were intimately involved in the planning movements of larger metropolitan cities.[159] In Kitchener-Waterloo, there were frequent references, before as well as after World War One, to the advantages to property-owners of planning and zoning in maintaining or raising property values by preventing the indiscriminate mixing of residential with other land uses, especially industrial. But the appeal was overwhelmingly to individual homeowners, and its success was the main factor in the public acceptance of the 1924 by-law. Owners of real property in the central business district were generally opposed to the corrective approach in the earlier phase of the planning movement. Owners of suburban properties, interested in developing small subdivisions, were also hostile to the street-opening proposals of the 1919 plan. Real estate agents and developers were not prominent in the local planning movement, as they are claimed to have been in larger, metropolitan cities.

Why did the Adams-Seymour plan of 1923-24 succeed when the Leavitt plan of 1913-14 and the Breithaupt plan of 1919 had failed? As with other municipalities considering city plans in 1913-14, the timing was wrong, as the urban land boom ended then and a crisis in municipal finance began which lasted nearly ten years; and the war distracted attention from urban problems. The 1912-14 phase was an essential preliminary to later success of the 1924 plan, however, in creating local awareness of the implications of city planning. The single most important factor in the passing of the 1924 by-law was the commitment of the outside planners, and especially of Seymour, to devising a practical and acceptable plan. Leavitt had paid only fleeting visits to Berlin and had clearly not known how limited were the powers of the municipal coun-

159. See, for example, Walter Van Nus, "Towards the City Efficient: The Theory and Practice of Zoning, 1919-1939," in Alan F.J. Artibise and Gilbert A. Stelter, eds. *The Usable Urban Past* (Toronto, 1979), pp. 226-46; John Weaver, "The Property Industry and Land Use Controls: The Vancouver Experience, 1910-1945," *Plan Canada*, 19 (Sept.-Dec. 1979), pp. 211-25.

cil. Seymour worked in Kitchener-Waterloo for well over a year and was
prepared to explain the purpose of the plan to the community and to
work out the legislative and administrative implications at the local and
provincial levels.

In the final reports for Kitchener and Waterloo, Seymour summed up
the requirements of a successful plan. It must, above all, be practical –
politically realistic and economically acceptable – in terms of costs,
municipal powers and public acceptance. Adams and Seymour seemed
personally to share many of Leavitt's aesthetic ideals on the dignity of
the new city hall surroundings and connections with the railway station,
and on the desirability of parkways and boulevards, but these personal
preferences were subordinated to suit locally perceived values and
needs.[160] This political realism may be contrasted with Leavitt's (and
Breithaupt's) approach in several ways. First was the shift in emphasis
from "corrective" to "preventive" planning, as it was impractical to
propose radical remodelling of the inner city or retroactive zoning of
land uses. Second, the plan should be considered, not as a fixed blueprint by which to control all future developments, but as "elastic and
capable of modification to meet unforeseen changes in conditions," for
"planning is an operation that never ceases."[161] Third, the single most
important requirement in making the plan effective was interesting the
public and winning its confidence and co-operation, which included
helping the local planning authorities with drafting by-laws and parlia-
mentary bills, and proving to businessmen that planning paid, "not only
in the conservation of life but also in the saving of money."[162] Seymour
declared:

> The final plan can only be successful if it has the support of a strong public
> opinion. It is a comparatively easy matter to prepare a plan showing extensive
> street widenings, openings and improvements, restricted areas, parks, park-
> ways and boulevards and a civic centre with monumental groups of public
> buildings, if no regard is paid to cost or public feeling – a plan that may be
> technically and artistically desirable but for the present impossible of being
> carried out . . . . The plan if successful must be based on present possibility . . .
> it must be a plan of such present practical value that public opinion will be the
> greatest single factor in making it effective now and assuring its success in
> future.[163]

Kitchener-Waterloo's experience of planning may be compared with
general trends in planning thought between 1912 and 1925. Others have

160. See, for example, Adams and Seymour, "Report on . . . Kitchener," 4 April 1924, pp. 4,
    17-18; Seymour, "Report on . . . Waterloo," 30 April 1924, 1-3.
161. "Report on . . . Kitchener", April 1924, p. 5.
162. "Kitchener Plan Becomes Law," Town Planning Institute of Canada *Journal* IV, 1
    (1925), p. 1.
163. H.L. Seymour, "Report on the Town Planning Survey of Waterloo May to August
    1923," September 1923 (Seymour Papers, Public Archives of Canada), p. 2.

considered the changing ideas about planning throughout Canada at this time, basing their generalizations mainly on the statements of professional planners or on the planning movements in larger, metropolitan cities.[164] Distinct phases have been identified, with that inspired by the ideal of the City Beautiful lasting till about 1913 and characterized by the architectural features of coherence, civic grandeur and visual variety, with the civic centre, parks and parkways prominent in the plans. After the interruption of World War One, the planning movement was dedicated to the "City Efficient," shaped more by the concerns of the municipal engineer, and purporting to be based on "scientific" principles. In practice, as Van Nus and Gunton have argued, planners in the 1920s settled for something far short of an optimally efficient urban organism.

Berlin-Waterloo became involved in planning and more general schemes of civic improvement relatively late, from mid-1912 when the City Beautiful movement was waning. Because of this late start and also because of the community's small size and sub-metropolitan, industrial functions, the local movement did not exemplify pure City Beautiful traits. Enthusiasm for civic grandeur and an imposing civic centre was certainly there, but beautification for its own sake was disavowed from the beginning. Mitchell and Leavitt spoke to approving audiences of primary importance of efficiency and functional order. Hodgetts stressed the fundamental need to plan for better housing and public health, as well as the value of starting to plan while towns were still small. All these were arguments more commonly quoted from planners after World War One. Leavitt's 1914 plan did have various hallmarks of City Beautiful plans but, given the small size of Berlin-Waterloo, it could also be interpreted as being similar to Garden Suburb plans. The Leavitt plan proved unacceptable because of its implications for private property rights, and because of the powerlessness of small municipalities to plan under provincial legislation of the time.

Planners of the 1920s have been implicitly condemned for their failure to adhere to purer planning ideas, and generally to bring in a new, perfect urban order. Van Nus has argued that planners forsook the ideal of the efficient city for "a more passive, managerial sort of planning which sought little more than the co-ordination of the desires and development policies of private interests." Planning became a "continuous administrative process rather than the gradual completion of a largely fixed design," with "elastic zoning in place of a workable long-range plan;" and planners compromised their principles for "political depend-

164. See Walter Van Nus, "The Fate of City Beautiful Thought in Canada 1893-1930," in Gilbert A. Stelter and Alan F.J. Artibise, eds., *The Canadian City: Essays in Urban History* (Toronto, 1977), pp. 162-85; Van Nus, "Towards the City Efficient," in Artibise and Stelter, *The Usable Urban Past*; and Thomas I. Gunton, "The Ideas and Policies of the Canadian Planning Profession, 1909-1931," in *ibid.*, pp. 177-95.

ence on local businessmen" and the property industry.[165] Evidence can be found in the case of Kitchener-Waterloo to support most of such generalizations. But detailed study of planning efforts in a particular time and place makes one more sympathetic to the planners. To begin with, planners were undoubtedly dependent on the invitation and fees of municipal councils. Further, planners were not engaged in a doctrinaire exercise of designing the perfect urban environment but in achieving public acceptance of legislation for some measure of urban planning, in the hope that more could be added later. And Kitchener-Waterloo was a test case, success in which would encourage other communities to follow suit. The planners' objective of an acceptable plan and the local constraints on their efforts explain their political realism and emphasis on public relations.

"Town planning is civil and political engineering, engineering technology and the psychology of suggestion," as "an Ottawa Planner" remarked of Seymour's plan of Waterloo. "It takes a great deal of patient education and persistence to 'put a plan across' and this is what must definitely occur if good is to come to the cause and encouragement to the planner."[166] Important as planning was as a science, it was still an art of the possible.

165. Van Nus, "Towards the City Efficient," pp. 226, 239.
166. Town Planning Institute of Canada, *Journal* III, 1 (1924), pp. 8-9.

1. Berlin's central business section viewed from the southwest, c. 1910. Victoria Park in the foreground was developed as a park from 1896. *Kitchener Public Library*.

2. Grand Trunk Railway tracks and industrial belt, Berlin c. 1906, as seen from the Margaret Avenue overbridge looking southwestwards. G.T.R. station left centre with tower (1897); Waterloo branch of G.T.R. swings off to the right. The three large factories which are visible all made furniture. *Kitchener Public Library*.

3. Berlin Market at rear of town hall, c. 1905, with Frederick Street to right looking down to the post office, King Street. The market building was replaced in 1907; but local demands for a new city hall in place of the plain structure of 1869, and a larger post office instead of the 1886 building, perhaps to be combined in a civic-federal square, were one element in the early civic planning movement. *Kitchener Public Library.*

4. King Street, Berlin c. 1910, main business street of Berlin, from just north of the Queen Street intersection to the Water Street corner. The Berlin and Waterloo Street Railway Company was municipalized in 1908 and the line double-tracked in 1909-10. *Kitchener Public Library.*

5. Residential area between business district on King Street to the south and the railway-industrial zone to the north, c. 1910; Ahrens Street runs from left to right across the picture. Hibner Place, the small triangular park created 1895, marks the junction between the two discordant street-systems of Berlin, one parallel to King Street and the other to the G.T.R. tracks. Most houses built in 1890s; cement sidewalks, curbs and gutters laid in 1907 and financed by frontage tax system. *Kitchener Public Library*.

# The Development and Beautification of an Industrial City: Maisonneuve, 1883-1918

PAUL-ANDRE LINTEAU

In the four decades following 1880, Canada underwent a major transformation, marked in particular by the acceleration of industrializtion and urbanization. In the province of Quebec, the population of the metropolitan area of Montreal grew from 177,000 in 1881 to 618,000 in 1921. Part of this urban growth overflowed the legal limits of the city itself and resulted in the establishment and expansion of several suburban municipalities. One of these was the city of Maisonneuve. Created in 1883, Maisonneuve became by the outbreak of the first World War a significant industrial centre – an experience shared by hundreds of other municipalities in North America during this period. In one respect, however, Maisonneuve was unique – its leaders conceived and implemented an impressive beautification program for which there are no Canadian equivalents. This program, which included a combination of features such as the erection of imposing buildings, boulevard and park development, and zoning, was undertaken in 1910 and was carried out during the next few years. In contrast to the haphazard city-building process in other centres – the product of thousands of individual decisions – the shaping of the urban landscape of Maisonneuve was generally the result of conscious decisions made by a small, influential elite.

## THE FRAMEWORK FOR DEVELOPMENT

The city of Maisonneuve was founded in 1883. Following the annexation of the municipality of Hochelaga to Montreal in 1883, the owners of land in the eastern section of Hochelaga obtained permission from the provincial government to establish a new municipality. Maison-

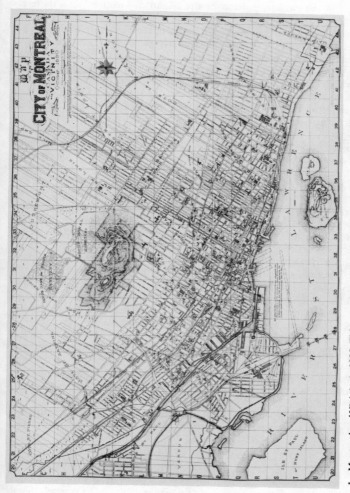

1. *Montreal and Vicinity in 1890.* This map shows the major suburban towns surrounding the city of Montreal. Maisonneuve is located at the right-hand lower corner. (*Public Archives Canada, C 85682*).

neuve was to exist independently until 1918 when, overcome by debts, it, too, was annexed to Montreal.[1]

Three factors explain the distinct development of Maisonneuve between 1883 and 1918. The first of these is location. Maisonneuve was situated in the eastern portion of the Island of Montreal and its territory, almost rectangular in shape, was bordered on the south by the St. Lawrence River, on the west by Bourbonniere Street, on the north by Laurier Street, and on the east by Viau Street. Several facts concerning this location were especially significant. Maisonneuve was on the edge of the river and in east Montreal. This meant that it was situated in an area where the population was predominantly francophone, and that it had easy access to port facilities. It also had access to the Canadian Northern Railway network and to the centre of Montreal, the latter thanks to an electric tramway which was extended to Maisonneuve in 1892.

Equally important, however, was the time frame in which Maisonneuve developed. The period following the establishment of the city was one of general economic prosperity for Canada. Together, rapid population gowth (fuelled by heavy immigration), the development of the Canadian west, and the initiation of several railway-building projects, created a strong demand for manufactured products and provoked a significant expansion in the production capacity of Canadian industry. Montreal, which at the time dominated the industrial production of the country, experienced a period of rapid growth, especially after 1900. New industries appeared, but above all there was a tremendous expansion of businesses that had been established prior to the turn of the century and which had to expand their plant facilities to meet new demands. And, since the central core of Montreal was already built up, many firms had to expand to the suburbs where land was available. Maisonneuve was an ideal location for such expansion.

The third and most important factor from the point of view of the city-building process in Maisonneuve was that the city's development was controlled by a small but influential elite. The city's founders and the prime agents in its development were a group of French-Canadian businessmen who, for the most part, were engaged in commerce, industry, and finance. They belonged to the middle-income bourgeoisie, a class whose social ascent had been rapid at the end of the nineteenth and the beginning of the twentieth century.[2] Two of these developers were

---

1. For more detailed information on the development of Maisonneuve, see Paul-André Linteau, *Maisonneuve, Comment des promoteurs fabriquent une ville, 1883* (Montreal: Boreal Express, 1981).

2. Paul-André Linteau, "Quelques réflexions autour de la bourgeoisie québecoise,1850-1914," *Revue d'histoire de l'Amerique française*, 30 (1976), pp. 55-66.

particularly noteworthy. Alphonse Desjardins was a lawyer and journalist as well as a politician (he was an influential member of the ultramontanist wing of the Conservative party and was later to become a member of Parliament, a senator and a federal cabinet minister). He was also an industrialist, having founded a brick works, and as a financier he was for twenty years president of the Jacques Cartier Bank, as well as being an administrator of other financial organizations. As early as 1874 he invested in the territory of what was to become Maisonneuve, and in subsequent years he along with other members of his family, was to acquire one of the large agricultural properties. Desjardins became deeply involved in the expansion of Maisonneuve and the development of his property. For several years his son Hubert sat on the municipal council, as either mayor or councillor. The other important promoter was a prosperous industrialist, Charles-Theodore Viau, founder of Viau & Son Biscuit Factory. In 1886 he acquired a large parcel of land in the eastern part of Maisonneuve. He also owned adjacent land in the neighbouring municipality of Longuepointe. During the course of the next decade he launched an ambitious scheme to urbanize his property and even thought of creating a distinct municipality called Viauville. Viau died in 1898, but his work was carried on by his heirs.

Other members of the promotional group involved in the development of Maisonneuve included the Letourneux family who were Montreal merchants, and Isaie Préfontaine, a businessman who was the brother of Raymond Préfontaine, a future mayor of Montreal and cabinet minister under Laurier. The only two anglophone members of the initial group were Wm. Bennett, a farmer, and James Morgan, owner of a large Montreal retail department store; these two men, and particularly Morgan, were much less involved in the operations of promotion than their francophone colleagues. At the beginning of the twentieth century a second generation of promoters appeared, often as members of joint stock companies; these included Senator Mitchell, Alexandre Michaud, Oscar Dufresne, and Mendoza Langlois.

The objective of these promoters was to realize a profit by the sale of land. What was needed was an influx of people in Maisonneuve. To achieve this, they attempted to attract industries whose workers would establish residence near the factory. By the intervention of the municipal council, the promoters established policies designed to attract industry. Between 1884 and 1915, almost without interruption, at least one of these great landowners sat on the municipal council as mayor or councillor. They were therefore in a position to orient municipal policies according to their interests. In the first place, the council had recourse to fiscal measures. One example illustrates the bias of the municipal directors. In 1884, the Jesuits, who owned a large property in Maisonneuve, asked to be exempt from paying taxes. The municipal council refused on

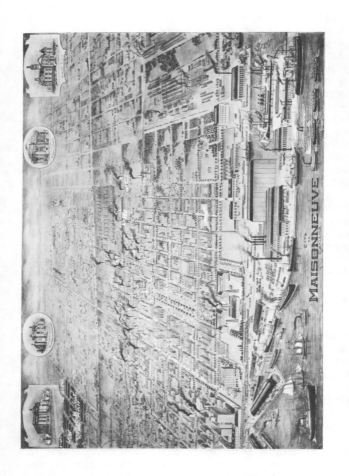

2. *Bird's eye view of Maisonneuve, on the eve of the First World War.* The mixed industrial and residential character of the city appears clearly. So does the new urban landscape shaped around Morgan Boulevard, with the new public buildings. The drawing is not perfectly accurate. The artist added planned buildings which were not built. Among these, the proposed scheme of Maisonneuve Park, north of Sherbrooke street, at the upper centre part of the drawing. (*Montreal City Archives*).

the grounds that the municipality was too poor.[3] Three months later the same council offered a twenty-five-year exemption from taxes to any enterprise that would establish a manufacturing plant in the municipality.[4] The Jesuits merited no consideration because their presence would not bring about any development. Between 1894 and 1900, the municipal council instituted another policy: the payment of grants for factory construction. Ten businesses benefited in this manner from municipal largesse. After 1900 there was a return to fiscal measures; namely, exemption from taxes, or a freeze on land assessment, for a given period.

These policies were successful in attracting several industries to Maisonneuve. By the first World War, the industrial structure of the city included several shoe manufacturers and a number of huge factories employing several hundred workers. These included the St. Lawrence Sugar Refinery, a paper dyeworks, the Viau Biscuit Factory, the Warden King and Son Foundry Works, the Canadian Spool Cotton Spinning Mill, and the naval dockyard of Canadian Vickers. Several firms of lesser importance were also scattered in various districts.

Maisonneuve became, therefore, a city with two well-defined roles: one industrial, the other residential. How was this dual vocation reflected in patterns of land use? Up to 1914, only the southern part of the city was occupied from the St. Lawrence River to the railway tracks, a distance of about six blocks. Industries were situated in two parallel bands, and in between was a residential zone. One part of the industry was concentrated on the edge of the St. Lawrence. In this category were the industries which depended on shipping, such as the sugar refinery whose primary product was imported by water. The majority of the industries, however, were located in the second zone to the north of Ontario Street on both sides of the railroad tracks. In both cases, the means of transportation was a major factor in location.

The variety of districts represented in Maisonneuve was reflected by the industrial architecture. In certain instances there were tall buildings, such as the St. Lawrence Sugar Refinery plant, whose oldest part dates from 1887, or the Canadian Spool Cotton complex, erected twenty years later in a more elaborate style. Farther along, the Warden King and Son foundry spread its installations over some ten acres, in a group of two-storey buildings. But the majority of factories were housed in typical brick buildings of three or four floors with simple lines and, occasionally, a pediment or other decorative element overhanging the principal entrance.

The working population was concentrated in a zone which extended from Notre-Dame Street to Ontario Street. The majority of the houses were constructed after 1900, when industrialization of the city took

3. Minutes of Municipal Council, October 1, 1884.
4. *Ibid.*, Jan. 7, 1885.

3. *Typical working-class housing in Maisonneuve.* (*Photo: Alain Laforest*).

4. *Morgan Boulevard's vista.* The impressive building of the Maisonneuve market stands at the head of the boulevard. The public bath is on the right. (*Photo: Alain Laforest*).

5. *The new fire station (1915).* This modern-style building has a design quite similar to Frank Lloyd Wright's Unity Temple. (*Photo: Alain Laforest*).

6. *The Château Dufresne.* This mansion, reproducing architectural features of the Petit Trianon, was built for the Dufresne brothers. The building contains two distinct houses. (*Photo: Gilles Rivest*).

place. Rows of two or three-storey houses each generally consisted of two to five apartments. Brick was the main material of construction. Many of these houses were decorated with the famous outside staircases typical of Montreal architecture. It must be emphasized, however, that the workers rarely owned their own homes. The assessment register of 1911 indicated that 90 per cent of the heads of households were tenants. These workers were thus little involved in the decisions concerning the direction of their city; these decisions were in the hands of the large landowners. In the centre of the residential district were two Roman Catholic churches, which constituted an essential feature for the new population. Moreover, it was Desjardins and Viau, the two most active landowners, who donated the land necessary to build the new churches. They considered the donation an investment which would facilitate the development of their domain.

In this residential area observers were attracted to one sector, Viauville, because of its originality. Here the promoter, Charles-Theodore Viau, left his imprint on the development he planned for his property. In all deeds of sale for land, he inserted a clause which obligated the buyer to construct a house with a stone facade and ten feet of sidewalk.[5] Various styles of greystone houses were constructed in this district. They gave a particular character to the Viauville area which can still be seen today. The population of Viauville had a large proportion of white-collar workers.

The *Montreal Atlas* for 1907 and 1914 indicates the stages of development.[6] By 1907 the land already built on was concentrated in the western sector, near the city of Montreal. Between 1907 and 1914 there was a marked growth and by 1914 the greatest part of the land located to the south of Boyce Street (now de Coubertin Street) was also built up.

By the eve of the First World War Maisonneuve had become one of the great industrial centres of Canada, and its municipal leaders enjoyed the accompanying recognition. In 1912, for example, in their publicity, they quoted federal statistics which put Maisonneuve in second place in the province and in fifth place in Canada for the value of the industrial production. It was with great pride that they dubbed their city the "Pittsburg [sic] of Canada."[7]

## A NEW URBAN AESTHETIC, OR "MAISONNEUVE, GARDEN OF MONTREAL"

It was this pride in the progress achieved that the municipal leaders wished to express in their grand plans for improving Maisonneuve. But,

5. *La Presse*, May 6, 1899, p. 16.

6. A. R. Pinsonneault, *Atlas of the Island and City of Montreal and Île Bizard: A compilation of the most recent cadastral plans from the book of reference* (1907); Charles Edward Goad, *Atlas of the City of Montreal and Vicinity in four volumes* (1912-1914).

7. *Le Devoir*, Nov. 16, 1912.

in order to understand the circumstances surrounding the birth of this mammoth project, it is essential to put it in the context of the years preceding the first World War – *La Belle Epoque*. It was in many ways the golden age of capitalism. Great industrialists and financiers showed off their wealth and constructed lavish residences. The large newspapers in Montreal such as *La Presse* gave accounts of the careers of these men, and wealth became a spectacle for those on the outside to watch. Prosperity engendered optimism. The leaders looked to the future with confidence. Prime Minister Laurier affirmed that the twentieth century belonged to Canada. There were some periodic crises but they were quickly resolved, and no one could see what could possibly put an end to the prosperity. Progress seemed to be limitless. This prosperity was obviously not shared by everyone, as Terry Copp showed in his study of the conditions of life among the Montreal working class,[8] but those in power were scarcely preoccupied with the disparities surrounding them.

The grandiose plans of Maisonneuve, then, were bathed in an atmosphere of euphoria which characterized the ruling classes, not only in Canada, but in all the industrialized countries or those moving in that direction. These grand designs are recorded in the North American trend to beautification of cities, which originated in the nineteenth century. The birth of the Park and Boulevard Movement in the 1870s was represented in the work of Frederick Law Olmstead, who planned the development of Central Park in New York and also of Mount Royal Park in Montreal. The representatives of this movement proposed the creation of towns in a network of vast green areas interlinked by tree-lined boulevards.[9] At the turn of the century, the City Beautiful Movement revised the emphasis of the previous century by enlarging its perspectives. To the boulevards and parks were added beautiful buildings and an entire urban structure, answerable to aesthetic criteria. The "White City," constructed for the Chicago Exposition in 1893, set the tone.[10] Plans for the beautification of cities were put forth just about everywhere, in Europe as well as in North America. Costly and grandiose, plans for beautification were submitted for several Canadian cities.[11]

The grandiose plans of Maisonneuve cannot be fully explained outside this general context, but there were some circumstances peculiar to Maisonneuve. The city was directed by a local bourgeoisie formed from three main elements: land speculators, who had a direct interest in the growth

8. Terry Copp, *The Anatomy of Poverty: The Condition of the Working Class in Montreal, 1897-1929* (Toronto: McClelland and Stewart, 1974).

9. Charles N. Glaab and A. Theodore Brown, *A History of Urban America* (New York: Macmillan, 1967), pp. 254-57.

10. *Ibid.*, pp. 259-63.

11. Walter Van Nus, "The Fate of City Beautiful Thought in Canada, 1893-1930," in G. A. Stelter and A.F.J. Artibise, eds., *The Canadian City: Essays in Urban History* (Toronto: McClelland and Stewart, 1977), pp. 162-85.

of the city; industrialists, who had set up both their factory and their residence in Maisonneuve and who wanted to be part of its development; and local merchants, whose volume of business could not help but benefit from the growth of the municipality. It was a social class on the upswing, a class for whom the possibilities of becoming wealthy were numerous. Members of this bourgeoisie were desirous of displaying their wealth and of asserting their presence in a permanent way.

Two figures dominated the council: the mayor, Alexandre Michaud, and the chairman of the finance committee, Oscar Dufresne. Alexandre Michaud was a former grain merchant who became a real estate agent and land speculator. Oscar Dufresne was a rich industrialist who managed one of the biggest shoe factories in Maisonneuve. In 1910, Marius Dufresne, Oscar's brother, was engaged as city engineer. He was to play an important role in the preparation of new projects. Michaud and Dufresne owed their personal wealth in large measure to the Canadian economic growth of the beginning of the century. They shared the optimism characteristic of the period. In 1912, *Le Devoir* reported that:

> Mayor Michaud declared that Maisonneuve which was now just at the beginning of its industrial life had the promise of a flourishing future. It had prosperity unequalled in the annals of the country. And that this was due to the policy of progress which the governing leaders of the municipality had always followed.[12]

Between 1910 and 1915 this ruling team outlined a beautification plan which was to make Maisonneuve one of the most beautiful municipalities on the island. The "Pittsburgh of Canada" was to become the "Garden of Montreal." The plan included four inter-related elements: the erection of imposing public buildings, the creation of a large park, the construction of wide boulevards, and zoning. Together, these programs would shape the urban landscape of Maisonneuve.

*Public Buildings*

During the first two decades of its existence, Maisonneuve had constructed only one building worthy of mention. Situated on Notre-Dame Street, it had been erected in 1888, then enlarged substantially in 1897, and housed the city hall, the police station, and the fire department. In 1906-7 a second fire hall was built on Ontario Street. It was a building of conventional style having no architectural pretensions. Then, in rapid succession over a period of five years, four new public buildings were started: the city hall, a market, public baths, and a fire station. Also significant was a new interest in architecture which had obvious aesthetic pretensions. The round of construction opened with the city hall. Its construction was imperative because the rapid growth of the city and municipal services had rendered the previous building quite inadequate.

12. *Le Devoir*, Oct. 3, 1912, p. 5.

The new building was to be situated on Ontario Street, between Pie IX Boulevard and Desjardins Street, and this northward movement was an indication of the physical expansion of Maisonneuve. Construction began in 1910 and continued the following year. A journalist wrote of it in 1915;

> Maisonneuve has had its City Hall since 1912. It stands as an edifice both severe and magnificent, with a facade adorned with elegant Greek columns, flanked by two superb candelabra and surrounded by a new lawn on Ontario Street. You enter through heavy bronze doors.[13]

The establishment of a public market, where farmers of the region could sell their produce, was a common practice in the Montreal area at the turn of the century. The municipal council decided to construct such a market in 1912. It was to occupy a large space on the north side of Ontario Street, opposite Morgan Boulevard. Construction took two years, with a public opening in September 1914. Here again no expense was spared and simple utilitarian considerations were largely transcended. The building was majestic with a huge entrance, a central dome, and four turrets. It was furnished with ultra-modern equipment and possessed a large hall suitable for huge assemblies. In fact, this hall was for a long time the privileged location for political assemblies in East Montreal. The cost of construction exceeded a quarter of a million dollars. The market square was marked by a monumental fountain which alone cost the taxpayers of Maisonneuve more than $20,000. The third of these grand structures housed a public bath and a gymnasium. The project was on the drawing board as early as 1911 but construction did not begin until 1914. Built at a cost of $215,000 it was inaugurated in May 1916. Its architecture, the choice of Marius Dufresne, marked a return to classicism. The building closely resembled New York's Grand Central Station, constructed between 1903 and 1913.

Finally, in 1914, the municipal council decided to rebuild the first fire station which dated from 1888. Work began in the summer of 1914 and the city took possession of the station at the end of the following year at a cost of $142,000. In this instance Marius Dufresne literally copied the plans for the Unity Temple which Frank Lloyd Wright had constructed in Oak Park, Illinois, in 1906. The comparison is striking if one examines the architect's drawings or the photographs of the buildings. This building contrasts with its predecessors by its more modern architecture, and probably reflects the personal taste of Marius Dufresne. This phenomenon of imitation underlines the importance of the North American networks of information on architecture. Through their journals and specialized publications, architects were rapidly informed of new designs. In this respect Quebec was not isolated and architects followed with interest what was going on in the rest of the world.

13. *La Patrie*, Dec. 4, 1915, p. 10.

*Park Development*

The new direction that the city fathers wanted to give to their city did not stop with the architecture of public buildings. They wished to modify the whole urban environment by introducing into it grandiose elements. The creation of Maisonneuve Park was a good illustration of this desire. The project began in 1910 when city council acquired 150 acres of land in the northeast section of the city with a view to creating a park. Matters remained stationary for sometime, but a grand scheme fostered by Oscar Dufresne took form in 1912. On the one hand it involved enlarging this park – to quadruple its area – in order to make a meeting place which would be at the disposal of all the inhabitants of the eastern part of the Island of Montreal. This park would thus become the equivalent in East Montreal of Mount Royal Park in the west. There was even thought of linking this park to Mount Royal by a grand boulevard which would also pass by Lafontaine Park.[14] On the other hand, Dufresne's scheme involved the construction of a cultural and sports complex on the site of the park. This aspect of the plan was inspired by the large European parks which included such features as hotels, cafes, casinos, race tracks, museums, libraries, zoos and botanical gardens. These features would, in many cases, generate revenue which would then be divided up in the following manner: half to the municipality, half to hospitals, schools and the charitable organizations of Maisonneuve.

The scheme, which was approved by the Quebec government, was to be administered by a commission separate from the municipality. The acquisition of the necessary land for the enlargment of the initial park was made between 1913 and 1916. It gave way to many land speculation deals referred to as the "Scandal of Maisonneuve Park." The technique of successive resales was practised systematically with the help of intermediaries. The majority of the city's large landowners managed to make a considerable profit from a transaction at one or another of its steps.[15] In the final analysis, the acquisition of the land cost Maisonneuve taxpayers about 6.5 million dollars.

In the meantime the Park Commission was formed, presided over by Oscar Dufresne. The commission entrusted Marius Dufresne with the preparation of a plan for development. A journalist described the project in the following manner:

> Access to the park will be provided by two monumental doors, one between Pie IX Boulevard and Sherbrooke St., and the other between First Avenue and Sherbrooke St. Besides a magnificent race track, more than a mile long which alone will be a big source of revenue for the city, this park will have amphitheatres with stone steps for hockey games, *balle au but*, lacrosse, etc., a motor-racing track, a race track, artificial lakes around which will be built elegant hotels, casinos with bandstands, wooded paths, cafés, etc. Then, on the educa-

14. *Le Devoir*, Jan. 21, 1914, p. 6.
15. Rodrique Langlois, *Le scandale du parc de Maisonneuve.*

tional side, the public will be able to take advantage of an art gallery with a museum and library, botanical gardens with an aquarium and a zoo. And during those hours when time hangs heavily, one can visit a "singing café," and on the more exotic side, there will be a Japanese café with gardens adorned with fairy-like waterfalls. In the middle of the central avenue the Maisonneuve monument will be erected.[16]

The First World War was to prevent the carrying out of this plan. Maisonneuve Park was, eventually, to realize its cultural and sports goals, but step by step over a long period of time. Today, three main complexes can be found on this site: a municipal golf course, developed in the 1920s, and more recently converted into a park; the Botanical Gardens created in the 1930s; and the installations constructed for the 1976 Olympic Games.

## Boulevards

The development of the city of Maisonneuve was completed by the creation of the grand boulevards: Pie IX Boulevard, Morgan Boulevard, and Sherbrooke Street. The most important was Pie IX. This artery has a particular character. It is wider than the other streets of the city (100 feet instead of 66), and several local middle-class families chose to establish their residence on this street. As early as 1907 the council decided to make it the line of communication between Maisonneuve and Rosemount, its neighbour to the north.[17] Employing the ideas of development in vogue at the time the engineer, Marius Dufresne, dreamed of making a grand boulevard lined with trees having a centre island of grass. Soon the project went beyond the boundaries of Maisonneuve and became an intercity boulevard running from the north to the south of the Island and even beyond by linking the Rivière des Prairies with a bridge. For that it was necessary to cross the territory of Saint Michel and Montreal North, two small municipalities which hardly had the means to effect such an undertaking. That did not present an obstacle, for Maisonneuve became the chief contractor. Maisonneuve expropriated the land, had the work done, and then charged the cost to the two municipalities.[18] The Quebec legislature gave its approval in 1912 and ratified Marius Dufresne's "two twenty-seven foot roadways and in the centre a twenty-six foot median."[19] First the southern section in the confines of Maisonneuve was developed, then the northern section in Saint Michel and Montreal North was begun in 1915.[20]

Sherbrooke Street was also seen to take on a special role. It occupied a magnificent site and the council decided to widen it with the hope that

16. *La Patrie*, Dec. 4, 1915, p. 10.
17. *Minutes of the Municipal Council*, April 17, 1907.
18. *Quebec Statutes*, 3 Geo. V (1912), Chap. 58.
19. *La Patrie*, Dec. 4, 1915, 10.
20. *Ibid.*; and *Minutes of the Municipal Council*, 1914 and 1915, *passim.*

some beautiful homes would subsequently be built there. It was, in fact, at the corner of Pie IX Boulevard that the Dufresne brothers had their lavish home built. The development of this part of the city remained nevertheless limited because of Montreal's refusal to open up Sherbrooke Street where it meets with Hochelaga.[21] Finally, Morgan Boulevard was developed in 1914-15, as a result of gifts of land made by James Morgan and Alexandre Michaud.[22] It opened a perspective on two of the new public buildings: the market and the baths. Its conception was linked to the models of the Paris World Fair of 1900 or of the "White City" where the grand boulevards were bordered by imposing buildings.

## Zoning

The creation of these grand boulevards provided the opportunity for the city to establish zoning to control construction. Thus on Pie IX the erection of factories was forbidden. Houses were to be situated twelve feet from the street and were to have a brick or stone front. Outside staircases, which had been numerous on Maisonneuve's other streets, were also forbidden.

> In the future no plant, factory or mill of any sort will be erected on Pie IX boulevard, from the St. Lawrence river to the river des Prairies except between Notre-Dame St. and St. Lawrence and from Ontario St. to the Canadian National and Canadian Pacific railway tracks. Lumber yards, coal depots and the manufacture of refrigerators are also forbidden. The only houses, stores and business establishments, which will be permitted to be built, must be exactly twelve feet from those lines previously ratified, they must be at least two stories high and be constructed of either stone or brick or of wood ornamented with stone or brick.
>
> It is forbidden to build staircases in front of any of the above mentioned buildings.[23]

This type of regulation was later to be extended to other streets of the city. The technique of zoning was quite new at the time, and in adopting it Maisonneuve was in the forefront in matters of development. Maisonneuve was also ahead of its time in having the surface of all the roads south of the railway paved with cement or asphalt.[24] Maisonneuve was therefore in the latest style, but it had to pay for it. In the words of a contemporary, it was a considerable and exaggerated expense given the municipal debt.[25]

21. *Le Devoir*, Oct. 2, 1913, p. 6
22. Sept. 30, 1914, Office of Notary, G. Ecrement, no. 3451; *Minutes of the Municipal Council*, July 29, Oct. 7, 1914.
23. *Quebec Statutes*, 3 Geo. V (1912), Chap. 58, Art. 5.
24. *Minutes of the Municipal Council*, 1911-1916, *passim*.
25. Interview with Dr. Edouard Desjardins, July, 1973.

The zoning regulation of Pie IX Boulevard marked a will to harmonize private dwellings with public development. Several members of the bourgeoisie had already established themselves on Pie IX Boulevard. Thus the most important move on the side of the private sector had been that of the Dufresne brothers. Not only had they cheerfully spent public funds but they also committed considerable personal investment. They had an extravagant mansion built, christened "Château Dufresne," whose architecture was inspired by the Petit Trianon. It is situated at the corner of Sherbrooke Street and Pie IX Boulevard, just beside the site of Maisonneuve Park. This building reproduced on an individual scale the dreams of grandeur which had been manifest in the public buildings.

CONCLUSION

At the end of 1915 the newspaper *La Patrie* published a special section called "The New Maisonneuve" which described the transformations that the city had undergone in the years preceding the First World War. Indeed the appearance of the city had been noticeably altered. The city fathers were proud of these changes and mentioned them in their publicity.

> With its grand boulevard of Pie IX Avenue extending from the St. Lawrence to Riviere des Prairies, Maisonneuve possesses the most beautiful avenue of the entire island of Montreal. Its public buildings ... are the most beautiful that we have on the whole island and they themselves constitute the finest monument which has been erected on the initiative of its government.[26]

In 1917 at a banquet held in Maisonneuve, a provincial minister, Jérémie Décarie, congratulated the municipal leaders:

> "I have seen good beautiful and solid things in Maisonneuve," said the Minister. And later congratulating the city on its progress: "I have faith," he added, "in the energy which is characteristic of our people, once they have undertaken a task. You wanted to do something artistic – and I am proud of that fact. Like our forefathers you wanted to build something which would last. You have succeeded in building something of lasting beauty."[27]

It was a policy of grandeur that the Michaud-Dufresne partnership had pursued in Maisonneuve; a policy that cost a great deal. The municipality had to borrow considerable sums of money. This situation might not have posed insurmountable problems had the city's growth continued. But the war came, putting a virtual end to urban growth for several years. Maisonneuve, laden with debts, was in serious difficulty. In 1916 and 1917 the municipality even had to borrow to pay the interest owed. The Quebec government in 1918 simply decreed that

26. *Minutes of Municipal Council*, May 26, 1917.
27. *Le Devoir*, May 28, 1917, p. 2.

Maisonneuve was to be annexed to Montreal. It marked the end of an adventure. Only the bills remained.

In 1917 Maisonneuve published an announcement which described the city as both the "Pittsburgh of Canada" and the "Garden of Montreal." These two seemingly contradictory descriptions summarized very well the history of the development of the suburban city. Behind these two expressions were outlined the objectives of the land promoters. In order to realize any profit on their lands they had had to initiate development and populate the city.They did this first by attracting industry. Then, as interest in quality gained momentum in North America, they wanted to beautify their city, so as to attract still more residents. But the beautification could have been accomplished in a more modest fashion. How does one explain this policy of grandeur? Obviously it was the work of opportunists. The French-Canadian businessmen, who brought about this policy, were self-made men. They showed off their newly acquired wealth. Their cultural vision was pretentious and they were satisfied with importing models. In this sense they rather resembled American businessmen of the same period who made just as much of a show of their wealth.

But there is more to it. This policy of grandeur was the affirmation of a social class, the bourgeoisie, and more specifically in this case, the French-Canadian bourgeoisie. This bourgeoisie had seen its strength increase considerably in this period of economic expansion. They wanted to reach the top and affirm their capabilities. This French-Canadian bourgeoisie was neither rural nor agricultural. It believed in progress and sought prosperity. And, as elsewhere in North America, during this, *La Belle Epoque*, it succumbed to an exaggerated optimism and believed too easily that prosperity would last forever. The workers, who constituted the majority of the population, had nothing to say; they were not consulted, but they had to pay the bill. The great buildings and Maisonneuve's other developments are the symbols and the reflection of an era; developments which gave to Maisonneuve a special dimension.

# From Land Assembly to Social Maturity:
# The Suburban Life of Westdale (Hamilton), Ontario, 1911-1951

JOHN C. WEAVER

With insights drawn from American and British scholarship, Canadian historians and historical geographers have been participating in what has become an international study of cities.[1] In Canada, as elsewhere, writing has tended to focus on the mid-nineteenth century, testing hypotheses relating to a shift from the pre-industrial pedestrian city to the

SOURCE: *Histoire sociale/Social History*, vol. 11 (1978), pp. 411-40. Reprinted by permission of the author and the University of Ottawa Press.

Research was funded by a McMaster University Arts Research Council Grant. The author is indebted to his research assistants Julie Backholm, Ron Elliott and Martin Lawlor. Lawlor's research on contractors and mortgage brokers was particularly original. Students in the urban history research seminar have provided essential studies on a variety of related issues. Their works have been cited throughout the study. David Gagan gave vital assistance.

1. See for example Peter Goheen, *Victorian Toronto, 1850 to 1900* (Chicago: University of Chicago, 1970) particularly Chapters 1 and 2. Another indicator of international exchange is found in a review of literature on housing in the second footnote of Michael J. Doucet, "Working Class Housing in a Small Nineteenth Century Canadian City. Hamilton, Ontario 1852-1881," *Essays in Canadian Working Class History*, ed. Gregory S. Kealy and Peter Warrian (Toronto: McClelland and Stewart, 1976). Michael Katz, *The People of Hamilton, Canada West: Family and Class in a Mid-Nineteenth-Century City* (Cambridge: Harvard University Press, 1975) is the most striking attempt to place Canadian social history into an international context.

industrial city with mass transit.[2] The decades from 1840 to 1880, therefore, emerge as a critical era for urban and social history. After the nineteenth-century watershed, there remain significant issues concerning "the city building process"[3] and major growth periods with waves of immigration between 1905 and 1913, 1925 and 1930, and again after World War II. Certain of the themes to be developed in twentieth-century urban history represent a continuation of earlier processes, but others indicate divergence. One matter of fundamental interest to all who study the city, one where twentieth-century urban history has an opportunity to advance several distinct trends, concerns the dwelling place. In Canada, planned suburbs with racial and social segregation, the evolution of the real estate agent, changes in land development and the building trades, and an expanding government role were initiated in the first half of the twentieth century. A national study of these concerns rests beyond the scope of the current article; rather it is reasonable to adopt an approach recommended by Michael Katz. "Only through analyzing the expression of the general through the particular," Katz reminds us, "can we construct stable and satisfactory explanations of social development."[4]

Westdale, the Hamilton community selected as a Canadian measure of twentieth-century urban trends, sheltered more than 1,700 households when completed. As a private enterprise, commercialism governed Westdale's construction, but like the best-planned North American suburbs of the era – the exclusive Country Club District of Kansas City or Vancouver's Shaughnessy Heights – it balanced aesthetic and environmental concerns with financial ones. This made it different from the many commuter suburbs, but its history still progressed in step with national urban circumstances. Developers, builders, and residents shared experiences with counterparts across Canada and the United States. The developers and builders of Westdale worked from a body of knowledge and from traditional practices that were by no means limited to their locale. The pitfalls and business practices of development and contracting were not unique to Westdale. The owners and tenants of the suburb sorted themselves out spatially in ways that conformed with attitudes and economic conditions that were continental in scope.

2. Though Gideon Sjoberg's *The Pre-industrial City, Past and Present* (New York: The Free Press, 1960) stimulated a considerable volume of case studies, the concept of a clear shift has been discredited. See for example, Tamara K. Hareven, "The Historical Study of the Family in Urban Society," *Journal of Urban History*, I (May 1975), 259-65; Herbert Gutman, "Work, Culture and Society in Industrializaing America, 1815-1919," *American Historical Review*, LXXVII (June 1973), 531-87.

3. For an excellent introduction to the concept of "the city-building process," see Roy Lubove, "The Urbanization Process: An Approach to Historical Research," *Journal of the American Institute of Planners*, XXXIII (Jan. 1967), 33-39.

4. Katz, *Hamilton*, p. 316.

THE DEVELOPERS

"Someone has said there is only one crop of land, but there is an endless crop of natives and every baby on the face of the earth makes every foot of land more valuable."[5] This article of faith, used by Westdale's developer, has been promoted by "the property industry," past and present. Critics of land developers, on the other hand, have maintained for generations that scarcity is a product of spectulators controlling supply. Yet, scarcity forms only one dimension of land value.[6] According to a solid body of historical research, particularly in the United Kingdom, successful land developers have affected land value by creative activity.[7] Discerning patterns of urban growth, both spatial and temporal, they have attempted to intepret and influence public taste. Developers have performed as instruments for drawing together political and legal acumen, capital-raising facilities, planning talents and the building trades. Value judgements about land developers, therefore, must be carefully constructed to permit areas of ambiguity. Westdale, for example, was a well-conceived community with a fine array of amenities. It also encompassed racial discrimination and "clever" tactics on the part of the developer. Nonetheless, it is difficult to indict the developer without condemning the prejudices and business practices of an era.

At the turn of the century, the thrust to Hamilton's expansion was being channelled by topography. By 1910, the escarpment to the south, Burlington Bay on the north, and a wide ravine in the west had turned land development eastward.[8] East-end surveys soon stood at considerable distance from downtown Hamilton. For that reason the level plateau which stood on the far side of the western ravine caught the eye of

---

5. Hamilton Public Library, Special Collections, F. Kent Hamilton, *Beauty Spots in Westdale* (c. 1928), p. 7.

6. N.H. Lithwick and Gilles Paquet, "The Economics of Urban Land Use," in *Urban Studies: A Canadian Perspective*, ed. Lithwick and Paquet (Toronto: Methuen, 1968).

7. I am indebted to Jean-Claude Robert of La Groupe de Recherche sur la société montréalise au 19e siècle for calling my attention to a pioneering study of land developers, Maurice Halbwach, *Les expropriations et le prix des terrains à Paris, 1860-1900* (Paris, 1909). Canadian studies of land development include the following: Jean-Claude Robert, "Un seigneur entrepreneur, Barthélemy Joliette, et la fondation du village d'industrie (Joliette), 1822-1850," *Revue d'histoire de l'Amérique française*, XXVI (déc. 1972), 375-95; Paul-André Linteau et Jean-Claude Robert, "Propriété foncière et société à Montréal: Une hypothèse," *Revue d'histoire de l'Amérique française*, XXVIII (juin 1974), 45-65; Peter Spurr, *Land and Urban Development: A Preliminary Study* (Toronto: James Lorimer, 1976). On the British historiographic trend see Richard Shannon, "The Genius of the Suburbs," *Times Literary Supplement*, 31 Dec. 1976, p. 1626.

8. The clearest indication is established by studying a map prepared by the Ramsay-Thomas Realty Company in 1913. The map presents newly registered and proposed surveys and provides a radius measure indicating distances from the city core. A brief account of the east-end expansion is found in Charles M. Johnston, *The Head of the Lake: A History of Wentworth County* (Hamilton, 1967), pp. 246-48 and Appendix E, "City Growth from the Year of Incorporation to 1914."

Toronto contractor, J.J. McKittrick, who, in 1911, began to promote a
100 acre plot, "Hamilton Gardens."[9] His venture lacked urban services
and he did not have the resources to secure them. Therefore, McKittrick
became associated with local partners whose careers had been meshed
with the development of Hamilton. Legal talent came from Sir John
Gibson, former Lieutenant-Governor of Ontario. Gibson and other
members of the McKittrick syndicate were connected with Hamilton's
pioneer utility firm, Cataract Power and Light and with the Hamilton
Street Railway. The Southam family, publishers of the *Spectator*,
acquired a major interest. Soon the new group had expanded the original 100 acres to 800 and successfully negotiated an agreement whereby
in January 1914 Hamilton annexed the survey.[10] The agreement set
down conditions among which was one that forced the syndicate to
construct and maintain a bridge. In return, the pre-annexation rural
assessments were frozen from 1914 to 1919. Even with entrepreneurial
talent and annexation, the endeavour proved unpromising. The 1913-15
recession and the war economy retarded property sales.

The cash flow anticipated by McKittrick Properties dried up for eight
years during which time its tax cushion expired and had to be renego-
tiated for a period extending to 1926. This provided some relief
although a furthur commitment entered into during the balmy days
before the war returned to plague the syndicate. To secure a key parcel
of 100 acres belonging to the Hamilton Cemetery Board, McKittrick
Properties purchased another site, traded in for the desired land and
included $40,000 compensation. Financial stringency forced the com-
pany to default on a compensation instalment. Eventually, the Board
agreed to a settlement, but not before a political move by the syndicate
was turned back by the Hamilton electorate. A "McKittrick man" ran
for mayor in 1916 and, despite the backing of the *Spectator*, he was
defeated.[11] The early years of the syndicate suggest that even a powerful
alliance among the civic elite could experience fiscal and political
embarrassment.[12]

Facing difficulties, the developers sought fresh management. One of
the investors, John Moodie, president of Eagle Knitwear of Hamilton,
invited his son-in-law, F. Kent Hamilton, to guide the company. A
Winnipeg lawyer, Hamilton had learned a great deal about the planning
and promotion of a suburb from his western experience. Upon his

9. Hamilton Public Library, Reference Room, Hamilton Collection, Scrapbook on Real
   Estate, "Action Requested to Promote Old Suburbs," undated clippings.
10. "Copy of Agreement among McKittrick Properties, the City of Hamilton and Ancaster
   Township presented to the Ontario Railway and Municipal Board, 26 January 1914,"
   *Hamilton City Council Minutes, 1914.*
11. Hamilton *Spectator*, Dec. and Jan. 1916.
12. This is somewhat in conflict with the impression of the Winnipeg business community
   as analysed by Alan Artibise, *Winnipeg, A Social History of Urban Growth, 1874- 1914*
   (Montreal: McGill-Queen's University Press, 1975).

arrival in 1918, Hamilton established a sales staff that grew to eight in good times; he also designed the publicity campaigns for the next seven years. Hamilton commissioned New York landscape architect Robert Anderson Pope to prepare a street plan.[13] Pope was one of many urban planners to have been influenced by German civic concepts. As early as 1910, he had recommended German-style urban decentralization, urging a shift away from urban systems which concentrated lines of transportation on the core city. Pope argued that this led to overcrowding and high residential land costs. New suburbs, more or less self-contained, promised a remedy.[14] When Pope designed Westdale, his plan included a central shopping district.

Kent Hamilton had a full understanding of the housing issues which had concerned North American reformers between 1900 and 1920, having read articles and attended lectures on most of the era's remedies for the housing crisis. In 1919 Hamilton was aware of English garden cities, limited dividend housing companies, tax incentives for builders and homebuyers, company housing, and even the wartime housing constructed by the United States government.[15] For a while, Hamilton considered supporting public housing. For example, he supported the creation of a Hamilton Town Planning Commission, apparently recognizing that studies by such a public body would necessarily benefit McKittrick Properties whose land was the last major tract near the central city and susceptible to an experiment in public housing. In the economic uncertainty of 1919, a sale, even to a public housing corporation, was to be welcomed. There was one further reason why Hamilton supported action on the housing problem. He had been advised that

. . . failure to take care of the returned soldiers, not only from the stand point of housing, but also from the stand point of opportunity for earning a decent livelihood, may result in a social uprising that in the end would be far more expensive to the city of Hamilton than a theoretical excess of providing adequate housing.[16]

The turbulence of 1919 frightened some civic leaders into repression but it moved others toward expedient consideration of reform. How-

13. F. Kent Hamilton Papers, William Lyon Somerville, Robert Anderson Pope and Desmond McDonaugh, *Report of Survey and Recommendations, McKittrick Properties* (1 Feb. 1919); Somerville to Hamilton, n.d.
14. Mel Scott, *American City Planning Since 1890* (Berkeley: University of California Press, 1971), pp. 96-99.
15. F. Kent Hamilton Papers, Somerville *et al.*, *Report on Survey and Recommendations, passim.* Also see Scrapbook 1 for various clippings kept by Hamilton on housing questions.
16. Somerville *et al.*, *Report of Survey and Recommendations*, p. 16. For similar fears as an impetus to housing reform, see John C. Weaver, "Reconstruction of the Richmond District in Halifax: A Canadian Episode in Public Housing and Town Planning, 1918-1921," *Plan Canada* (March 1976), pp. 36-47.

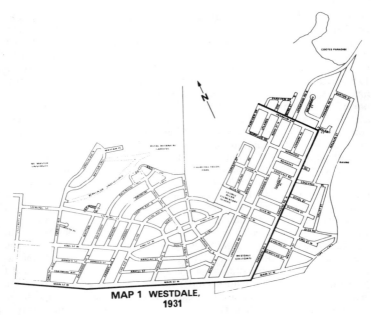

**MAP 1   WESTDALE,
1931**

ever, with the return of social stability, public housing proved a will-o-the-wisp. Indicative of a more conservative approach was a week-long Better Homes Exhibit in the Hamilton Armouries sponsored by Kent Hamilton and the newly-created Hamilton Real Estate Board. Their aim was "to educate the average renter into the method and means of ownership."[17] As a tangible move in that direction, the first Westdale surveys came onto the market as workingmen's parcels arranged on a grid layout with thirty-foot frontages. The areas set aside for workingmen had a lower potential land value than elsewhere in Westdale, for along the opposite slope of the ravine the city maintained a garbage dump. As late as 1928, a syndicate official would write that these lands were unattractive: "I do not expect that they ever will be very desirable."[18] One direct incentive hastened the surveying of workingmen's lots. The Ontario Housing Act of 1919, applying federal funds, assisted the raising of mortgages for modest six room homes with minimum standards costing under $4,000. In fact, the "Hamilton A-1 Plan", pro-

17. Hamilton Real Estate Board, Kent Hamilton Scrapbooks, Scrapbook 2, p. 11, clipping dated June 1922.

18. I am indebted to Mr. St. Clair Balfour and his Secretary, Mrs. M. Shano, for providing copies of Southam correspondence relating to Westdale. Southam Papers, J.P. Mills, Secretary, Westdale Properties to F.I. Ker, 19 Nov. 1928.

viding a home for $3,850 was a government-approved brick dwelling seeing considerable service in Westdale (see Illustration 1).[19]

From 1920 to 1926, the syndicate faced lean times. Local conditions formed an element in a national interlude of slow residential construction after the boost provided by the Housing Act. Moderate improvement occurred in 1922, but construction suffered a setback in the next two years. It joined business in a subsequent steady improvement which continued unbroken until 1928, a peak year for residential building.[20] As conditions improved, the syndicate responded, shifting tactics, preparing surveys that would appeal to affluent homebuyers (see Table 1). Still, a prosperous clientele brought no sudden windfall. Twenty years after the intital land assembly, ten years after the registration of the first survey, only half the residential lots had been build upon. While some of the 830 vacant properties which appeared on the 1931 assessment rolls were held by contractors, the property developers retained 570. Free bus trips for inspection tours, gold certificates buried on a few lots, the opening of a model home, a contest to name Westdale, and the barnstorming of Jack V. Elliott in his *Canuck* aeroplane could not stir sufficient interest. McKittrick Properties went into bankruptcy.

With the prospect of the tax assessment freeze being lifted in 1926 and the suburban bridge, which the syndicate financed, now serving a major highway, the developers requested an amended agreement with the city. In fact, there were serious cash flow problems while fixed annual charges had to be met. To meet expenses, the syndicate turned to an expedient. Since land had been mortgaged to finance the intital assembly, some additional collateral had to be discovered. The company borrowed against contractor's agreements-to-purchase and loans came from the major shareholders. As contractors repaid the company, the loans were retired.[21] The sticky cash flow became a critical worry, for with time running out on the assessment freeze, the Corporation of the City of Hamilton and the Ontario Municipal Board refused once again to amend the 1913 agreement. Caught in a pinch that required dramatic proof to the city that relief was required, the Board of Directors met in June 1926, refused to pay a minor bill, and precipitated bankruptcy proceedings.[22]

Two major unsecured creditors were affected by the bankruptcy: those shareholders who had loaned funds against the agreements-to-purchase, and the city which was owed $200,000 in taxes arrears and

19. Bureau of Municipal Affairs, *Report re Housing for 1919* (Ontario Sessional Papers, 1920).
20. Canada Dominion Bureau of Statistics, *Census Monograph No. 8 Housing in Canada (A Study Based on the Census of 1931 and Supplementary Data)*, p. 104.
21. Supreme Court of Ontario, *in Bankruptcy, Re: McKittrick Properties*, Affidavit of Frank Ernest Roberts, 19 March 1928, p. 6.
22. S.C.O., *in Bankruptcy, Re: McKittrick Properties*, Petition of Charles Delamere Magee, 29 June 1926.

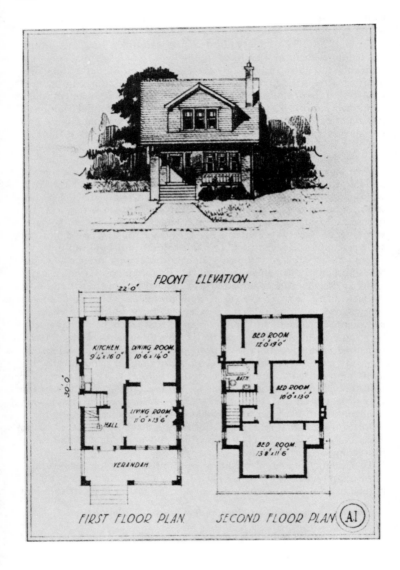

FRONT ELEVATION.

FIRST FLOOR PLAN.          SECOND FLOOR PLAN.

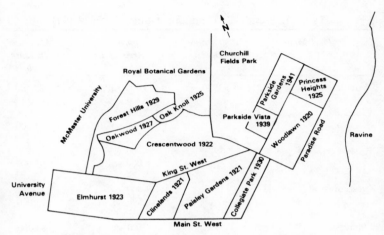

**MAP 2 WESTDALE'S COMPONENT SURVEYS & DATES OF REGISTRATION (see Table 1)**

Table 1: WESTDALE'S COMPONENT SURVEYS, 1920-1941

| Plan Name | Registration Date | Approximate Area of Lot | Minimum Building Values |
|---|---|---|---|
| Woodlawn | 1920 | 3,000 sq. ft. | $2,500 |
| Clinelands | 1921 | 3,000 to 4,500 sq. ft. | $3,000 |
| Paisley Gardens | 1921 | 4,000 to 5,000 sq. ft. | $3,000 |
| Crescentwood | 1921 | 4,000 to 5,500 sq. ft. | $3,000 |
| Elmhurst | 1923 | 4,500 sq. ft. | $3,000 |
| Princess Heights | 1925 | 3,000 to 3,500 sq. ft. | none |
| Oak Knoll | 1925 | 7,500 to 10,000 sq. ft. | $7,000 |
| Oak Wood | 1927 | 6,000 to 15,000 sq. ft. | $8,000 |
| Forest Hills | 1929 | 7,000 to 10,000 sq. ft. | $8,000 |
| Collegiate Park | 1930 | 3,000 to 3,300 sq. ft. | $3,000 |
| Parkside Vista | 1939 | 3,000 to 3,200 sq. ft. | $2,500 |
| Parkside Gardens | 1941 | 3,300 to 3,600 sq. ft. | $2,500 |

Sources: Survey Maps and Sample Property Instruments, Wentworth County Courthouse.

Table 1: Westdale's Component Surveys, 1920-1941

| Plan Name | Registration Date | Approximate Area of Lot | Minimum Building Values |
|---|---|---|---|
| Woodlawn | 1920 | 3,000 sq. ft. | $2,500 |
| Clinelands | 1921 | 3,000 to 4,500 sq. ft. | $3,000 |
| Paisley Gardens | 1921 | 4,000 to 5,000 sq. ft. | $3,000 |
| Crescentwood | 1921 | 4,000 5,500 sq. ft. | $3,000 |
| Elmhurst | 1923 | 4,500 sq. ft. | $3,000 |
| Princess Heights | 1925 | 3,000 to 3,500 sq. ft. | none |
| Oak Knoll | 1925 | 7,500 to 10,000 sq. ft. | $7,000 |
| Oak Wood | 1927 | 6,000 to 15,000 sq. ft. | $8,000 |
| Forest Hills | 1929 | 7,000 to 10,000 sq. ft. | $8,000 |
| Collegiate Park | 1930 | 3,000 to 3,300 sq. ft. | $3,000 |
| Parkside Vista | 1939 | 3,000 to 3,200 sq. ft. | $2,500 |
| Parkside Gardens | 1941 | 3,300 to 3,600 sq. ft. | $2,500 |

SOURCES: Survey Maps and Sample Property Instruments, Wentworth County Courthouse.

service charges.[23] As real as the financial crisis was, certain features suggest that this bankruptcy had tactical dimensions. Shareholders paid to the full all creditors, except the city.[24] It soon dawned on the mayor and Board of Control that their obduracy had drastic implications. Future tax revenue, the city's credit rating, and Hamilton's reputation as the "Ambitious City" were jeopardized. The episode had a familiar ring. A major private endeavour had become so closely identified with commitments to expansion that a government hard-line was precluded despite the hard-nosed posturing to convince voters that City Hall was not "soft on" developers. The Public Trustee appointed by syndicate shareholders expressed the situation well enough. "The city asked us to come, as a matter of fact, because they were worried about the situation. . . . I told them that there was now no money to pay taxes. . . and out of that arose the suggestion that they might take some of our land for parks, to clear up those arrears of taxes.[25] The city subsequently took 377 acres of ravine land for park purposes and erased the tax debt. That the rugged land was unsuited for development while a park bolstered land values made the agreement a rewarding one for the developer.[26] With the debt lifted and the Ontario Municipal Board approving a relaxed agreement that released the developers from maintenance of the bridge, a newly chartered company composed of shareholders in the old syndicate sprang to life and bought out McKittrick Properties. Except for the sacrificed park land and a legal bill of $7,573, the manoeuvre had cost nothing but it had succeeded in forcing the better terms which the syndicate had sought for five years.[27]

Just as the new company took shape, it fell heir to a boon secured by the old syndicate. Commencing late in 1921, Kent Hamilton had begun courting McMaster University which had been considering a move from its Bloor Street location in Toronto.[28] After lengthy negotiations and generous contributory pledges to McMaster from J.R. Moodie, William Southam, G.H. Levy, and Sir John Gibson – all members of the original and new land companies – McMaster located in Westdale.[29] Imme-

23. *Ibid.*, Report of the Official Receiver, 28 July 1926.

24. *Ibid.*, Debtor, filed 6 July 1926.

25. S.C.O., Depositions of Frank E. Roberts, taken before John Bruce, Special Examiner, 25 Oct. 1927.

26. Hamilton Public Library, Special Collections, Scrapbooks, West-End and Westdale Scrapbook (1926-28), *passim*.

27. Memorandum on Board of Directors' Meeting, Westdale Properties, 1 Nov. 1928.

28. Charles M. Johnston, *McMaster University*, Vol. 1, *The Toronto Years* (Toronto: University of Toronto Press, 1970), pp. 204-36.

29. Baptist Archives, McMaster Univesity, Removal to Hamilton/Hamilton Chamber of Commerce File, F. Kent Hamilton to F.P. Healey, Secretary Chamber of Commerce, 3 Oct. 1921; W.W. McMaster to Chancellor A.L. McCremmon, 21 Feb. 1922. For the contributions, see New McMaster File, W.J. Westaway to Chancellor H.P. Whidden, 13 June 1928. Altogether, those associated with Westdale Properties had pledged $171,000.

diately, property in two of Westdale's surveys was "expected to be more attractive than anything at present on the market as a result of the McMaster University location."[30] The developers also calculated that the University would carry the burden of expense for a major water main, that it would purchase electricity from their power company, and that it would lure "a colony of Professors."[31] For all of the apparent ingenuity and sinister cunning, the syndicate was neither a cohesive nor an instant success. Kent Hamilton sued for payment of commissions owing during the bankuptcy; the Southams found certain legal charges of their syndicate partners, Gibson and Levy, far too high; as for professors, most were too impecunious to buy Westdale land.[32] The Depression ruined whatever expectations had been raised by the new agreement and the enticing of McMaster. Only in the building boom after World War II did vacant fields disappear.

## THE BUILDERS

The creation of Westdale preceded the era of totally integrated property development. In a sense, Westdale emerged at a mid-point in the evolution of the "property industry." It had the benefit of sophisticated town planning but, unlike recent packaged suburbs, construction was not a branch of the developers' activities. Given this absence of corporate integration, Westdale's completion drew upon an array of traditional crafts and specialists whose relationships tended, for all their diversity, to be intimate and coordinated. Mortgage brokers, contractors and sub-contractors worked separately, but their ability to deviate from the developer's vision of the community was constrained.

By the early twentieth century it was assumed in real estate circles that unregulated growth, non-conforming buildings (gas stations, tenements, corner stores), and certain ethnic groups offended middle-class home buyers. No longer were building lots sold by auction, a system which minimized the developer's control. Conformity was now deemed impor-

30. Southam Papers, J.P. Mills, Secretary, Westdale Properties to F.I. Ker, Hamilton *Spectator*, 19 Nov. 1928.

31. Baptist Archives, new McMaster File, W.J. Westaway to Chancellor H.P. Whidden, Re.: Agreement with Westdale Properties, 30 Oct. and 6 Dec. 1928. The power issue was discussed in Westaway to Whidden 13 June 1928. Westaway destroyed his copy "as I do not want it to appear in our correspondence." The housing of Professors was raised in Westaway to Whidden, 12 Oct. 1929. Westaway to Whidden, 7 Apr. 1930.

32. Kent Hamilton's litigation appears in S.C.O. *in Bankruptcy, In the Matter of the Estate of McKittrick's Limited* and *In the Matter of the Claim of F. Kent Hamilton*, copy of Depositions of F. Kent Hamilton, 29 Sept. 1927 and 11 Oct. 1927. The conflict among the syndicate members is cited in Southam Papers, Memorandum on Board of Directors' Meeting, Westdale Properties, 1 Nov. 1928. The state of the professors' purchasing power is contained in Baptist Archives, New McMaster File, Westaway to Whidden, 7 Apr. 1980.

tant; its implementation with so many individual actors led to the evolution of the real estate agent and to the appearance of planning and restrictive covenants. The latter had wide application in Hamilton. Those enforced to the industrial east-end specified a minimum building value and brick construction,[33] but Westdale's covenants went further and contained two categories: structural and ethnic. The first defined the minimum dwelling value and the building materials. Each of the component surveys in Westdale had a specific set of standards (see Table 1). In more exclusive areas the developer retained the right to approve "the location, plan and specifications, exterior elevation and type of construction." As for the second feature, this was a typical clause: "None of the lands described. . . shall be used, occupied by or let or sold to Negroes, Asiatics, Bulgarians, Austrians, Russians, Serbs, Rumanians, Turks, Armenians, whether British subjects or not, or foreign-born Italians, Greeks or Jews."[34] Developers' brochures emphasized that Westdale was "restricted."[35] Regulation in the early years was enforced. A real estate agent warned a contractor not to sell to an interested Italian green grocer: "Tom, we don't want people like that in here."[36]

Builders, dependent upon credit and a sound reputation with developers, lacked the security to risk breaking covenants. Relying on quick sales, they dared not risk architectural innovations. Most of Westdale's builders operated as family combinations with design facilities. They depended upon standard blueprints of demonstrated popularity. This meant that, for all of Westdale's distinctive planning, the bulk of its housing stock resembled much of contemporary Hamilton. The appearance of large custom-built homes in the 1940s altered styles somewhat, for even smaller homes constructed in these years attempted to incorporate imitations of expensive flourishes: diamond-shaped window panes, stonework, bay windows, and wood and stucco finishing on the second storey.

The builders themselves were a mixed group. For many men anxious to improve themselves, an opportune route out of the labour pool was to become a small-scale builder. Like street peddling or the corner store, contracting provided a few urban labourers with access to an independent occupational ladder. A few, such as the Hamilton-based builder

---

33. Anna Chiota, "Somerset Park, 1910-1960: A Quantitative Study of an Industrial Suburb," McMaster University, 1977 (Mimeograhed). Michael Doucet found few convenants between 1847 and 1881 and those examined pertained to building materials. See Michael Doucet, "Building the Victorian City: The Process of Land Development in Hamilton, Ontario, 1847-1881" (Ph.D. Dissertation, University of Toronto, 1977). Doucet also provides an excellent discussion of the land auction system.

34. Wentworth County Courthouse, Land Registry Office, property instrument 327114 (Hamilton) is cited, but its racial clause was standard throughout Westdale.

35. *Beauty Sports in Westdale* (c. 1928), p. 22.

36. Interview with contractor Thomas Casey, 24 Sept. 1976.

Michael Pigott, carried success beyond their original locality. Some insight into the hustle and flexibility of builders is provided by merely considering the fragments of information conveyed in city directories, advertisements, and assessment rolls. One family, the Theakers, demonstrated a few characteristics of the approximately thirty builders involved in Westdale.[37] The men worked on homes in the east-end and Westdale, using the same plans. Gladys kept the books and worked for a mortgage broker. The family lived within walking distance of Westdale. Occasionally, such family-builders formed transitory partnerships, for the building trades were fluid in their association. J. Vickers and his son held twenty Westdale lots in 1931, but the father also retained several with plasterer Fred Beldham in the east-end. These were small father and son operations, but certain family-builders were virtual dynasties. The Mills family had participated in home construction and property development since the mid-nineteenth century.[38] The Armstrongs had depth and experience, having constructed by 1930 some 600 dwellings in Hamilton, and with William C. Armstrong as realty and financial agent and William D. Armstrong heading the architectural department, they were the most ambitious domestic contractors in the city.[39] Thomas "Carpenter" Jutter and his son Charles were builders, but Jutter senior, Mayor of Hamilton and later a Member of the Ontario Legislature, was a signficant local politician. The father, in his role as mayor, cut the opening ribbon on a Westdale economy home in 1924; the son was erecting dwellings in Westdale as late as 1948.[40] Whatever their scale of operation, the builders had a local commitment. They depended on local reputation for credit and buyers. They were not working for large suburban builders with an impersonal corporate name that one finds today. Building techniques have advanced since the 1920s, but there was something of value in having contractors mindful of their community's esteem. One builder, Thomas Casey, could identify the homes that he had constructed fifty years later. Having done so, he proceeded to rhapsodize about the quality and distinctive character of his brickwork.

There was no typical builder in Westdale, but the operations of Thomas Casey, who constructed seventy homes between 1920 and 1932, and between 1945 and 1955, suggest in greater detail the practices used by

37. The information on builders was generated from the 1931 assessment file, the City Directories, real estate advertisements between 1920 and 1930, and interviews with a contractor of the era, Thomas Casey. Now 97, Casey had excellent recall and his memory of purchases was cross-checked with the land registry office and found accurate. Although Victorian London was hardly comparable to Hamilton in the 1920s, the latter's family contractors and credit system seems remarkably similar to that outlined in H.S. Dyos and D.A. Reeder, "Slums and Suburbs," in The Victorian City, ed. H.S. Dyos and Michael Wolff, vol. I (London: Routledge and Kegan Paul, 1973), pp. 378-79.
38. Hamilton Spectator, 15 July 1926.
39. Hamilton Spectator, real estate section, 30 Apr. 1926.
40. Kent Hamilton Scrapbooks, Scrapbook 2, Spectator, 17 Nov. 1924.

those with few assets save work and ambition. Raised in Ireland in County Cork, Casey had laboured in Liverpool before emigrating to Canada in 1914. Between 1914 and 1920, he found employment on Toronto construction projects, on a Caledonia, Ontario, farm, and in Hamilton and later Pittsburgh steel mills. During the winter and spring of 1919 he built a worker's cottage near Dominion Foundries and Steel, hewing out a basement with pick and shovel. Casey thus illustrated the truth that any man with the ambition and the rudimentary tools could style himself as a contractor. He continued to construct small dwellings in the slow growth of the 1920s, but by his peak year, 1930, Casey employed eight carpenters and was working on fifteen substantial homes in Westdale. Like the land developers who increased the size of their lots in the prosperous late 1920s, builders began to fashion homes for the swelling numbers of middle-class purchasers (see Table 1). Reduced to one house a year from 1933 to 1939, Casey left Westdale for Burlington, which had adopted a Depression policy of inducements to attract builders.[41] He returned to Westdale in the mid-1940s when the Central Mortgage and Housing Corporation Act (effective 1 January 1945) once more stimulated construction.

Casey's transient career and his lack of apprenticeship in the building trades present an occasional view of labour history, but it is his building practices which merit attention. Casey normally selected a parcel of adjacent lots in a survey and signed an agreement to purchase. He could not afford clear title. He then put in a number of basements simultaneously and started to erect the frame for one dwelling. Other work he subcontracted to "the seven trades": masons, lathers, plasterers, electricians, plumbers, roofers, and tinsmiths. By staggering construction along a row of basements, Casey provided an even sequence of work. More importantly, the stages in home construction were paced by his line of credit.

The half-dozen major office buildings in downtown Hamilton housed a large and yet almost invisible capital market. Banks and trust companies provided building loans, but at least thirty non-institutional mortgage brokers advertised in real estate columns during the 1920s. In some instances, barristers functioned as brokers, handling estate funds or working for clients interested in sound investments. During the 1920s mortgage interest varied from 6 to 6½ per cent. Casey preferred borrowing through these barristers. Institutional lenders demanded quarterly instalment payments which included principal as well as interest, but private lenders accepted payments of interest and a lump sum for principal when a house was sold. For the builder, loans came in "three draws." When the roof went on, the lender's agent drove out to inspect

---

41. Interview with Hughes Cleaver, former M.P. and developer cited in Stephen White, "The Business Community in Burlington's Development, 1920-1939," McMaster University, 1977 (Mimeographed).

the dwelling. If he approved, Casey could make his initial draught. When the white finishing plaster dried, he was entitled to a second. With interior trim and fixtures in place, he could make a final draught. Each draught helped to finance subcontracting work on adjacent houses. The chain effect meant that while Casey sensed a declining demand in 1930 and 1931, he could not trim back on building activities. Having signed agreements to purchase lots along one side of a block and having put in basements, he could only continue and hope for the best. He was locked into the completion of what he had begun in a year of incredible optimism, a fact which helps to explain the general over-building in Westdale in 1930 and 1931. What was true for Westdale may well have applied elsewhere, for the national collapse in residential construction did not arrive until 1932.[42] As for Casey, he had to let several of his completed homes. In addition, he exchanged one of the new dwellings for a cheaper east-end dwelling, believing that in hard times it would sell more readily.

Several major property owners had no interest in construction, but instead held lots for future capital gain. In 1931, the largest investor, with twenty-two lots, was a doctor whose brother managed the real estate branch of National Trust and held office as president of the Hamilton Real Estate Board in 1930. The degree of speculation and the background of participants is difficult to establish, in part because of a complex set of "straw companies." Kent Hamilton is a case in point. Besides his connection with McKittrick Properties, he managed several companies which dealt in property: Blackstone Realty Securities Limited and Gorban Land Comapny Limited. Hamilton and similar property entrepreneurs placed some of their activities (at a remove) in order to limit personal liability, but the numerous corporate labels imply additional motives. They at least suggest the involvement of different combinations of shadow investors and the desire for anonymity.

In many instances, those who dealt in Westdale real estate had assorted interests across the city and engaged in other facets of the housing industry. Kent Hamilton, for example, also functioned as a mortgage broker. Forty Investors, the largest owner of Westdale commercial property was managed by W.C. Thompson, real estate agent and mortgage broker. His other operations included Forty Associates, Hamilton Home Builders Limited, Hamilton Improvement Company, Traders Realty, and Thompson and Thompson Realty. Less diverse agents understandably located their offices close to complimentary services. Realtor Norman Ellis owned three Westdale lots in 1931, and to arrange financing for builders or prospective home owners he had only to go next door to the Hamilton Finance Corporation. Realtor J.W. Hamilton, who held three lots in 1931, sold building lots to contractors. At the same time, he served as secretary of a major wholesale lumber

42. Canada, Central Mortgage and Housing Corporation, *Housing in Canada, A Factual Summary*, vol. I (Oct. 1946), p. 47.

company with offices on the same office-tower floor as his real estate agency. These and a host of comparable connections helped to guarantee Westdale's disciplined development. The realty agents and investors who owned lots were not likely to sell to contractors who might introduce structural or ethnic non-conformity for fear of depressing the value of remaining properties. Moreover, it appears that speculation had few amateurs. Those who owned several lots were involved in real estate or the building industry (see Table 3).

Just as Westdale's development preceded the era of integrated development and yet represented a progression from the relatively modest land assembling of the nineteenth century, the financial arrangements illustrate a transition. The home buyer could pay interest in two or four payments annually, repaying the principal at the end of an agreed interval. However, in the mid 1920s, what "Carpenter" Jutter billed as a "modern technique" came into use. After a down payment of $500, the purchaser would pay $42 a month, a blending of principal and interest.

## THE COMMUNITY, 1931[43]

Instalment plans notwithstanding, the Westdale of 1931 stood incomplete, frozen midway in its settlement by the economic depression. Of 1,734 lots, 48 per cent lay vacant, while 46 per cent had dwellings with residents and 6 per cent had buildings under construction or were unoccupied. There were many signs of the community's raw state. Children raised in Westdale during the 1930s and the war years would recall the opportunities for play in vacant fields and hollows. For several years Anglicans held services in the basement of their unfinished church. Altogether, some 760 families resided in the incomplete suburb, 556 (72 per cent) as owner-occupants, and 205 (28 per cent) as tenants either in rented homes, over commercial establishments or in the five apartment buildings. The ratio of homeowners to tenant households was quite different from the whole city where 52 per cent of households rented, one of many indicators of Westdale's situation as a distinct community within the larger urban setting.

Settled to a large extent between 1925 and 1930, the Westdale of 1931 had a relatively young population, a reflection of the affluent 1920s and the developer's advertising which lauded the suburb as a proper nesting ground. Completion of an elementary school in 1927 and Westdale

43. Unless otherwise stated, data for this section was derived from the Assessment rolls using S.P.S.S. A COBOL programme was used to sort and list names of occupants and owners alphabetically. This proved useful in establishing the names of owners of more than one property and in linking "persisters" from the 1931 to 1951 files. Subjective observations were collected from a number of interviews, including the following: Philip Barrs, Diane Turner, Sheila Scott, Thomas Casey, Mrs. F. Kent Hamilton, and Gordon Hamilton.

Table 2: Urban Canada, Hamilton and Westdale, Homeowners and Tenants, 1931 and 1951

|  | 1931 | | 1951 | |
|  | Owners | Tenants | Owners | Tenants |
|  | % No. | % No. | % No. | % No. |
| Westdale | 72 (556) | 28 (217) | 82.5 (1,408) | 17.5 (301) |
| Hamilton | 48 (17,876) | 52 (19,341) | 65.8 (36,090) | 34.2 (19,250) |
| Urban Canada* | 37.6 (226,136) | 62.4 (375,445) | 46.1 (390,930) | 53.9 (457,710) |

SOURCE: Westdale file and Census of Canada

*cities over 30,000.

Table 3: Builders and Speculators: Owners of Two or More
Vacant Lots, 1931

| | No. of Lots | Owner's Occupation | Comments and Names Mentioned in Text |
|---|---|---|---|
| 1. | 132 | Manufacturer | Associated with McKittrick Syndicate |
| 2. | 25 | Doctor | Brother was President of Real Estate Board |
| 3. | 18 | Contractor | |
| 4. | 16 | Contractor | Thomas Casey |
| 5. | 15 | Contractor | Bryers and Son |
| 6. | 12 | Contractor | Budd and Son |
| 7. | 11 | Contractor | Vickers |
| 8. | 10 | Contractor | |
| 9. | 10 | Contractor | |
| 10. | 10 | Real Estate Agent | |
| 11. | | Widow | |
| 12. | 7 | Real Estate Agent | |
| 13. | 7 | Contractor | |
| 14. | 6 | Contractor | Theaker Family |
| 15. | 6 | Contractor | Jutten and Son |
| 16. | 4 | Contractor | Armstrong Family |
| 17. | 3 | Contractor | |
| 18. | 3 | Confectionary | |
| 19. | | Manufacturer | |
| 20. | 3 | Real Estate Agent | Norman Ellis |
| 21. | 3 | Contractor | |
| 22. | 3 | Contractor | |
| 23. | 3 | Contractor | |
| 24. | 3 | Widow | |
| 25. | 3 | Contractor | |
| 26. | 3 | Real Estate Agent | J.W. Hamilton |
| 27. | 3 | Radio Repairman | |
| 28. | 2 | Contractor | Most of the owners of two lots appear to have been planning to build homes on a double lot or to build on one and sell the other. |
| 29. | 2 | Locomotive Engineer | |
| 30. | 2 | Manufacturer | |
| 31. | 2 | Railway Superintendant | |
| 32. | 2 | Vice President of Manufacturing Company | |
| 33. | 2 | Widow | |
| 34. | 2 | Real Estate Agent | |
| 35. | 2 | Widow | |
| 36. | 2 | Merchant | |
| 37. | 2 | Office Clerk | |
| 38. | 2 | Widow | |
| 39. | 3 | Manufacturer | |

Source: Westdale files using a COBOL 6 SORT/MERGE Programme.

Collegiate in 1930, touted as the finest institution in the region, reinforced its attraction. With the arrival of McMaster University, a child could progress from kindergarten through university within a mile's radius. The suburb was overwhelmingly Protestant with 95 per cent of the households affiliated with Protestant churches (see Table 4). Only thirty-three Roman Catholic and five Jewish families resided in Westdale. Restrictive covenants account for part of this resounding Protestant character, but nothing prohibited native-born Catholics or Jews from moving to the suburb. Some further considerations discouraged non-Protestants and attracted Protestants. The developers only encouraged Protestant churches to locate in Westdale. Syndicate partner John Moodie figured prominently in Hamilton's very influential Presbyterian community: "he was. . . behind a movement to build a nice Presbyterian church in Westdale."[44] The campaign to attract McMaster University, a Baptist institution, fitted the Protestant design. Possibly it was coincidental that the winner of the contest to name the suburb was an esteemed Anglican canon, but the *Spectator* made good use of the canon's name and his praise for Westdale's sylvan splendor.[45] It seems plausible to consider these events along with covenants and advertising as signals that Westdale provided an escape from what was regarded as a mounting alien presence elsewhere in the city. The period of heaviest construction, 1925 to 1930, coincided with what would be peak years for eastern and southern European arrivals in Hamilton until after 1945. Hence, in the decision to reside in Westdale, more than simple proximity to work or attractive situation was involved. Conventional nativist and sectarian biases of the 1920s were combined with that powerful determinant of middle-class behaviour: parental desire to secure decent education and "the right" neighbourhood influences. Catering to this combination of aspirations and prejudices, the developer added to the land value. Ironically, Westdale Collegiate, designed as part of the suburb's self-contained image, became an instrument of contact between its children and those from quite different backgrounds. Built in advance of the completed community, the collegiate attracted pupils from Hamilton's working-class north end as well as rural Ancaster.[46]

Religious distinction was a consideration in measuring the contrast between Westdale and the city. A common and simple statistical device, the index of dissimilarity yields a rough measure of that contrast. If the index number was 0, then religious affiliations would have been distributed in an indentical fashion both in Westdale and the city of Hamilton. If the index number was 1, then the two areas would have been com-

44. Baptist Archives, New McMaster File, W.J. Westaway to Chancellor H.P. Whidden, 10 Oct. 1930.
45. Hamilton Real Estate Board, Kent Hamilton Scrapbooks, Scrapbook 2, clippings from *Spectator*, Apr. 1923.
46. Interview with Diane Turner, a former resident raised in Westdale, 20 Nov. 1976.

Table 4: Religious Affiliation of Hamilton and Westdale Residents, 1931 and 1951

| | 1931 | | 1951 | |
|---|---|---|---|---|
| Denomination | Hamilton | Westdale | Hamilton | Westdale |
| United Church | 20.6 | 35.4 | 23.2 | 35.5 |
| Anglican | 29.4 | 25.6 | 25.1 | 21.9 |
| Presbyterian | 17.1 | 18.7 | 12.4 | 11.7 |
| Baptist | 5.2 | 6.9 | 5.1 | 7.4 |
| Lutheran | 1.2 | 1.1 | 1.8 | 0.8 |
| Salvation Army | 0.7 | 0.3 | 0.5 | 0.4 |
| Roman Catholic | 18.5 | 4.5 | 23.0 | 7.6 |
| Jewish | 1.7 | 0.7 | 1.5 | 8.3 |
| Other or Unspecified | 5.6 | 6.8 | 7.4 | 6.5 |
| | 100.0 | 100.0 | 100.0 | 100.0 |

1931 Index of dissimilarity = .193
1951 Index of dissimilarity = .211

SOURCES: The percentages for Hamilton were based on Census returns which included all individuals; percentages for Westdale were based on the affiliation of household heads.

pletely distinct in terms of religious affiliation. With the large areas and numerous households involved, a high score close to 1 was unlikely.[47] Even so, the index score of .193 seems low, but it does indicate segregation. Nonetheless, comparable studies of urban segregation treat such a score as marginal.[48] Indeed, another variable in Westdale had a more impressive score. Occupation ranked on a scale devised by Bernard Blishen[49] or classified by economic function (vertical and horizontal classification respectively) showed a contrast between Westdale and

47. To calculate an index of dissimilarity, the percentages of every category of a variable within a designated area (for example Westdale) are prepared. The process is repeated for another area (Hamilton). The result is a two column frequency distribution expressed in percentages (see Table 4). The difference in percentages in each row is noted and the sum taken of the positive differences is the index of dissimilarity. For more information see Charles M. Dollar and Richard J. Jensen, *Historians' Guide to Statistics: Quantitative Analysis and Historical Research* (New York: Holt, Rinehart and Winston, 1971), p. 125; Sam B. Warner, Jr., *The Private City: Philadelphia in Three Periods of its Growth* (Philadelphia: University of Pennsylvania Press, 1968), pp. 13-14.
48. Kenneth L. Kusmer, *A Ghetto Takes Shape: Black Cleveland, 1870-1930* (Urbana: University of Illinois Press, 1976), pp. 44-46.
49. Bernard R. Blishen, "The Construction and Use of an Occupational Class Scale," *Canadian Journal of Economics and Political Science*, XXIV (Nov. 1958). Given the valid criticisms against occupation as a surrogate for status, occupations were also coded for horizontal classification. See Katz, *The People of Hamilton, Canada West*, pp. 51-52.

Hamilton with index of dissimilarity scores of .383 and .365 (see Tables
5 and 6). Professionals on the vertical scale accounted for 40.1 percent
of Westdale's household heads while only 14.1 per cent of Hamilton's
male labour force had a corresponding rank. Aside from rank, the
employment characteristics expressed in terms of economic sectors
reveal other vivid distinctions. If Hamilton was a lunch bucket city,
Westdale was a white-collar suburb. Manufacturing and construction
provided jobs for 57.1 per cent of Hamilton's male work force. In
Westdale, only 25.3 per cent of household heads were so employed,
while 57.0 per cent worked in trade and commerce, finance, and as
clerks and professionals (see Table 6). As a further measure of Westdale's social character, comparison can be made with a contemporary
land development near Hamilton's industrial district. Completed in the
early 1920s, Somerset Park had a 10 per cent professional component in
1931; 90 per cent of household heads were skilled or unskilled labourers,
most working in the metal and electrical industries (see Table 7).

Table 5: Vertical Classification of Occupation in Westdale and
Hamilton, *1931* and *1951*

| Rank | Hamilton | Westdale | Hamilton | Westdale |
|---|---|---|---|---|
| Elite professions | 1.6 | 4.5 | 1.6 | 5.1 |
| Professions and management | 12.5 | 35.6 | 15.3 | 48.3 |
| White collar — semi-skilled | 2.0 | 13.6 | 3.5 | 8.1 |
| Blue-collar– foreman level | 8.2 | 8.9 | 6.8 | 7.2 |
| Skilled labour | 33.1 | 26.4 | 39.8 | 24.9 |
| Semi-skilled | 17.0 | 9.1 | 18.2 | 4.7 |
| Unskilled | 25.6 | 1.9 | 14.8 | 1.7 |
| | 100.0 | 100.0 | 100.0 | 100.0 |

1931 Index of dissimilarity between Westdale and Hamilton = .383
1951 Index of dissimilarity = .429

SOURCES: The percentages for Hamilton were based on census returns for all
employed males; percentages for Westdale were based on occupation of household heads. Since census data did not combine work force distribution with
numbers of widows, widowers and pensioners, the later cases were excluded
from the Westdale computation.

Despite Westdale's contrasts with Hamilton proper, it displayed
internal variety within its Protestant boundaries. There were indentifia-
ble physical and social regions. We have noted how in the year immediately after 1918, the first portions of Westdale to be developed were

working-class surveys: Woodlawn and Elmhurst. In addition to a site bordered by a garbage dump, the main survey (Woodlawn) was near the City Isolation Hospital and a brickyard. As Kent Hamilton expressed it: "that brickyard. . . was like a boil on the thumb so far as our property was concerned."[50] House assessments in the area ranged from $1,900 to $2,500 in 1931 when the mean for the whole Westdale suburb was roughly $2,700.

The separate spatial and structural features of the workingman's areas were matched by social distinctions (see Table 8). Typical occupations of household heads included railway firemen and brakemen, mechanics, machinists, moulders, truck drivers, printers, bookkeepers, office clerks, and sales and shipping clerks. Though the United Church had the largest following of any denomination in Westdale, in the workingmen's sections the Church of England prevailed. This fact, along with occupational traits, gave these areas a social profile similar to the city. It is also worth noting that the daily routines of these families remained detached from the more affluent. Located on a grid at the

Table 6: Horizontal Classification of Employment in Hamilton and Westdale, *1931* and *1951*

|  | Hamilton | Westdale | Hamilton | Westdale |
|---|---|---|---|---|
| Agriculture, forestry and mining | 1.4 | 0.8 | 0.6 | 0.1 |
| Manufacturing | 47.4 | 19.9 | 46.8 | 20.4 |
| Utilities | 2.1 | 1.0 | 3.0 | 0.7 |
| Construction | 9.7 | 5.4 | 8.7 | 3.3 |
| Transportation/ Communication | 11.2 | 10.2 | 8.4 | 5.5 |
| Trade/Commerce | 11.3 | 17.9 | 6.2 | 33.5 |
| Finance | 1.6 | 12.0 | 1.4 | 6.5 |
| Clerical | 5.2 | 10.1 | 8.8 | 8.0 |
| Professional Service | 4.8 | 17.2 | 6.4 | 18.0 |
| Personal Service | 3.8 | 2.9 | 6.9 | 2.0 |
| Public Service | 1.2 | 2.3 | 2.4 | 1.9 |
| Recreational Service | 0.3 | 0.3 | 0.4 | 0.1 |
|  | 100.0 | 100.0 | 100.0 | 100.0 |

Index of dissimilarity 1931 = .365
Index of dissimilarity 1951 = .376

SOURCE: See the published census of Canada, 1931 and 1951 for the lists of occupations appearing under the classfication headings.

50. S.C.O., *in Bankruptcy*, Depositions of F. Kent Hamilton, p. 11.

Table 7: Vertical Classification of Occupation in Westdale and the Workingman's Neighbourhood of Somerset Park, 1931 and 1951

|  | 1931 | | 1951 | |
|  | Westdale | Somerset | Westdale | Somerset |
|---|---|---|---|---|
| Elite professional | 4.2 | 0.8 | 4.2 | 0.0 |
| Professionals and management | 33.2 | 6.3 | 39.8 | 5.3 |
| White-collar — semi skilled | 12.7 | 3.1 | 6.7 | 4.1 |
| Blue-collar — foreman level | 8.3 | 8.7 | 6.3 | 4.1 |
| Skilled labour | 24.7 | 44.2 | 20.7 | 33.0 |
| Semi-skilled | 8.5 | 17.3 | 3.9 | 14.9 |
| Unskilled | 1.8 | 10.2 | 1.4 | 26.4 |
| Widowed, widowers, retired | 6.6 | 9.4 | 17.0 | 12.2 |
|  | 100.0 | 100.0 | 100.0 | 100.0 |

1931 Index of dissimilarity = .399
1951 Index of dissimilarity = .434

SOURCE: Westdale and files and Anna Chiota, "Somerset Part, 1910-1960: A Quantitative Study of an Industrial Suburb" McMaster University, 1977 (Mimeographed).

eastern extremity of Westdale, the workingmen's homes were positioned so that residents did not have occasion to travel through abruptly different social areas to shop, reach places of employment, or attend school. Major thoroughfares, like King and Main, as well as important feeder streets, like Longwood and Stirling, spanned Westdale but did not erode patterns of segregation supported by the location of parks, business districts, and school property (see Map 3).

The most isolated and exclusive surveys – Oak Knoll, Oakwood and Forest Hills – clung to a narrow, ravine-indented fringe at the western extremity. Deep lots fronted on secluded curving avenues. Several humble avenues were elevated to crescent status as they entered the district. High minimum building values were written into the covenants, many of which inserted specifications for building materials and established broad architectural guidelines. Therefore, like the eastern sector, this neighbourhood evolved an architectural style but with a pronounced difference. Interpretations of English manors, stone as well as stucco and wood, dominated with some sharp variety provided by several examples of the austere international style. A tasteful variety here contrasted with what would become cheap eclecticism a half mile to the east in the workingmen's district. Building assessments reflected the differ-

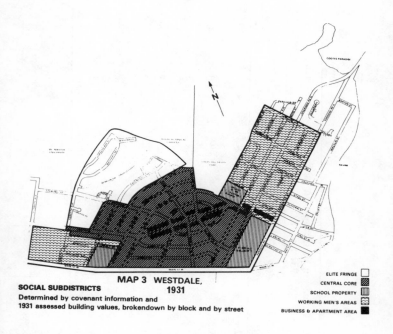

**MAP 3   WESTDALE, 1931**

**SOCIAL SUBDISTRICTS**
Determined by covenant information and
1931 assessed building values, brokendown by block and by street

ELITE FRINGE ☐
CENTRAL CORE ▨
SCHOOL PROPERTY ▥
WORKING MEN'S AREAS ▩
BUSINESS & APARTMENT AREA ■

ence, extending from $4,000 to $5,700 or more than twice that for the workingmen's homes.

Needless to say, the "old" wealthy families of Hamilton did not move into the exclusive portion of Westdale. Hamilton's elite avenues remained settled among the Victorian estates which had backed onto "Hamilton mountain" in the south central part of the city. By a process of "in-filling," spacious homes were constructed within an already prestigious area which had housed two Lieutenant-Governors of Ontario.[51] By contrast, Westdale's unfinished appearance and distinctly middle-class image, established by the developers, meant that the exclusive portion of the suburb attracted not old money, but affluent young professionals. At least 55 per cent of household heads had a professional ranking. Merchants, corporate executives, and senior educators were frequently encountered. United Church affiliation soared to 45 per cent as compared with 35 per cent for all Westdale and 21 per cent for Hamilton. In an industrial centre like Hamilton with its immigrants, labourers, aspir-

---

51. Terry Naylor, "Ravenscliffe: A Hamilton Elite District," McMaster University, 1978 (Mimeographed).

Table 8: Vertical Classification of Occupation in Westdale Subdistricts, 1931 and 1951

| | Working-class Area | | Central Core | | Elite Fringe | |
|---|---|---|---|---|---|---|
| | 1931 | 1951 | 1931 | 1951 | 1931 | 1951 |
| Elite professions | 2.4 | 1.9 | 4.2 | 4.8 | 5.8 | 7.0 |
| Professions and management | 17.5 | 30.9 | 19.0 | 36.6 | 54.6 | 56.7 |
| White-collar — semi-skilled | 5.4 | 6.9 | 16.9 | 8.4 | 14.4 | 16.5 |
| Blue-collar — foreman level | 10.2 | 8.3 | 12.0 | 5.0 | 7.9 | 4.9 |
| Skilled labour | 41.7 | 28.9 | 29.0 | 18.4 | 9.4 | 8.4 |
| Semi-skilled | 11.4 | 5.2 | 7.0 | 5.5 | 4.3 | 0.7 |
| Unskilled | 3.0 | 1.7 | 4.2 | 2.0 | 0.0 | 0.0 |
| Widows, widowers, pensioners | 8.4 | 16.2 | 7.7 | 19.3 | 3.5 | 5.8 |
| | 100.0 | 100.0 | 100.0 | 100.0 | 100.0 | 100.0 |

SOURCE: Assessment rolls.

ing middle class and old elite, notions of class had recognized boundaries. While the old elite generally shunned even the best sections of Westdale, these same areas must have seemed quite remote to residents of the working-class east-end of Hamilton. What they saw was inaccessibility and the affluence of the elite fringe. One experience, though not conclusive, is suggestive of an outsider's sense of turf, religion and class which, as an impressionable youth, seemed to threaten his life's goals.

At high school he had been considered presumptuous for wishing to become a teacher. It was pointed out that he had attended neither West-dale Collegiate nor Westdale United Church. The images of neighbour-hood, although they are not perfect assessments of society, have a crude accuracy. Wealthy and poor had carved out, by their feelings and actions, a position for Westdale's exlusive area in a hierarchy of Hamil-ton neighbourhoods.

The central blocks of Westdale were made up of surveys with moder-ate convenant restrictions on building values and unpretentious survey names: Crescentwood, Clinelands, and Paisley Gardens. The streets here did not consist of a simple grid pattern like the workingmen's areas, but neither were they protected from the noise and commotion of through traffic like the elite fringe. Geographically and figuratively, the centre was truly mid-way. Contractors responded most frequently to the developer's characterization of the property by constructing the 2½ sto-rey "square-plan" homes which had proven popular across Hamilton. Slight variations in the style of brick, porch, and window details did little to break the monotony or, as some preferred, the continuity. The centre evolved as something of a compromise between the simplicity apparent in workingmen's portions and the elegance arrayed along the fringe, though it inclined toward the former and included a number of the "Hamilton A-1 Plan" houses (see Illustration 1). Since Westdale builders responded to the prosperity of 1925-29 with frantic construction of middle-class dwellings, this central portion had a number of blocks completed by 1931. After the 1930s it continued to have more architectu-ral cohesion than any other neighbourhood in the development. For example, while Casey built the whole side of one block, his rival, Vickers, built a row behind him using similar materials, blueprints, and embellishments.

Occupational ranking paralleled the compromise structural features by indicating a middle-class mix of blue and white collar. Households headed by professionals accounted for 23 per cent of the total. In the skilled-labour tier, the portion was 29 per cent (see Table 8). The heart of Westdale housed white-collar professionals such as accountants, clergymen, teachers, retail managers, and commercial travellers. It also sheltered manufacturing foremen, railway conductors, station employees, and assorted clerks.

What of the tenants? Some 205 of the Westdale households rented. Approximately fifty resided in five apartment buildings;[52] these had been provided for in Pope's plan, and so as not to jar homeowner tastes, they were located within the business district. A dozen families rented apartments above commercial establishments. The bulk of Westdale's tenants were scattered throughout the suburb, to some extent reflecting

---

52. The apartment dwelling made its most dramatic showing in Hamilton in 1924 when 48 were constructed. Kevin Braybrook, "The First Apartment Dwellers in the City of Hamilton: A Social Profile," McMaster University, 1977 (Mimeographed).

Table 9: Occupational Rank, Age and Household Size: Homeowners, Home Tenants, Apartment Dwellers, 1931

| | *Rank by % of Household Heads* | | | |
| | *All Households* | *Homeowners* | *Home Tenants* | *Apartment Dwellers* |
|---|---|---|---|---|
| (Total cases) | (761) | (556) | (143) | (62) |
| Elite professions | 4.2 | 4.2 | 2.8 | 4.9 |
| Professions and management | 33.2 | 29.1 | 41.8 | 36.1 |
| White-collar — semi-skilled | 12.7 | 12.7 | 14.9 | 14.8 |
| Blue-collar — foreman level | 8.3 | 9.2 | 5.7 | 8.2 |
| Skilled labour | 24.7 | 27.3 | 20.6 | 18.0 |
| Semi-skilled | 8.5 | 8.3 | 7.8 | 8.2 |
| Unskilled | 1.8 | 2.0 | 1.4 | 1.6 |
| Widows, widowers, pensioners | 6.6 | 7.2 | 5.0 | 8.2 |
| | 100.0 | 100.0 | 100.0 | 100.0 |

| | *Age in Years of Household Head* | | | |
|---|---|---|---|---|
| Mean | 40.0 | 41.0 | 38.4 | 31.8 |
| Median | 38.7 | 40.0 | 37.3 | 29.8 |
| Mode | 32.0 | 42.0 | 37.0 | 28.0 |

| | *Household Size by Number* | | | |
|---|---|---|---|---|
| Mean | 3.6 | 3.7 | 3.7 | 2.4 |
| Median | 3.4 | 3.5 | 3.6 | 2.2 |
| Mode | 3.0 | 3.0 | 3.0 | 2.0 |

the contractors' overbuilding in the early months of the Depression. Four to five years younger than homeowners and with a comparatively small families, the heads of tenant households actually had a slightly higher occupational ranking than homeowners. Generally, Westdale's tenants appear to have been upwardly mobile, small, young families who took advantage of the abundant good rental accommodations. Dividing tenants into apartment dwellers and tenants renting homes does not alter this impression of upwardly mobile residents. What did separate those who rented homes from those who rented apartments was their stage in the life cycle. As is shown in Table 9, the residents of apartments were considerably younger and had far smaller households than home tenants. Young couples and singles, quite naturally, could accept the limited space and lack of yard inherent in an apartment situation.

Family and household in Westdale were virtually the same. As a middle-class suburb with certain social attitudes, the practice of taking in boarders appears to have been shunned. What American researchers have observed for the end of the nineteenth century seems applicable in Westdale: "family boarding had lost. . . its middle class respectability."[53] Assessment rolls and city directories, despite their limitations as accurate sources for such details, indicate very little boarding and "doubling up" of families. The same sources, however, do indicate that nearly 10 per cent of Somerset Park's homes contained two families in 1931. The added income or economy, therefore, do not appear to have been important considerations in Westdale. The records revealed only half a dozen households which contained either in-laws or boarders. A few households headed by a bachelor or spinster included a relative or friend. Though the records did not list or describe all of household arrangements in suitable detail, it is reasonable to conclude that in household composition, as in occupation and religion, Westdale's deviation from homogeneity and middle-class standards registered in small degrees, if at all.

THE COMMUNITY, 1951

By definition, suburbs stand apart from the city, but eventually lines of distinction fade with land-use transformations and the overlapping of new surveys. However, twenty years after its first period of growth, Westdale retained a separate character. The Chedoke ravine and the McKittrick bridge had preserved the suburb against assimilation by preventing the extension of Hamilton's grid sheet system. The unique street layout of Westdale itself countered fusion with the new post-war suburbs around Westdale. All the same, the community was not quite what it had been. Age, occupational composition, and denominational features had altered, revealing new social forces at work in Hamilton and across Canada.

The national housing boom that swept Westdale into its second period of construction, eliminating the vacant tracts, came partly through government-stimulated activity. Efforts to encourage residential building began with the Dominion Housing Act of 1935. When

53. For a survey of the boarding phenomenon see John Modell and Tamara K. Hareven, "Urbanization and the Malleable Household: An Examination of Boarding and Lodging in American Families," in *Family and Kin in Urban Communities, 1700-1930*, ed. Tamara K. Harevan (New York: New Viewpoint, 1977, 164-83; Dominion Bureau of Statistics, *Census Monograph No. 7, The Canadian Family (A Study Based on the Census of 1931 and Supplementary Data)*, pp. 71-72; Chiota, "Somerset Park," Table 24, p. 26.

revised as the National Housing Act of 1938, fewer than 5,000 new units across Canada had benefited from the federal loans. War curtailed operations, but not the involvement of Ottawa in shelter matters. A Wartime Housing Corporation created by Order in Council concentrated on the provision of temporary quarters for immigrants coming into the areas of war industry activity. The affected urban land was not of prime suburban quality. In Hamilton, the areas were scattered. The site nearest to Westdale was over one mile away. In the planning for postwar reconstruction, a revised National Housing Act of 1944 prepared the way for accelerated suburban lending. Frequently reorganized and expanded, the program joined postwar prosperity in providing a sustained house-building effort that produced 500,000 dwelling units across Canada from the beginning of 1945 to the end of 1951. Abstract books in the Land Registry Office indicate considerable movement in Westdale real estate from 1945 to 1951. During 1951, NHA loans were approved for 800 housing units in Hamilton.[54] Their presence in Westdale was reflected in the fact that many mortgages were held by some of the thirty government-approved institutional lenders and "His Majesty the King, by Minister of Finance." By 1951, home ownership in the nation, Hamilton, and Westdale had risen substanially above 1931 levels (see Table 2). The incomplete Westdale of depression and wartime rapidly filled with NHA homeowners.

The 1931 households had had relatively young heads, but in 1951 the average age of households heads tended to be ten years above that of 1931. The initial families had aged and new homeowners had been delayed in their entry into the housing market by depression as well as by the dislocations of life and the construction shortages caused by war. Most striking as a measure of maturation, households headed by pensioners, widowers, and widows increased from 6.6. per cent to 16.3 per cent. The major component consisted of widows of whom there were 150 or 8.5 per cent of household heads. These widows were not impoverished. Of 196 houses rented in Westdale, 72 (36.7 per cent) were owned by widows. Seventeen widows owned two dwellings each, living in one while letting the other. Elizabeth Groh illustrated the condition. During the 1920s, her carpenter husband had built and let Westdale homes. In 1951, his widow let two dwellings while residing in a third. Not all of the property-owning widows had been Westdale residents like Mrs. Groh. Twenty appear not to have been former Westdale residents. Some might have purchased homes for rental income and future capital gains. Several appear to have been holding titles to homes in which married sons resided. Widowhood, being more apparent on older streets, was not

54. Canada, Central Mortgage and Housing Corporation, *Housing in Canada, A Factual Summary*, Vol. 6 (Fourth quarter, 1951), p. 11; Canada, Central Mortgage and Housing Corporation, *Annual Report to Minister for Resources and Development for the Year 1951*, p. 51.

evenly distributed across the community. Along one of these early streets, South Oval, 24 out of 66 homes were owned by widows.

Unfortunately, cross-sectional analysis prevents a full understanding of widowhood, so it is difficult to estimate how many Westdale wives were widowed between 1931 and 1951. It was probably greater than the 150 found residing there in 1951 since some might have moved away. On the other hand, few widows had moved into Westdale. By tracing through City Directories, it was established that all but 14 of the 150 had resided there before the death of their husbands. With an average age of 59.6, it is unlikely that resident widows had lost husbands in the war. Other explanations must be sought for a 5 to 1 ratio of widows to widowers. The answer seems to be the fact that for our time period, across Ontario, male deaths out numbered female deaths in all age cohorts from birth to 70. The greatest differential appeared in the 50 to 59 age cohort. Significantly, the ratio of male to female heart disease victims in the age group 40 to 59 was 3 to 1.[55] Given the age composition of Westdale household heads in 1931, and of widows in 1951, it seems plausible that the suburb had had a number of male heart-attack victims throughout the 1940s and early 1950s.

In comparison with 1931, there had been a slight upward trend in the occupational ranking of Westdale's household heads by 1951 (see Table 5). Very little of this can be attributed to the occupational gains of the older residents. Of the 556 homeowners in 1931, 190 or 34 per cent could be located at the same address in 1951. Many had retired. Of the remaining 143 still in the labour force, it can be said that as a group they had realized a few advances in occupational rank with some rising from skilled labour into management (see Table 10). More essential to the higher occupational tone of Westdale were 900 new households. Most were beneficiaries of wartime prosperity and the government housing measures. The largest single increase in occupation involved retail merchants. Though there were 33 in 1931, there were nearly 100 more in 1951. As Westdale's occupational profile was enhanced, the measure of its dissimilarity with the city and with the east-end workingman's survey, Somerset Park, broadened moderately (see Tables 5 and 7).

The newcomers who effected the shift in occupational distributions also altered denominational traits (see Table 4). A Jewish migration into Westdale had begun in 1944, the year in which Ontario passed its Racial Discrimination Act. Although it required considerable litigation by the Canadian Jewish Congress and civil liberties activists, the racial ingre-

---

55. Ontario Department of Health, *Annual Reports*. See "Chief Causes of Death by Age and Sex" for 1948, 1949, 1950, 1951. This observation is supported by Robert D. Rutherford, *The Changing Sex Differential in Mortality* (Westport, Connecticut: Greenwood Press, 1975).

Table 10: Occupational Ranking of Employed Household Heads Who "Persisted" for Twenty Years*

|  | Persisters in 1931 | Persisters in 1951 |
|---|---|---|
| Elite professions | 6.3 | 6.3 |
| Professions and management | 28.7 | 41.2 |
| White-collar — semi-skilled | 9.0 | 9.1 |
| Blue-collar, foreman level | 7.6 | 9.1 |
| Skilled labour | 34.6 | 24.5 |
| Semi-skilled | 10.2 | 6.3 |
| Unskilled | 3.6 | 3.5 |
|  | 100.0 | 100.0 |

* The address, religion, name and age provided a sure linkage. In no case did the age cited in 1931 vary by more than two years from the twenty year differential anticipated in 1951.

dients in restrictive covenants were finally eliminated by 1951.[56] Interestingly, the Jewish population in Westdale concentrated on four of the suburb's thirty streets. Indeed, along one block of Bond Street, sixteen of thirty household heads were Jewish retailers. All four streets had been underdeveloped prior to the mid-1940s. As lots here had come onto the market, there had been a convergence of events: the challenge to covenants, a housing boom, and the fact that the Jewish community, newly affluent, began to move to the west-end seeking middle-class amenities.[57] Along with United Church adherents, Jews were now over-represented in Westdale when compared with their proportion of the city-wide population. Roman Catholics alone remained under-represented (See Table 4) and churchless.

Westdale's completion as a prime and distinctive development was fully realized by 1951, but as a large district whose formative planning in the 1920s had allowed for workers dwellings, it retained elements of the social mix present in 1931. Overall, however, the proportion of professionals and managers rose while the proportion of skilled and semi-skilled labourers declined. The greatest shifts appeared in the central core and workingmen's surveys where Westdale's image of convenience and prestige dispelled the compunctions of the 1920s about raw and unattractive lots in such surveys as Woodlawn and Princess Heights. Measured against the new suburbs laid out in Stoney Creek, Hamilton mountain, and on land to the west of Westdale, the district had appeal.

56. Alan Burnside Harvey, ed., *The Ontario Reports, Cases Determined in the Supreme Court of Ontario, 1945* (Toronto: Carswell, 1945), pp. 778-80; *Canada Law Reports: The Supreme and Exchequer Courts of Canada*, Part I (Ottawa: King's Printer, 1915), pp. 64-80.
57. Louis Greenspan, "The Governance of the Jewish Community of Hamilton," (Paper prepared for the Center for Jewish Community Studies, 1974) p. 6.

In contrast to rising status among homeowners, rental accomodations now housed a population with a lower occupational profile than in 1931. This condition most likely reflected the quite different prospects faced by young families in 1931 and 1951. In the depressed housing market of 1930 and 1931, a few young upwardly mobile families had found abudant rental bargains until financial circumstances brightened. Their 1951 counterparts seeking shelter had the advantages of economic boom and government incentives for home ownership, points conveyed in the rise of city-wide ownership from 48 per cent (1931) to 65 per cent (1951). The average age of 1951 tenants was still lower than that of homeowning neighbours. However, while small young families continued to rent apartments and homes in Westdale, a group composed of the retired and the widowed accounted for one tenant in five. The ratio was somewhat higher among apartment dwellers (see Table 11). In sum, the social attributes of Westdale in 1951 had come to reflect not only the developers' activites of the 1920s, but now they also bore the impression of aging and the significant domestic ramifications of the war.

\*     \*     \*

This cross-sectional analysis of a community has spanned two decades, over depression, recovery, and boom. Conclusions drawn from this admittedly limited process suggest a number of hypotheses pertaining to city-building and to urban society in the twentieth century. Land value was a complex function of natural settings, provision of services, and entrepreneurial activity. These gave tone to the property and that enhanced its value. To create a scarce commodity was not a policy on the part of the developers as demand remained below expectations for nearly thirty years. The creation of a suburban community was neither smooth nor sudden. It required forty years as well as a variety of individual skills and government aid. Morevoer, Westdale's history testifies to the importance of examining local events in the light of national and, at times, international forces.[58] The planning and promotional tactics of developers drew on concepts that were not strictly local. Builders followed practices in contruction, sub-contracting, and credit that has wide North American appreciation.[59] The local building cycles in Westdale had no independent significance. They coincided precisely with national trends in residential construction. In another important matter, racial discrimination, the Westdale or Hamilton experience did not differ from the North American norm.

58. Gilbert A. Stelter, "A Sense of Time and Place: The Historian's Approach to Canada's Urban Past," *The Canadian City: Essays in Urban History*, ed. Stelter and Alan F.J. Artibise (Toronto: McClelland and Stewart, 1977), p. 435.
59. Sam Bass Warner, Jr., *Streetcar Suburbs, The Process of Growth in Boston, 1870-1900* (Cambridge: Howard University Press and the M.I.T. Press, 1969), pp. 117-52.

Table 11: Occupational Rank, Age and Household Size: Homeowners,
Home Tenants, Apartment Dwellers, 1951

|  | All Households | Homeowners | Home Tenants | Apartment Dwellers |
|---|---|---|---|---|
| (Total cases) | (1,709) | (1,408) | (196) | (105) |
| Elite professions | 4.2 | 4.5 | 2.6 | 3.8 |
| Professions and management | 39.8 | 41.6 | 34.3 | 26.6 |
| White-collar — semi-skilled | 6.7 | 7.0 | 6.2 | 4.8 |
| Blue-collar — foreman level | 6.3 | 6.3 | 5.1 | 7.6 |
| Skilled labour | 20.7 | 19.3 | 25.1 | 30.5 |
| Semi-skilled | 3.9 | 3.6 | 6.7 | 1.9 |
| Unskilled | 1.4 | 1.4 | 1.5 | 1.9 |
| Widows, widowers, pensioners | 17.0 | 16.3 | 18.5 | 22.9 |
|  | 100.0 | 100.0 | 100.0 | 100.0 |

Age in Years of Household Head

|  | All Households | Homeowners | Home Tenants | Apartment Dwellers |
|---|---|---|---|---|
| Mean | 50.4 | 51.4 | 45.1 | 47.5 |
| Median | 50.4 | 51.3 | 42.0 | 44.4 |
| Mode | 52.0 | 52.0 | 33.0 | 40.0 |

Household Size by Number

|  | All Households | Homeowners | Home Tenants | Apartment Dwellers |
|---|---|---|---|---|
| Mean | 3.3 | 3.4 | 3.2 | 2.4 |
| Median | 3.2 | 3.2 | 3.0 | 2.2 |
| Mode | 2.0 | 2.0 | 2.0 | 2.0 |

Peter Goheen has observed that "the element of residential choice which was introduced into the city with the. . . . street railway apparently served to rationalize the distribution of families within the city."[60] Westdale supports the generalization. The McKittrick bridge and a streetcar line – at one time operated by some of the same businessmen who had taken up an interest in the land syndicate – made possible this essentially middle-class suburb. And yet, while Westdale demonstrated the "rationalize[d] distribution of families within the city," it embraced a measure of diversity and experienced some modest social transformations. The developers of Westdale originally had attempted to build exclusiveness into their community, but they also had to accede to

60. Goheen, *Victorian Toronto, 1850 to 1900*, p. 200.

market forces, government inducements, and legal decisions. The dynamic tension produced when a protestant, middle-class design was challenged eventually generated a community where managers and truck drivers, Christians and Jews, resided. Admittedly, they did not dwell in close proximity. Over time, the suburb had matured, progressing from a raw incomplete community with young families and a degree of ethnic isolation into a tree-shaded area familiar with aging and some religious diversity. However unique it might have been in its scale and physical layout, Westdale was so much a part of broad economic and social events that it can serve as one point of departure for comprehending the mundane processes shaping the contemporary Canadian city. Aside from what it demonstrates about the complexity of making a suburb, Westdale offers some reflections on North American social change. More mature than their 1931 counterparts, the post-war newcomers were to reside in a finished suburb during prosperous years. Thus Westdale became a place where people could aspire to an affluence almost within reach. As well as greater economic security, the arrival of Jews heralded national shifts in which the urban middle class of Canada would become different in outlook and background from its depression counterpart.

# Politics, Space, and Trolleys: Mass Transit in Early Twentieth-Century Toronto

MICHAEL J. DOUCET

During the last half of the nineteenth century several developments in the field of urban mass transit helped to alter both the spatial structure and the way of life in cities in Europe and North America. The most important and widespread of these innovations was the electric streetcar which evolved during the 1880s. More efficient and cheaper to operate than its predecessors – omnibuses, stage coaches, and horsecars – the electric streetcar meant lower fares and a greatly expanded ridership in most places where it was introduced. Moreover, the greater speed of the trolley (roughly twice that of earlier vehicles) increased the opportunities both for more distant suburban residences and for leisurely holiday and Sunday excursions to outlying recreational facilities. Wherever it was adopted, the electric streetcar fostered urban physical expansion. On a more humane level the trolley broadened the awareness space or zone of familiarity of many city dwellers. Yet, if the advantages of electric traction were virtually universal, attitudes towards the operation and control of this important utility were quite different throughout the western world. As John McKay has noted, in France and Germany street railways were in private hands but were regulated by strict government controls relating to levels of service, fare structure, and the expansion of facilities. Moreover, in most instances the ownership of the utility reverted to the municipality upon the expiration of the franchise. In Great Britain, on the other hand, municipal ownership was the norm from a very early stage. Finally, in the United States street railways were operated by private entrepreneurs under essentially laissez faire condi-

SOURCE: This article is substantially revised from that published in the *Urban History Review*, no. 3-77 (February 1978), pp. 3-33, entitled "Mass Transit and the Failure of Private Ownership: The Case of Toronto." Reprinted by permission of the author and the editors of the *Urban History Review*.

tions, with system expansion frequently linked to real estate speculation. This lack of regulation often proved to be particularly vexatious upon the expiration of the private franchise, since it was seldom clear how the municipality could gain control of the utility or how much compensation the private entrepreneur should receive.[1]

The history of mass transit operations in Toronto reveals a rather interesting intermixture of the attitudes towards traction ownership that had been evolving elsewhere. Nor did attitudes remain constant over time, and the shifting perceptions of the role of the public sector in supplying mass transportation in the Ontario capital provide a major focus for this paper. In 1861, for example, when the first horse-drawn streetcars, or horsecars, were introduced to the city, Toronto followed the example set in U.S. cities. The horsecars were seen not as a public utility but as a novelty, and the rights to their operation were handed over to private entrepreneurs in the form of a thirty-year monopolistic franchise. The city did regulate the fare structure (five cents) and some aspects of the operation of the horsecars (for example, hours of service) and it did tax the company's revenues, but in typical North American fashion the question of what was to happen when the franchise expired was not clearly spelled out. Indeed, there was considerable bickering over the cash value of the utility as the first franchise drew to a close. By 1891 mass transit had become an essential part of life to many Torontonians. Some even advocated municipal ownership of the utility, and the horsecars were operated by the municipality while the civic leaders deliberated upon a course of action. Public ownership, however, proved to be too radical a step for most of Toronto's leading politicians and businessmen in 1891. In spite of the poor service provided by the private company during the last years of its franchise, most civic leaders still felt that transit had to be operated on strict business principles.[2] Torontoni-

1. John P. McKay, *Tramways and Trolleys: The Rise of Urban Mass Transport in Europe* (Princeton, 1976). Other important studies include G. Smerk, "The Streetcar: Shaper of American Cities, *Traffic Quarterly*, XXI (1967), 569-84; G.R. Taylor, "The Beginnings of Mass Transit in Urban America," in J.F. Richardson, ed., *The American City: Historical Studies* (Waltham, 1972), pp. 125-57; Donald F. Davis, "Mass Transit and Private Ownership: An Alternative Perspective on the Case of Toronto," *Urban History Review*, No. 3-78 (1979), 60-98; and Charles W. Cheape, *Moving the Masses: Urban Public Transit in New York, Boston, and Philadelphia, 1880-1912* (Cambridge, Mass., 1980).

2. See Christopher Armstrong and H.V. Nelles, "The Un-Bluing of Toronto and the Revenge of the Methodist Bicycle Company: the Fight over Sunday Street Cars, 1891-1898," paper presented to the annual meeting of the Canadian Historical Association, Kingston, 1973. This article has been expanded recently into a book, *The Revenge of the Methodist Bicycle Company: Sunday Streetcars and Municipal Reform in Toronto, 1888-1897* (Toronto, 1977). For other discussions of the early transit history of Toronto, see Harold Moras, "A Study of the Development of the Public Transit System in the City of Toronto from 1861 to 1921" (B.A. thesis, University of Toronto, 1970) and Louis H. Pursley, *Street Railways of Toronto, 1861-1921, Interurbans*, special volume No.25 (Los Angeles, 1958).

ans still had to learn about the drawbacks of private ownership through experience. Another three decades under this system was a more than sufficient period for the education of the riding public.

The second street railway franchise, also for thirty years, was let to William Mackenzie and his associates in the Toronto Railway Company (TRC). Mackenzie was probably the greatest Canadian steam and electric railway magnate of his time, with interests that included the vast Canadian Northern Railway (later to become the Canadian National Railway) and other rail lines, along with street railway and electric power companies in such diverse places as Nova Scotia, Manitoba, Alberta, Mexico, and Brazil. In the TRC, as in most of his ventures, Mackenzie was joined by contractor Donald Mann.[3] A third important TRC official was Robert John Fleming, a former mayor of Toronto, who joined the firm in 1905 as general manager.[4] It was often the "astute and affable" R.J., overseer of the day-to-day operations of the TRC, who bore the brunt of the criticism levelled against the company.[5]

According to a recent analysis by Christopher Armstrong and H.V. Nelles, the primary reason for the granting of the second franchise was economic.[6] Thirty years of service by Toronto's first private operator, the Toronto Street Railway Company, had left the city with a system that was badly in need of both modernization and expansion. More specifically, to keep up with the latest developments in North American street railway technology, the Toronto utility had to be electrified.[7] This meant a large outlay of capital for new equipment which most felt would have been well beyond the means of the municipality. It, therefore, called for tenders for a new franchise and made the responsibilities for refurbishing the street railway a part of the agreement between the City and the successful applicant.[8]

3. *An Encyclopaedia of Canadian Biography*, II (Montreal and Toronto, 1905), 21-22; John F. Due, *The Intercity electric railway Industry in Canada* (Toronto, 1966), pp. 14, 82-87 and Pursley, *Street Railways*, p. 153.

4. J.E. Middleton, *The Municipality of Toronto. A History*, 3 vols. (Toronto, 1923), II, 54.

5. This description is found in Pursley, *Street Railways*, p. 140.

6. For a discussion of the reasons behind the rejection of public ownership at this time, see Armstrong and Nelles, "The Un-Bluing of Toronto," pp. 28-31.

7. In spite of the fact that the first North American commercial electric railway began as an annual experiment at the Canadian National Exhibition in 1883 (see Mike Filey, Richard Howard and Helmut Weyerstrahs, *Passengers Must Not Ride on Fenders* [Toronto 1974], p. 20), Toronto lagged behind many North American cities in the adoption of the new power source. For details of this, see W.G. Ross, "Development of Street Railways in Canada," *The Canadian Magazine*, XVIII (January 1902), 276 and Harold C. Passer, *The Electrical Manufacturers 1875-1900* (Cambridge, Mass., 1953), pp. 216-55. On European developments, see McKay, *Tramways and Trolleys*, pp. 35 - 83.

8. See *The Charter of the Toronto Railway Company Together with Subsequent Statutes, Agreements and Judgments Relating to the Said Company and the Corporation of the City of Toronto From April 14th 1892 to December 8th 1905* (Toronto, 1906) and Statutes of Ontario, 1892. The legislation establishing the TRC was 55 Vic., c. 99.

For a time, the Toronto Railway Company made a conscientious effort to serve the citizens of Toronto. The car lines were completely electrified by 1894, and, between 1892 and 1910, the number of miles of single track increased by about 74 per cent, from 68.5 miles to just under 120 miles. Even though the population of the city had increased by about 102 per cent during this period, the growth of the street railway had been encouraging (Table 1). Furthermore, the fleet of electric street cars operated by the company had increased dramatically and at a rate that far outstripped population growth. In 1894 there had been 0.61 electric street cars for every one thousand Torontonians; by 1910, this figure had more than tripled to 1.91 per thousand of population.

The TRC did not go unrewarded for its early efforts. Passengers increased by 473 per cent between 1892 and 1910 and the number of trolley trips per capita grew from 113 in 1891 to 320 in 1910, a figure that far outdistanced patronage in European and most U.S. cities at that date.[9] Even more important from the private company's view, net earnings had grown by a hefty 830 per cent. By 1910, the net annual earnings of the TRC had passed the $2 million mark for the first time and the company was able to pay out more than $0.5 million in dividends to its shareholders. Two important trends emerge from these figures; namely, Torontonians were becoming increasingly dependent upon the street railway and the TRC had become an extremely lucrative property.

After 1910 the expansion of Toronto's street railway system was hardly spectacular. Indeed, in the interval between 1910 and 1920 only about 20 miles of single track were added to the network, a 17 per cent increase; and of these, 8.7 miles were actually constructed by the City. The TRC, in fact, laid no new trackage of its own after 1915. Population increased by 50 per cent in Toronto between 1910 and 1920; yet, in spite of the fact that public dependence on streetcars had increased to the point where every man, woman, and child in the city took an average of almost 385 trips per annum on the system, the ratio of electric cars to population had declined to 1.46 per thousand by the latter year. The fleet of streetcars itself remained stationary in size after 1911, with virtually all additions being offset by losses of 200 cars due to fires in the King Street car barns in 1912 and again in 1916. By the end of the TRC's franchise each of its streetcars was transporting an average of 220,000 passengers per year, double the load carried per car in 1900. On the other hand, in spite of a deficit of $420,000 in its last full year of operation, the net earnings of the TRC for the decade from 1911 to 1920, $21.02 million, actually exceeded those for the first 19 years of the franchise, $18.28 million, by almost 15 per cent. But if the company surpassed the expectations of its shareholders, it did not keep pace with the growth of the city. By the latter years of the franchise the extent of

---

9. McKay, *Tramways and Trolleys*, pp. 194-98. Only New York, with 330 trips per capita, was ahead of Toronto in 1910.

Table 1: Average Financial Statistics, Trackage, and Rolling Stock
Toronto Railway Company, 1892-1920

| Period | Pop. (000's) | Passengers (millions) | Gross earnings | Net earnings | Payments to city | Dividends paid | Miles of single track | Streetcars in service | Passengers per car (millions) |
|---|---|---|---|---|---|---|---|---|
| | | | in millions of dollars | | | | | | |
| 1892-1895 | 172 | 21.6 | 0.92 | 0.38 | n.a. | n.a. | 86.1 | 96 | 0.23 |
| 1896-1900 | 188 | 29.1 | 1.22 | 0.62 | n.a. | n.a. | 92.2 | 263 | 0.11 |
| 1901-1905 | 220 | 53.1 | 2.17 | 0.96 | 0.3 | 0.35 | 96.8 | 475 | 0.11 |
| 1906-1910 | 296 | 91.8 | 3.71 | 1.77 | 0.56 | 0.52 | 109.8 | 614 | 0.15 |
| 1911-1915 | 434 | 140.6 | 5.63 | 2.55 | 1.01 | 0.86 | 133.0 | 764 | 0.18 |
| 1916-1920 | 487 | 170.8 | 6.79 | 1.65 | 1.31 | 0.48 | 139.0 | 767 | 0.22 |

SOURCES: Calculated from figures in Louis H. Pursley, *Street Railways of Toronto, 1861-1921: Interurbans*, Special Vol. No. 25 (Los Angeles, 1958), 4 and 144; *The Railway and Shipping World*, II (January 1899), 16; and the Toronto Railway Company, *Annual Reports* (1905, 1906, 1908-1920).

Table 2: Per Capita Electric Railway Statistics
for Selected Canadian Cities, 1915

| City | Miles of track/ 000 Pop. | Passengers | Gross revenue | Net revenue |
|------|------|------|------|------|
| Calgary | 1.25 | 322 | $14.0 | $1.9 |
| Edmonton | 1.25 | 382 | 18.8 | -5.7 |
| Halifax | 0.26 | 157 | 10.8 | 5.1 |
| Hamilton | 0.30 | 172 | 6.8 | 1.1 |
| London | 0.56 | 225 | 8.3 | 1.7 |
| Montreal | 0.26 | 326 | 13.9 | 3.1 |
| Ottawa | 0.29 | 291 | 12.3 | 3.5 |
| Quebec | 0.25 | 143 | 6.1 | 2.4 |
| Regina | 1.02 | 130 | 6.2 | -3.4 |
| TORONTO | 0.17 | 389 | 15.6 | 3.0 |
| Winnipeg[1] | 0.80 | 394 | 15.9 | 6.5 |
| Vancouver[1] | 2.42 | 461 | 46.1 | 10.5 |

SOURCE: Calculated from figures in *Canadian Railway and Marine World*, XIX (June 1916), 241. Population figures based on the *Fifth Census of Canada*, I (Ottawa, 1912), 535-45.

Note: 1. Figures include extensive radial railway operations, serving outlying areas, as well as intra-urban street railway facilities. Per capita figures for these cities are, therefore, inflated because the population figures for the catchment areas of the inter-urban lines are not included in the census figures for the respective cities.

the TRC's streetcar system, as measured by miles of track per thousand of population, did not compare favourably with the pattern in other Canadian cities (Table 2).

One reason for this stagnation was the relationship between the TRC and the City, which was poor almost from the beginning of the franchise, a pattern that McKay found typical in most U.S. cities and in British cities prior to public ownership.[10] The granting of the second Toronto franchise had itself been shrouded in controversy and from a very early date, the TRC, like many other North American transit firms, had attempted to have the contract altered to suit its needs and, generally, to increase profits, usually at the expense of service quality.[11] Legal battles between the City and the TRC were numerous, with individual cases often being highly protracted affairs. These only served to inhibit

10. *Ibid.*, pp. 91-95.
11. Armstrong and Nelles, "The Un-Bluing of Toronto," p. 4.

the expansion of the system. One such case concerned the passenger loads being carried by the cars of the TRC. As early as 1899, the City was before the courts to accuse the street railway operators of over-crowding their cars.[12] With the creation by the province of the Ontario Railway and Municipal Borad (ORMB) in May of 1906 to regulate the operation of steam and electric railways and to hear appeals concerning municipal affairs, the operation of public utility companies, and labour conflicts, a permanent battle ground was established for all such disputes.[13] It came to be well utilized by both parties.

The City quickly raised the overcrowding issue before the newly created ORMB. In May of 1907 the board ruled that the TRC had indeed overcrowded its streetcars. The solution to the problem, board members felt, lay both in the provision of 100 new cars and the construction of 10 to 15 miles of new lines to relieve congestion in the downtown area, where many of the TRC's lines converged.[14] Further-more, the ORMB judgement admonished the combatants to be more wary of the public's interest:

> The City and the Company are co-partners in the enterprise of giving rapid transit to the people. The City contributes the right-of-way and the Company provides the capital, plant and labor. Both profit by the business and both owe a duty to the public in return for what the people pay. There is every reason why partners should agree rather than be at cross purposes.[15]

Unfortunately, these words had little impact on either party.

It took a number of years for the ORMB's orders to be carried out. The issue of track extensions proved to be particularly complex. A 1904 TRC policy had confined all track extensions made after that date to the boundaries of the City of Toronto as they had existed in 1891, when the franchise had been granted. To combat this and to assert its control over the streets the City, which had brought the problem of inadequate service before the ORMB, paradoxically attempted to prevent the company from making any extensions at all. On the instructions of the City Council, the city engineer refused to approve a series of new lines proposed by the TRC for the downtown area during the first half of 1908. Once again the so-called "transit partners" came before the ORMB. In December of 1908, the board ruled that the street railway company had the right to select the streets upon which new lines were to be constructed and that, by trying to prevent the TRC from making its desired extensions, the City was in breach of the contract between the two

---

12. "Toronto Railway Litigation," *Railway and Shipping World*, II (March 1899), 92.

13. S.J. McLean, "Ontario Railway and Municipal Board," *Railway and Marine World*, IX (September 1906), 505 and 507.

14. *Appeal Book in the Matter of an Application before the Ontario Railway and Municipal Board between the Toronto Railway Company and the Corporation of Toronto* (Toronto, 1909), p. 70, Ontario Archives.

15. *Ibid.*, p. 69.

parties.[16] Undaunted, the City appealed to both the Supreme Court of Canada and to the Judicial Committee of the Privy Council in London, without success. The decision of the latter body, rendered in September of 1909, was unequivocal: the City had neither the right to prevent the TRC from making extensions within the 1891 city limits, nor the right to force the company to make them in the outlying districts.[17] Years had been wasted in this legal battle and, as a result, it was not until 1912 that the TRC was able to complete its desired downtown improvements.[18] Delays of this nature were to become increasingly characteristic as the years passed.

None of the judicial decisions of this period did anything to provide improved transit facilities for the residents of Toronto's burgeoning suburbs. Since the City had annexed a number of outlying districts after 1891 on the understanding that street railway service would be extended into these areas within a reasonable time, the City found itself in a real quandary (Figure 1).[19] People in the suburbs wanted a direct link with the city centre, but the TRC would not comply. The short-term solution for the City was to build its own car lines (the Toronto Civic Railway) into some of the recently annexed areas. The first of these lines was begun in 1911, and by 1915 some 17 miles of civic double–track had been laid along thoroughfares such as St. Clair, the Danforth, Coxwell, Gerrard, and Kingston Road (Figure 2). While some areas were served for the first time by this civic venture, it was merely a stop-gap measure since the city's car lines were neither integrated with the much more extensive TRC network (separate fares were charged on each system) nor, in most instances, with each other. Only the TRC cars could get people into and out of the downtown area. In parts of the suburbs some sort of transit service was provided by two so-called inter-urban radial railway companies: the Toronto and York Railway which operated three widely separated lines along the Lakeshore, Kingston Road, and Yonge Street (the latter was known as the Metropolitan Line and went as far north as Sutton on Lake Simcoe); and the Toronto Suburban Railway which operated a series of interconnected lines in the west end along Davenport, Dundas, and Weston Road (the latter portion extended as far as Guelph by 1917).[20] Both radial companies had been

16. *Ibid.*, pp. 106-7.

17. *Report of the Proceedings in the Privy Council.* "On Appeal from the Court of Appeal for Ontario in the Matter of an Application before the Ontario Railway and Municipal Board between the Corporationof the City of Toronto and the Toronto Railway Company" (Toronto, 1910).

18. Pursley, *Street Railways*, p. 28.

19. Figure 1 is reproduced, with permission, from J. Lemon and J. Simmons, "A Guide to Data on 19th Century Toronto," Dept. of Geography, University of Toronto, n.d., mimeo.

20. For a discussion of the operations of these radials see Due, *The Intercity*, pp. 82-87; James V. Salmon, *Rails from the Junction* (Toronto, n.d.) and Ted Wickson, "The Radial Railways on North Yonge Street," *Upper Canada Railway Society Newsletter*, No. 326 (March/April, 1973), 44-58.

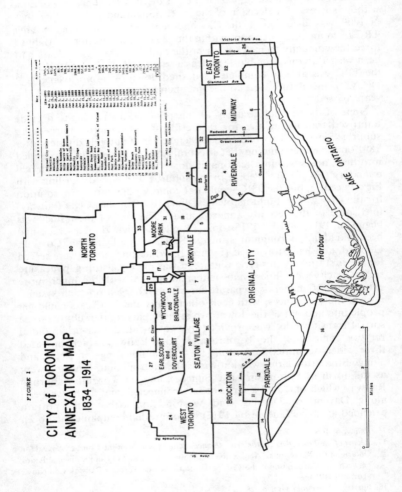

FIGURE 1

CITY of TORONTO
ANNEXATION MAP
1834-1914

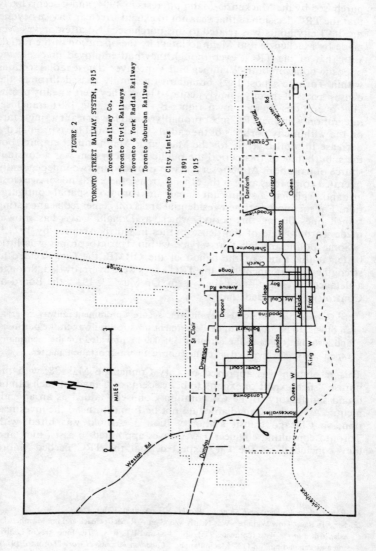

FIGURE 2

TORONTO STREET RAILWAY SYSTEM, 1915

— Toronto Railway Co.
—·—· Toronto Civic Railways
········ Toronto & York Radial Railway
—— Toronto Suburban Railway

Toronto City Limits
———— 1891
········ 1915

purchased by the Mackenzie-Mann interests in 1904, and it seems likely that the TRC decision of that year not to extend streetcar service beyond the 1891 city limits was related to this purchase.[21] No attempt was ever made by Mackenzie and Mann to integrate the operation of the radials with that of the TRC,, even though they had employed precisely the opposite policy in both Winnipeg and Vancouver. Those radial sections within Toronto's post-1891 boundaries generally passed through the densest parts of the city's newly acquired land. They were already profitable operations in their own right. By virtue of the 1891 franchise, integration with the TRC most probably would have meant a single fare on the within-city portions of the radials which would have produced a decrease in total revenue for the Mackenzie-Mann transit empire. For the suburbanite of this period the public transit system was a constant source of irritation. At best, service was provided by two independently operated companies and, at the other extreme, there was no service close at hand. This had a significant impact on Toronto's spatial configuration. For the most part new residential areas only emerged near existing transit lines during this period, which, as Donald Davis has shown, made Toronto one of the densest cities in North America by 1921.[22]

Some politicians felt that service within Toronto proper was little better at this time. At the behest of the ORMB, the City decided to strengthen its argument that the TRC's cars were overcrowded, through a detailed investigation of the system. On May 23, 1912, the Board of Control recommended that

> the Corporation Counsel be permitted to secure a prominent railway expert as an official to be attached to his department for street railway traffic purposes; and also that funds to the extent of $35,000 be provided for the securing of expert assistance required in connection with transportation matters.

This resolution was approved by City Council on May 28, with the phrase "such expert or experts to be residents of Canada if such can be found satisfactory to the Corporation Counsel," added as an amendment.[23] A suitable Canadian could not be found to head the investigation, so Chicago transit authority Bion J. Arnold was hired, with Toronto consulting engineer J.W. Moyes appointed to assist him, and they concluded, as the City expected, that the TRC needed "more

---

21. Reported in *Railway and Shipping World*, VII (August 1904), 269.
22. See J.E. Rae, "How Winnipeg Was Nearly Won," in A.R. McCormack and Ian Macpherson, eds., *Cities in the West* (Ottawa, 1975), pp. 74-87 and P.E. Roy, "The Fine Arts of Lobbying and Persuading," in D.S. Macmillan, ed., *Canadian Business History* (Toronto, 1972), pp. 237-54. Davis, "Mass Transit and Private Ownership," pp. 81-83.
23. *Toronto City Council Minutes, 1912* (hereafter, *Council Minutes*), Appendix C, 9 and Appendix A, p. 741.

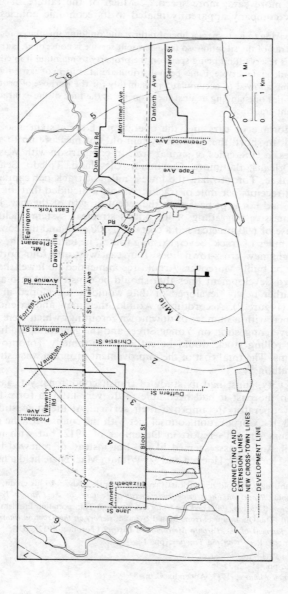

Figure 3  Bion Arnold's Proposed Extensions of the Toronto Transit System, 1912.

tracks, more cars, more speed."[24] Much of the failure of the street railway company apparently related to its economic policies:

> it is paying to its owners a liberal return, and building up a large reserve fund. The crux of the situation, so far as present service is concerned, is that instead of this reserve fund being spent on the property to maintain it in condition to give adequate service, it has been distributed in the form of extra dividends, or is being allowed to accumulate, presumably for the purpose of protecting the securities outstanding against the property at the time of the expiration of the present franchise.[25]

The shortage of cars and the operation of the system also troubled Arnold. In short, the Toronto system did not measure up to street car facilities in U.S. cities of similar size. In comparison with Washington, New Orleans, Baltimore, Cincinnati, Cleveland, Toledo, Seattle, and Milwaukee, Toronto had the lowest miles of track per capita and the highest receipts per mile of track.[26] Arnold concluded that the Toronto system had not kept pace with the recent growth of the city. He called for a massive upgrading of Toronto's streetcar system, including the purchase of 600 streetcars at a cost of $7,500 each and the construction of 127 miles of track. Seven existing routes, he felt, should be extended and eight new cross-town lines (four east-west and four north-south) should be built. Moreover, Arnold recommended no fewer than eleven so-called development lines that should be constructed into areas that were either newly developed or that would be developed in the near future (Figure 3). According to Arnold, these new and extended lines, combined with a number of downtown reroutings which were designed to relieve congestion on Yonge Street and the general upgrading of the TRC's rolling stock would be sufficient to remedy Toronto's transit problems. The total cost of the improvement program was estimated at $8.8 million.[27]

The City, thus, had its study of transit conditions; yet, as we have seen, in the past it had been able to do very little to force the street railway operators to improve their service. At the suggestion of the ORMB, the Corporation counsel met with the counsel for the company to discuss Arnold's report in December of 1912, but with no visible result.[28] An impasse seemed likely, but instead of renewed litigation, Mayor Horatio C. Hocken and Sir William Mackenzie held a number of

24. By this date, Arnold had at least two major transit analyses to his credit. See Bion J. Arnold, *Report on Engineering and Operating Features of the Chicago Transportation Problem* (Chicago, 1902) *and Report on the Pittsburgh* Transportation Problem (Pittsburgh, 1910). The Quotation is from Bion J. Arnold, *Report on the Traction Improvement and Development of the Toronto Metropolitan District* (Toronto, 1912), p. 3.
25. Arnold, *Report on the Traction*, pp. 1-2.
26. *Ibid.*, p. 13.
27. *Ibid.*, pp. 15-22, 35-36.
28. *Council Minutes, 1913*, Appendix C, p. 8.

meetings in an attempt to arrive at a mutually acceptable solution. And their solution proved to be quite simple: Mackenzie, for reasons that remain unclear, agreed to let the City buy up his Toronto-area utility holdings, including the TRC, the Toronto Electric Light Company, and the within-city portions of his radial railways.[29] It only remained to fix the price of sale. By October that price had been finalized – $22 million for the street railway properties and $8 million for the electric utility.[30] In presenting these figures to Council, Mayor Hocken took time to rationalize the idea behind the proposed purchase:

> We find ourselves with a monopoly of the surface transportation in the hands of a company enfranchised twenty-two years ago, upon such terms as make it impossible for the municipal administration to economically or satisfactorily provide transportation for nearly one-half of the present area of the City. This company has a franchise terminating eight years hence, with no inducement to maintain an efficient service, but with every reason from the standpoint of the shareholders to deteriorate both the service and the equipment. I believe that I am within the mark when I say that hardly a single thoughtful citizen expects from the present company reasonable consideration of their needs during the remaining period of the franchise.[31]

To be on the safe side, the mayor sent the valuations of the Mackenzie-Mann properties to Toronto chartered accountant John Mackay for appraisal. In a report dated November 17, 1913, Mackay stated that he had

> no hesitation in advising the completion of the transaction on the conditions stated herein, nor, in my judgment, is there any need for the Ratepayers or the Light and Power Consumers of the City entertaining the slightest doubt or fear as to the outcome, provided the basic conditions stated are observed.[32]

The way now seemed clear for a speedy public takeover of the Mackenzie-Mann utilities, but continued political infighting and the uncertain economic climate that accompanied the onset of the First World War would combine to create more delays. At least one politician had some doubts about the entire purchase scheme. Enter Thomas L. Church, lawyer and member of the Board of Control. By 1913 Tommy Church could be categorized as a veteran of Toronto civic politics, a classic example of a politician who had worked his way up

---

29. *Mayor's Message, Reports, and Correspondence in Relation to the Acquisition of the Assets and Franchise Rights of the Toronto Railway Company (and Subsidiary Lines) and the Toronto Electric Light Co.* (Toronto, 1913), pp. 46-57, Toronto Public Library.

30. *Council Minutes, 1913*, Appendix C, pp. 110-15.

31. *Ibid.*, p. 108.

32. John Mackay, *Interim Report on proposed Street Railway and Electric Light Purchase* (Toronto, 1913), p. 10, Toronto Public Library.

through the ranks.[33] He emerged as the chief opponent of the Mackenzie-Mann transit empire in general, and of the purchase deal in particular. In June of 1913 he moved in Council that the negotiations between Mackenzie and the Mayor be abandoned. This motion was easily defeated. When the cost of the public takeover was revealed in October, Church again took the offensive. Feeling that the public should be asked to approve such a large outlay of civic funds, he moved on November 11 that the specific details of the purchase of the various utilities be set before the electorate. This motion was approved, but in an amended form that called only for a vote on the idea of the purchase, not on its specific details.[34] The plebiscite, however, could not be prepared in time for the forthcoming municipal elections, but Church was returned to office. In January of 1914 he moved that the City Council abandon the purchase negotiations, procure a judgment from the ORMB for a better general transit service, and build more civic lines.[35] This motion, in various forms, was defeated by Council on January 12, January 26, and February 9. The purchase question then remained dormant until June, when Church moved that the purchase and all the details surrounding it be put to a public vote at the next civic election, January 1, 1915.[36] This motion carried unanimously.[37] Tommy Church had succeeded in his efforts to prevent a snap vote by Council on the Hocken-Mackenzie agreement. Both sides would now have to prepare for a plebiscite on the issue. But the matter never got to the ballot paper. The outbreak of the First World War intervened and made the financing for the scheme highly uncertain. On September 21, 1914, Mayor Hocken reported to Council that

> I have ... consulted with our Treasurer and taken counsel with other financiers as to the probabilities of the duration of the war, and the possibility of floating bonds for such a purpose at the conclusion of hostilities. I am advised that, even should the war end in six or nine months, the financial market will be in such a condition that it will be impossible to secure the funds necessary except at an exorbitant rate of interest.... In view of all the circumstances, therefore, I suggest to the Council that the negotiations for the acquisition of these properties be discontinued.[38]

33. *The Canadian Who's Who* (Toronto, 1948), p. 169, notes that Church served as school trustee from 1899-1904, alderman from 1905-1909 and as a member of the Board of Control from 1910 to 1914.
34. *Council Minutes, 1913*, Minutes, pp. 492-94.
35. *Council Minutes, 1914*, Minutes, p. 5.
36. Even Mayor Hocken agreed that little had happened during this interval. He attributed this to the fact that Mackenzie had been deeply involved in negotiations with the Dominion Government over the purchase of the Canadian Northern Railway. See *Council Minutes, 1914*, Appendix C, p. 115.
37. *Ibid.*, Minutes, p. 290.
38. *Ibid.*, Appendix C, pp. 115-16.

Council approved the mayor's suggestion on October 5th and the purchase deal episode drew to a close.[39] More than a year had been consumed in negotiating the agreement. Quite naturally, few improvements had been made to the transit system during this interval, and when the negotiations were terminated, the joint prospects of war and renewed bickering over the details of the street railway franchise did not bode well for a rapid solution to Toronto's transportation difficulties.

In the meantime developments had taken place on other fronts. After the judicial setbacks of 1909, the City had waited until 1911 before it again did battle with the TRC in the courts. In November of that year, the Corporation came before the ORMB to present a list of both alleged franchise defaults by the company and defects in the service provided. At the head of the list was the overcrowding issue, and the City used the findings of Arnold's report to substantiate its case. Since this study had dealt with an area that was considerably larger than the corporate limits of Toronto, the ORMB decided to commission its own investigation into transit conditions within the city proper. For this purpose, the services of C.R. Barnes, a man with twenty years experience as the Electric Railway Expert of the New York State Public Service Commission, were engaged. Barnes presented his report to the ORMB on May 15, 1914.[40] Basically, his conclusions corresponded to those of Arnold, and the members of the ORMB reacted favourably to the findings of the Barnes Report turning its seventeen specific recommendations into direct orders to the company.[41] For the most part, the TRC either ignored the Board's orders completely or else employed the excuse that due to the war the company could not obtain the necessary resources for it to comply with the ORMB's directives. Yet company profits remained respectable even during the war years (Table 1). In 1917, the City requested that the ORMB force the TRC to comply with the board's original orders. The company again refused to provide more cars and in 1918, the ORMB levied a fine of $24,000 against the TRC. This was a small price to pay compared to the total price of the 200 streetcars, which would have cost about $7,500 each, that the Board had ordered built.[42] Once again prolonged litigation had failed to bring relief to the transit-riding public.

While all of this had been going on in the courts, some interesting

39. *Ibid.*, Minutes, p. 425.

40. "Ontario Railway and Municipal Board Order Re Toronto Railway," *Canadian Railway and Marine World*, XVII (December 1914), p. 349 and ORMB, *Ninth Annual Report* (Toronto, 1915), p. 21. The final report was C.R. Barnes *et al.*, *Report to the Ontario Railway and Municipal Board on a Survey of Traffic Requirements in the City of Toronto* (Toronto, 1914). This came to be known simply as the Barnes Report.

41. *Barnes Report*, pp. 196-98. the extensions were to be on: 1) Wilton (later Dundas Street) and Pape Avenues; 2) Terauley (later Bay); 3) Bloor Street West; 4) Harbord Street; and 5) Dupont Street. See also ORMB, *Tenth Annual Report* (Toronto, 1916), pp. 16-19.

42. *Canadian Rrailway and Marine World*, XXI (April 1918), p. 163 and (September 1918), p. 398.

372                                                        MICHAEL J. DOUCET

events had taken place at City Hall. Following the failure of the purchase deal, Horatio Hocken had decided not to seek re-election for the
1915 term. One of his staunchest Council allies, Controller Jesse O.
McCarthy, immediately went after the soon-to-be-vacated mayor's
chair. His opponent was none other than Tommy Church, and the
handling of the purchase deal emerged as a very important campaign
issue. The *Star* and the *Globe* both opposed Church's candidacy, claiming he was the only person the Conservatives could get to run against
the Liberal's man, McCarthy. But federal and provincial political affiliations were not to be the basic issue in this municipal contest. Church
was enthusiastically supported by the *Evening Telegram* and the reasons
behind this paper's support were summarized in a pre-election headline.

> CHURCH'S RECORD AT CITY HALL SHOWS WHY YOU SHOULD VOTE FOR
> HIM . . . FOUGHT THE PURCHASE DEAL . . . CONTROLLER CHURCH LED IN
> STRUGGLE AGAINST THE MACKENZIE-MANN FORCES AND WON.[43]

To the *Tely's* delight, Tommy Church soundly defeated McCarthy, out-
polling him 26,042 to 19,573.[44] The day after his victory, the new mayor
summarized one of the key reasons for his victory in an exclusive *Telegram* interview:

> the people have given me an excellent expression of their opinion of the nefe-
> rious street railway company and have shown for all time to come that they
> will have no patience with such dealings with the street railway company to
> which they have been subjected to for the past two years.[45]

There would be no deals between the Mackenzie-Mann company and
the new mayor and since Church occupied that office until 1921, no real
peace on Toronto's transit scene until after the public takeover of that
year.

From the outset, Tommy Church gave every indication that he would
be a man of action, willing to take the transit bull by the horns. In his
inaugural address to Council he noted that

> it is true that the purchase agreement and negotiations are of the past, but the
> problem of the street railway and its inefficient and insufficient service is still
> with us and must be solved. There is only one possible policy now before the
> people, and that is fast traction.[46]

The new mayor was determined to expose once and for all the inadequate service provided by the Mackenzie-Mann traction companies.

43. *Evening Telegram*, December 28, 1914, p. 12.
44. "Mayor Church Mayor by 6,469," *Evening Telegram*, January 2, 1915, p. 13.
45. "By His Own Fireside Good News Found T.L. Church," *Evening Telegram*, January 2,
    1915, p. 9.
46. *Council Minutes, 1915*, Appendix C, p. 4.

One of his first acts was to set up a committee composed of R.C. Harris, Commissioner of Works, F.A. Gaby, chief engineer of the Hydro Electric Power Commission of Ontario, and E.L. Cousins, chief Engineer of the Toronto Harbour Commission, to investigate transit conditions in Toronto. In their report these civil servants noted that in 1915 more than 31,000 people resided beyond the 1891 city limits and that many of the city's newer areas were entirely without car service. They felt that the goal of Toronto's public transit system should be to provide a level of service such that all citizens living within a six-mile radius of King and Yonge streets could be transported to that junction within thirty-five minutes. Harris, Gaby, and Cousins also urged the City to quickly announce its intention to take over the streetcar system immediately upon the expiration of the TRC's franchise in 1921.[47] The transit utility increasingly was coming to be viewed as a necessary part of Toronto life, something that could no longer be left to the whims of private capital.[48] As the mayor's transit investigators noted:

> The future growth and development of the City of Toronto will be largely dependent upon the provision of adequate transportation facilities, properly located.
>
> The extension of these facilities should and usually does, precede the population, but in Toronto of late years, the conditions have been reversed.[49]

Mayor Church, himself, could not have written a more convincing indictment of the TRC.

Like its two predecessors, the report of the Civic Transportation Committee had very little impact upon streetcar operations in the Toronto area. The ordinary transit rider still remained poorly served. Nor did any of the court action initiated during Church's term of office do much to alleviate the city's transportation problems. TRC officials were willing to admit that overcrowding existed, but they did very little to come to grips with the problem. For the most part, the company's attitude was summarized by General Manager R.J. Fleming's cavalier comment that passengers could "take the next car."[50]

While the complex courtroom drama surrounding the overcrowding question was certainly protracted, and undoubtedly annoyed and inconvenienced many people, it did not arouse nearly as much public interest as did the question of the provision of car service in the recently annexed North Toronto district (Figure 1). There, the only service was provided along Yonge Street by the Metropolitan Branch of Mackenzie and Mann's Toronto and York Radial Railway. This company had received its franchise from York County in 1885 and, due to subsequent changes

---

47. Toronto Civic Transportation Committee, *Report on Radial Entrances and Rapid Transit for the City of Toronto*, 2 vols. (Toronto, 1915).

48. Armstrong and Nelles, "The Un-Bluing of Toronto, pp. 20-31 and 44.

49. Toronto Civic Transportation Committee, *Report*, p. 4.

50. "Just Take Next Car," *Evening Telegram*, February 6, 1915, p. 8.

in the boundary of Toronto, a quarter-mile portion of its trackage came to lie within the 1891 limits of the city. The franchise on this section (Figure 2), situated between the C.P.R. tracks (just south of Summerhill Avenue) and the 1891 northern city limit, Farnham Avenue, was due to expire on June 25, 1915.[51] Mayor Church saw this as an opportunity to take the offensive against the Mackenzie-Mann empire by asserting the right of the City to control the use of the streets. The mayor and his supporters, witnessed by what was termed a "great turnout of people," made absolutely certain that the franchise would terminate at the appointed hour. A newspaper headline succinctly described their night's work: "City Cuts Metropolitan Line. Tore Up Tracks at Midnight." A 1,320-foot gap, extending up the steep Gallows Hill, now existed between the termini of the TRC and the Toronto and York Radial Railway.[52] Citizens of North Toronto not only had to pay two car fares to get downtown, they had to walk a quarter of a mile to do it! Tommy Church assured the citizens that the inconvenience would be temporary. Once again, those supposed champions of the transit riders, the civic leaders, proved to be inconsistent. Bickering over who would run the cars over this section, and under what terms, took so long that individual citizens were forced to take the matter into their own hands. At least two solutions emerged.

The first citizen-inspired remedy attempted to provide an alternative transportation service. Along with other North American cities that were being poorly served by mass transit, Toronto became caught up in the jitney craze of 1914-15. Jitneys were simply private automobiles that were used by their owners to transport passengers along pseudo-routes for the payment of a five–cent fare. They had first appeared in Los Angeles sometime in 1914, and within a short time the idea had reached Toronto and other Canadian cities.[53] Yonge Street was by far the most popular Toronto jitney route, especially after Tommy Church's midnight theatrics, and, at one point, more than 700 of the vehicles plied that thoroughfare.[54]

In spite of the noticeable relief that jitneys brought to the Yonge Street gap and other parts of the city, civic officials greeted the cars with mixed feelings. The police commission, with the support of City Council, enacted a very stiff set of regulations to control the operation of

51. Wickson, "Radial Railways," p. 51 and "Buy Out Metropolitan Railway Some Say," *Evening Telegram*, June 23, 1915, p. 6.
52. *Evening Telegram*, June 26, 1915, p. 8. This hill forms part of the ancient shoreline of glacial Lake Iroquois.
53. "The Jitneys are Coming," *Star*, January 29, 1915, p. 8 and "Jitney Automobile Operation in Canada," *Canadian Railway and Marien World*, XVIII (June 1915), p. 230. See also P.E. Roy, "Regulating the British Columbia Electric Railway: the First Public Utilities Commission in British Columbia," *BC Studies*, 11 (Fall 1971), pp. 3-20.
54. "Only Eleven Jitneys Now on Yonge Stret," *Globe*, November 24, 1915, p. 9.

these vehicles.[55] The daily press strongly attacked such measures, with the *Globe* noting that

> people have put the jitney sign on their cars coming to work in the morning and again on their way home in the evening. They have actually carried passengers, much to their own profit and the satisfaction and comfort of the passengers carried. It has been a mutual advantage which in the eyes of a certain class of legislators is a contemptible, detestable, pernicious thing, to be supressed by every lawful means. When no lawful means can be found the law must be changed to provide some. If the city law-makers had only been wise enough to do nothing with regard to the jitneys for a few weeks the street railway managers would have seen the necessity of meeting the public convenience and would have wisely abandoned the policy of stinting the service and forcing the nuisance of overcrowding. Competition, the great regulator, would work a transformation.[56]

The "great regulator," however, was not to be allowed to take its proper course. Attempts by the jitney operators to establish a formal bus service in the city were rejected by City Council, which feared that the TRC would sue the City if competition was permitted to violate the street railway company's monopoly.[57] Moreover, by virtue of the franchise agreement between the TRC and the City, the former was compelled to pay a specified portion of its gross revenues to the latter.[58] The City thus had a vested interest in protecting the monopolistic feature of the street railway franchise. In the end, unfair legislation, timidity on the party of civic leaders, and the poor cold-weather performance of the jitneys themselves combined to ensure their virtual disappearance by the middle of autumn.[59] They were, however, the harbingers of a more flexible

55. The *Evening Telegram*, October 23, 1915, p. 17, in an article entitled "Jitneys are Dropping Off," reported that the restrictions were: 1) all passengers must have a seat; 2) maximum load set at seven; 3) no smoking in the cars; 4) all owners must take out a $1,000 insurance policy; 5) each year owners must purchase jitney, cab, and chauffeur licences. Needless to say, the net effect of these regulations was to greatly reduce the profitability of jitney operation.

56. "Street Traffic Relief," *Globe*, June 25, 1915, p. 4.

57. A number of different schemes were proposed. By far the most ambitious of these was put forward by one A.D. McBride, who intended to spend $250,000 to establish bus lines on a number of city streets including Yonge, King, Queen and Dundas Streets and Avenue, Kingston and Lake Shore Roads. For details of this plan, see "Controllers Fear Motor Bus Franchise," *Star*, June 15, p. 2.

58. *The Charter of the TRC*, p. 16, prescribed that the company pay to the city the following percentages of gross revenues: 8% of the first $1,000,000; 10% of the next 500,000; 12% of the next $500,000; 15% of the next $1,000,000; 20% of all gross revenues in excess of $3,000,000.

During its franchise, the TRC paid the city more than $14 million according to this schedule. Fully 45 per cent of this amount was paid out after 1914. The rigidity of this formula and a fixed fare combined over the length of the franchise to make the profit picture in later years less impressive than it might have been under unregulated market conditions.

59. "Only Eleven Jitneys Now on Yonge Street," *Globe*, November 24, 1915, p. 9.

and less capital-intensive form of public transit, the bus, whose impact would soon be felt dramatically in many cities.

Even before the failure of the jitneys, however, the disadvantaged citizens of Toronto's north end were seeking other remedies to their transportation woes. In this regard, the North Toronto Ratepayers' Association emerged as an active and vocal pressure group. At first, it was content to send deputations to City Hall to protest the Yonge Street situation, but eventually talk of secession from Toronto was heard at the meetings of the Association.[60] This movement died out early in 1916 after the TRC had complied with the ORMB directive to provide service along the Yonge Street gap.[61] It remains, however, as an example of the extreme lengths to which some citizens were willing to go in their search for transit relief.

The lengthy overcrowding case, three major investigations (all critical of the TRC), and the Yonge Street affair had all served to point out the shortcomings of Toronto's transit system. Increasingly, as in the earlier case of hydro-electric power, people had begun to question the wisdom of private ownership of a utility designed to serve the general public. The *Star* noted that

> private ownership has in nothing so clearly revealed its motive as in the trolley business. It is out to make money. In the trolley business private ownership is not content to take the fat with the lean – it wants to make money not only on the annual service, but it gets down to minute details and aims to give only such service as will make money every day, every hour, and on each journey

60. For an account of these activities see "Get Control of Yonge Street," *Evening Telegram*, August 11, 1915, p. 11 and "Deputation to City Hall," *Evening Telegram*, September 1, 1915, p. 6. A full record of the activities of this organization can be found in the "Minutes of the North Toronto Ratepayers' Association," MS, Ellis Family Papers, Ontario Archives. For a discussion, see Daniel J. Brock, "The Genesis and Demise of a Secession Movement within a Twentieth Century Metropolitan Centre: A Case Study of Toronto, Canada," unpublished term paper, Department of Geography, University of Toronto, April, 1971; "North Toronto Nears End of Its Patience," *Globe*, September 27, 1915, p. 7; "North Toronto Folk Talk of Seceding," *Globe*, October 11, 1915, p. 7 and "Un Uprising in the North," letter to the editor, *Star*, December 29, 1915, p. 5. The minutes of the North Toronto Ratepayers' Association, reveal that a motion to secede was discussed at meetings held by that organization on September 21 and 25 and October 9, 1915. No action, however, appears to have been taken on the motion.

61. According to "Black Eye for the City. Railway Gets Yonge Street," *Evening Telegram*, September 11, 1915, p. 15, the order to the TRC was issued on that day, with the company given until December 1, 1915, to put the extension into service. Typically, this was not Done until early in 1916. During the interval, an estimated ten to fifteen thousand people daily trudged along the Yonge Street gap.

that every numbered car makes. A car shed is like a livery stable. The vehicle stays in the barn unless its use is paid for.... The trolley is still indispensable, but private ownership of it has been a failure.[62]

Hope for the future came to be associated with the public takeover of the transit system. This issue of public ownership was finalized by a plebiscite held in 1918. On January 1, by a ratio of 11 to 1, Torontonians voted in favour of the public takeover of the street railway network upon the expiration of the TRC's franchise in 1921.[63] The TRC waited more than a year to respond. In an obvious last-ditch attempt to save its franchise, the company proposed a scheme called "Service at Cost." Under this plan, the private owners would continue to own the system, but, as in France and Germany, the directions as to its operations and expansion would come from the City. Fares would be geared to pay the costs of operations and improvements, and to give the owners a return of 6 per cent on their investment.[64] The citizens, however, had had their fill of private ownership. There was to be no retreating from the 1918 decision and by 1920 the Toronto Transportation Commission had been established by an act of the provincial Legislature to oversee the shift to public ownership.[65] In September of 1921, Toronto's street railway became a public utility.

Much work was needed to get Toronto's transit system into shape in the years after the public take-over. Among other things, the Toronto Transportation Commission scrapped more than 400 of the obsolete TRC cars, purchased 575 new cars, totally rebuilt 57 miles of single track (more than one-third of the system), and extended services into the newer areas of the city by building new lines and consolidating the existing radials with the rest of the system (Figure 4). In addition, feeder bus lines were inaugurated in those areas where the population density was not sufficient to support a car line. There were even some brief experiments with electric trolley buses in the north end.[66] Funds for

62. "Trolleys and Jitneys," *Star*, April 28, 1915, p. 6. For a thorough discussion of the public electric power movement in Ontario, see H.V. Nelles, *The Politics of Development: Forests, Mines, and Hydro-Electric Power in Ontario, 1849-1941* (Toronto, 1974), pp. 237-306.
63. *Evening Telegram*, January 1, 1918, p. 9 ("One Great Shout of Aye!"), reported that the actual vote was 39,979 in favour of public ownership and 3,769 opposed.
64. Toronto Railway Company, *Public Service Topics*, 6 (April 19, 1919), 1. This publication was a four-page newsletter that was issued every two weeks by the company. Extant copies (Ontario Archives) exist only for the March 29 to June 7 portion of 1919.
65. Toronto Transportation Commission, *Wheels of Progress*, 5th ed. (Toronto, 1953), pp. 120-21. The relevant legislation was contained in An Act respecting the City of Toronto (10-11 Geo. V., C. 144).
66. Toronto Transportation Commission, *Wheels of Progress*, pp. 35-64. See also John F. Bromley, *T.T.C. '28* (Toronto, 1968), pp. 3-8 and Robert McMann, "All About TTC's Track Rehabilitation Program, 1921-1924," *Upper Canada Railway Society Newsletter*, 308 (September 1971), pp. 131-38.

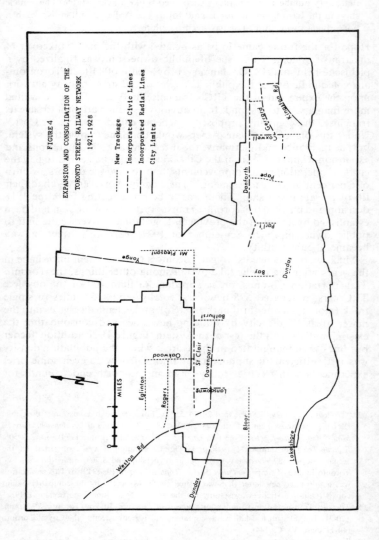

FIGURE 4

EXPANSION AND CONSOLIDATION OF THE
TORONTO STREET RAILWAY NETWORK
1921-1928

............ New Trackage
—·—·— Incorporated Civic Lines
———— Incorporated Radial Lines
———— City Limits

those improvements were raised through the issuance of municipal debentures, a process authorized by the legislation that created the TTC.

The Toronto transit problems of the early twentieth century were the result of a number of interrelated factors. Most certainly they were precipitated by the refusal of the operators of the private company to extend service beyond the 1891 city limits and to do anything, in spite of considerable judicial, legislative, and public pressure, to improve the quantity and quality of the service offered on existing routes. No doubt the war made improvements difficult, but the service had begun to fail the public long before 1914. And all the time that the transit riders suffered, the owners of the TRC made a fortune. As one editorial writer put it.

> the Toronto Street [*sic*] Railway Company has fought the city on every clause of its contract that was designed to ensure the rights of the passengers and the public generally, and the result is that if the strap-hanging population of Toronto could find other means of conveyance they would not worry very much about the loss of business to the company.[67]

But if most of the blame must fall on the private operators of the street railway system, at least they were true to the goals of capitalism. Civic officials, on the other hand, did not always act with such consistency. The purchase deal, for example, was probably not in the best interests of the citizens. Certainly the way in which its proponents tried to speed its implementation was somewhat irregular, given the vast sum involved. Tommy Church carefully guarded the public purse in this matter, but he showed little regard for the plight of the ordinary transit rider when he carried out his late-night raid on the Metropolitan's tracks. Indeed, Mayor Church, for reasons that remain unclear, seems to have let his personal feelings about Mackenzie and Mann influence his actions on a number of occasions. This prompted Chairman McIntyre of the ORMB to accuse Toronto's civic leaders of "quibbling and chicanery" and to suggest that they

> usually opposed anything tending to aid in a solution to the transportation problem, that they frequently took unreasonable attitudes in contesting these attempts at a solution, railing at the board when it failed to adopt the same attitude and that they threw the whole blame on the board or the railway company for delays for which they were chiefly responsible.[68]

Clearly, the street railway was not the only exploiter of the public interest.

Streetcars still ply a number of routes in central Toronto today. In fact the TTC is in the process of receiving a shipment of about 200 new,

67. "The Jitneys," *Star*, February 4, 1915, p. 6.
68. "City Officials Stupid Before Railway Board," *Globe*, November 12, 1915, p. 7.

Canadian-built streetcars, and a new high-speed suburban line (in the Borough of Scarborough) is in the final planning stages. This, however, does not make the city unique (though the condition and cleanliness of its streetcars might), as a handful of U.S. cities have retained their streetcar systems; and some, such as San Francisco, are experimenting with new vehicles as well. What makes Toronto unusual is the overall vitality of her public transit system. Both ridership and the system have continued to grow. In fact it was the only North American transit system to increase its total patronage between 1946 and 1980 and the TTC added an extensive, if expensive, subway system to its facilities during this period. The events described in this paper contributed to this in no small way. As Donald Davis argues, "avarice, poverty, monopoly, geography, and technological backwardness together laid the foundations for Toronto's success in urban transit." The TRC was at one and the same time the cause of both the inadequate system before 1921 and the superb one that followed 1921. "Its legacy was two-fold: first, a consensus ... in favour of municipal ownership that made it easier for the TTC to win popular approval; and second, a densely populated city with a high riding habit that made mass transit uniquely viable in Toronto."[69] The years leading up to 1921 had been difficult ones for Toronto's transit riders. They expected a good transit system after the public takeover, and the TTC responded in kind. By 1931, $50 million had been expended in the pursuit of this goal.[70] Indeed, compared to other Canadian cities, the growth of Toronto's street railway network was outstanding during this period (Table 3). (Such growth, of course, was possible in large part because of the limited system and poor level of service provided under the TRC.) But both the stake and public interest in mass transit have remained high in Toronto to this day; and in this era of ever-escalating energy costs, the TTC's system remains the envy of most other major North American centres. Seventy years ago many critics suggested that Toronto was poorly served by her mass transit system; today, analysts such as John Sewell would argue that the city's transit network is too extensive for its own good.[71]

69. Davis, "Mass Transit and Private Ownership," pp. 93-95.

70. Toronto Transportation Commission, *Ten Years of Progressive Public Service 1921-1931* (Toronto, 1931), n.p. See also John F. Bromley and Jack May, *Fifty Years of Progressive Transit: A History of the Toronto Transitn Commission* (New York, 1973). For some comparisons see Herbert W. Blake, *The Era of Streetcars and Interurbans in Winnipeg*, 1881 to 1955 (Winnipeg, 1974); R.D. Tennant, "Capital Traction: An Outline History of the Street Railway System of Ottawa," *Upper Canada Railway Society Newsletter*, 273 (October 1968), pp. 118-22 and S.I. Westland, "Steel City [Hamilton] Traction," *Upper Canada Railway Society Newsletter*, 304 (May 1971), 69-74.

71. John Sewell, "Mass Transit in Canada: A Primer," *City Magazine* III (May/June, 1978), 51.

Table 3: Street Railway Expansion in Selected Canadian Cities,
1922-1931

| City | Miles of Track[1] 1922 | 1931 | % Increase 1921-31 | % Population[2] Increase |
|------|------|------|------|------|
| Calgary | 66.5 | 77.0 | 15.8 | 32.3 |
| Edmonton | 33.3 | 54.5 | 63.7 | 34.6 |
| Halifax | 12.6 | 24.0 | 90.5 | 1.5 |
| Hamilton | 17.4 | 34.3 | 97.1 | 36.3 |
| London | 27.5 | 29.4 | 6.9 | 16.7 |
| Montreal | 142.8 | 283.7 | 98.7 | 32.3 |
| Ottawa | 26.6 | 52.4 | 97.0 | 17.6 |
| Quebec | 20.5 | 33.5 | 63.4 | 37.2 |
| Regina | 25.6 | 28.6 | 11.7 | 54.5 |
| TORONTO | 93.5 | 222.6 | 138.1 | 20.9 |
| Winnipeg | 63.6 | 106.4 | 67.3 | 22.2 |
| Vancouver | 245.8 | 302.3 | 23.0 | 51.1 |

NOTES: 1. *Canadian Railway and Marine World*, XXVI (December 1923), 595
and *Canadian Railway and Marine World*, XXXV (October 1935), 525,
2. Calculated from figures in the *Seventh Census of Canada*, 1931
(Ottawa, 1933), III, 8-13.

# "C.P.R. Town": The City-Building Process in Vancouver, 1860-1914

NORBERT MACDONALD

Vancouver's urban landscape is the outcome of numerous forces and decisions. Topography, climate, technology, the political system and population all played their part, as did the decisions of thousands of citizens, politicians and businessmen. The relative importance of these determinants varied depending upon the particular time or district, and no single agency controlled all development. In the first big surge of growth of the 1880s, the Canadian Pacific Railway played a critical role. But as Vancouver grew in size and complexity, especially in the decade before the First World War, a whole series of agencies such as city council, street car companies, real estate firms, businessmen, immigrants and national economic trends shaped and modified the original imprint provided by the CPR.

I

Even before the arrival of the first train in 1887, a series of events and decisions significantly influenced the future urban landscape. Victoria was founded in 1843, but it was not until the Fraser River gold rush of 1858-59 that attention was drawn to the mainland of British Columbia. As part of the early defence arrangements, a detachment of Royal Engineers constructed some rough roads from New Westminster on the Fraser River to Burrard Inlet about ten miles away. At the same time the Royal Navy surveyed the entire Burrard Inlet, English Bay region and set aside a number of military, Indian and townsite reserves.[1] In all,

---

1. F.W. Howay, "Early Settlement on Burrard Inlet," *British Columbia Historical Quarterly*, 1, no. 2 (1937), p. 101; M.A. Ormsby, *British Columbia: A History* (Toronto, 1958), pp. 189, 193.

these reserves amounted to some nine square miles, and would play an important role in Vancouver's development. About half of these lands soon moved into private hands, but the rest remained in the public domain. They determined the location of the original settlement, provided the land for the future university, and still make up the heart of Vancouver's extensive park system.

Except for these reserved lands, the entire area was opened for preemption (see Map 1). In 1862, John Morton, William Hailstone, and Sam Brighouse, dubbed the "Three Greenhorn Englishmen," took the necessary steps and obtained a crown grant to 540 acres in what is now Vancouver's West End. They paid $555.73. While there was no great rush to follow their example, crown grants were issued periodically throughout the 1860s and 1870s. Most of the grants were for land along the inlet, or in the False Creek region. They averaged 160 acres, but on occasion were much larger.[2]

While these landowners were very conscious of the potential value of their holdings, the real focus of interest and activity on Burrard Inlet was two lumber mills which were established in the 1860s. The original mill on the north side of the inlet started in 1863, and two years later Captain Edward Stamp established a mill on the south side of the inlet.[3] Operations expanded rapidly and in 1869 alone at least forty-five vessels left the inlet carrying lumber to California, Latin America, England and Australia.[4]

By the early 1870s a pattern of activity and settlement was established on Burrard Inlet that would not change significantly in the next dozen years. The entire area was geared almost solely to the production and export of lumber. Besides the main concentration at Moody's mill on the North shore, and Stamp's on the south, scattered camps and settlements provided logs, lumber, specialized timbers and handmade shingles.

In the early years, Moodyville on the north shore was clearly the largest settlement, but as the 1870s unfolded, the settlement on the south shore gradually inched ahead. Most of the men who worked in Stamp's mill lived in bunkhouses of shacks that they had thrown together. But after "Gassy Jack" Deighton opened a small hotel and saloon about half a mile to the west of the mill, people gravitated to the boardinghouses, hotels, homes, and stores that gradually accumulated in that

---

2. The activities of H.P.P. Crease provide an example of the kinds of profits possible. While Attorney General of B.C. in 1863, he obtained a Crown grant for 165 acres in District Lot 182 and paid $310.40. Two years later, he purchased the adjoining 165 acres in District Lot 183 from T. Ronaldson. In 1877, while a judge of the Supreme Court of British Columbia, he sold the entire parcel to Dr. I.W. Powell of Victoria for $3,500. See B.C. Department of Lands, Registry Office, and Townely MSS, Vancouver City Archives.

3. F.W. Howay, "Early Shipping in Burrard Inlet, 1863-1870," *British Columbia Historical Quarterly*, 1, no. 1 (1937), p. 4.

4. *Ibid.*, p. 20.

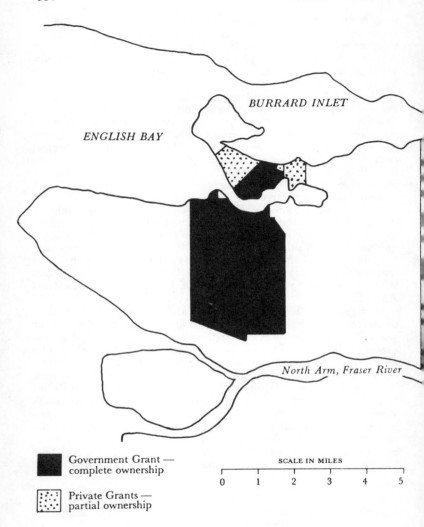

BURRARD INLET

ENGLISH BAY

North Arm, Fraser River

◼ Government Grant —
    complete ownership

⠿ Private Grants —
    partial ownership

SCALE IN MILES

0    1    2    3    4    5

Map 1 .        C.P.R. Land Grants in the Burrard Inlet, English Bay Region,
                1886.
SOURCE: Adapted from B.C. Electric Railway Company papers, 1907, and
Harland Bartholomew, *A Plan for the City of Vancouver*, (1929). Reprinted by
permission of *BC Studies*.

area. Whether referred to as Coal Harbour, Gastown or Granville, this
settlement was the nucleus of the future Vancouver.

While lumber production rose steadily, population growth was modest. By 1884 the entire Burrard Inlet area had about 900 residents, and
Granville was still only a small lumbering village. Its cluster of buildings
now extended four hundred yards along the shore, and the trees had
been cleared back some two or three hundred yards, possibly twenty
acres in all. Its total population was approximately 300.[5]

The announcement by the CPR in 1885 that the terminus of the
railway would be "in the immediate vicinity of Coal Harbour and English Bay" brought about an abrupt change in Granville's fortunes.[6] Up
to this time it had been understood that the main line would terminate
at Port Moody, some twelve miles to the east. But with this uncertainty
removed, conditions changed dramatically, and the previous trickle of
migrants to the area soon became a steady stream of enthusiastic
settlers.

The year 1886 was especially memorable. In April the City of Vancouver was officially incorporated (see Map 2). Just two months later a
devastating fire destroyed most of the city, and some twenty persons
died. Yet migrants continued to pour in and recovery was rapid. CPR
trains reached Port Moody in July, and from there it was only two hours
by steamer to Vancouver. Survey crews, road gangs, home builders and
real estate agents were busy throughout the year, and by Christmas the
city's population had reached 2,000 – more than four times what it had
been when the year began.

It was much the same in 1887. The big event of the year was the
arrival of the first CPR train on May 23. It pulled in about one o'clock
in the afternoon and virtually the entire city turned out. Businesses were
closed, city council adjourned its meeting, ships in the harbour were
decorated, and the fire brigade and city band led a parade of hundreds
to the station. After a welcoming address by Mayor McLean, Harry
Abbott of the CPR spoke. He mentioned the many difficulties that the
CPR had overcome in order to arrive in Vancouver, but added "Here
we are, and here we will remain." The crowd loved it. They were equally
excited on June 13 when the CPR liner *Abyssinia* arrived in Vancouver
harbour from the Orient. This marked the beginning of a regularly
scheduled steamship service to China and Japan and was taken as one
more piece of evidence that Vancouver was destined not only for
national but, indeed, international stature.[7]

5. R.T. Williams, ed., *British Columbia Directory 1884/85.*
6. For illustrations and a more detailed discussion of the role of the CPR see my article,
   "The Canadian Pacific Railway and Vancouver's Development to 1900," *BC Studies,*
   35 (Autumn 1977), pp. 3-35.
7. *Daily News Advertiser* and *Vancouver World,* 23-26 May, 13-16 June 1887, give detailed
   treatments. See also W. Kaye Lamb, "The Pioneer Days of the Trans-Pacific Service,"
   *British Columbia Historical Quarterly,* 1, no. 3 (1937), pp. 149-60.

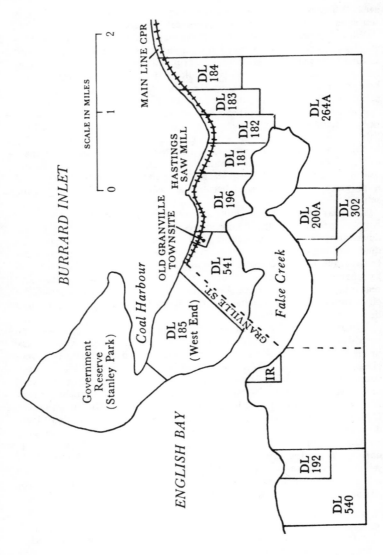

Map 2    Vancouver as Incorporated in 1886 Showing Earlier District
         Lots.
         SOURCE: N. MacDonald, "The Canadian Pacific Railway."

In a less dramatic way and at a less hectic pace, the entire period from 1886 to 1892 followed the pattern of 1886-1887. During this boom period the city absorbed 12,000 migrants. City officials supervised the clearing and grading of sixty miles of streets, provided schools and teachers for 2,000 students, established police and fire departments, and developed water lines, sewers, parks and hospitals. The city's business-men expanded the output of lumber and shingles, carried out construction valued at $4 million, built a thirteen-mile electric street railway system, created a variety of foundries, machine shops and small manu-facturing plants, and installed up-to-date lighting and telephone systems. Year by year the little lumbering village of Granville faded into the background as a virtually new city was created.[8] While a great variety of individuals and institutions contributed to this process, the CPR was of fundamental importance.

First and foremost, the CPR ended Vancouver's isolation and pro-vided a quick, convenient means of moving there. The very existence of its rail and steamship facilities also shaped a whole set of expectations about the city's future. Thousands of migrants reasoned that with such facilities Vancouver was destined for inevitable prosperity, and through-out the 1880s they flooded in from Eastern Canada, Britain, the United States, and the Orient. By 1891 the federal census showed that Vancouv-er's population had soared to 13,709.[9]

Beyond contributing to population growth, the CPR played a major role in determining the city's street layout and general land-use patterns. The government of British Columbia had agreed in 1884 that in return for the extension of the mainline from Port Moody to Coal Harbour and English Bay, it would grant the CPR some 6,000 acres of land in the vicinity of the new terminus.[10] On 13 February 1886, two significant crown grants were issued to Donald A. Smith and Richard B. Angus as trustees of the CPR.[11] The first, amounting to some 480 acres, granted the former government reserve on Coal Harbour to the two trustees. This grant included thirty-nine specific lots (about eight acres) in the Granville townsite. The second, a great tract of untouched forest south of False Creek, amounted to 5,795 acres (see Map 1).

Private owners in the city also made donations. They realized that

8. For a detailed analysis of this period, see Robert A.J. McDonald, "City-Building in the Canadian West: A Case Study of Economic Growth in Early Vancouver, 1886-1893," *BC Studies*, 43 (Autumn 1979), pp. 3-28.
9. For detailed statistics, see N. MacDonald, "Population Growth and Change in Seattle Vancouver, 1880-1960," *Pacific Historical Review*, 39 (1970), pp. 279-321.
10. B.C. Sessional Papers, 1885, pp. 129-36, 385-86, provides the main details of the discus-sion between William Smithe of B.C. and William Van Horne of CPR. Both the government and the railway were anxious to have the main line extended to Coal Harbour, and while the CPR originally sought some 11,000 acres, they later accepted the offer of 6,000 acres. See also *Victoria Colonist*, 14 January 1885.
11. B.C. Legislative Assembly, *Journals*, 1888, XL-XLI.

real estate values would soar if Vancouver got the transcontinental railroad. While it might have been painful to make voluntary contributions
to an already large and powerful organization, they undoubtedly saw it
as an appropriate move that would help assure future gains. These
scattered grants lying east and west of the Granville townsite amounted
to about 175 acres.[12]

The system of public and private grants to encourage railroad construction was a familiar practice in both American and Canadian cities,
yet it is noteworthy that the CPR received much more land in Vancouver than in other western Canadian cities. Winnipeg, for example,
contributed a city-built bridge, a $200,000 cash bonus, approximately
thirty acres of land for station and shops, and a permanent exemption
from taxation on railway property.[13] In Regina and Calgary the CPR
obtained only the alternate sections of land that were bestowed under
the original federal charter. Had the Canadian government originally
fixed on Coal Harbour rather than Port Moody as the terminus of the
CPR, the standard federal land grants would have applied all the way to
Vancouver. But the choice of Port Moody, and later extension to Vancouver, meant that special arrangements were made with the government of British Columbia. The CPR undoubtedly was aware of the steps
taken by various American cities on the Pacific coast. Thus, San Diego
offered the Santa Fe Railroad some 15,000 acres, while Seattle in 1873
offered the Northern Pacific Railroad a package consisting of 7,500
town lots, 3,000 acres of land, $50,000 cash, $200,000 in bonds as well as
use of the city's waterfront.[14]

Regardless of the rationale for these grants, there is no doubt that the
CPR's acquisition of 6,458 acres – more than ten square miles – in the
heart of the future city marked the most significant land transaction in
Vancouver's entire history. The CPR immediately became the largest

12. This estimate was made from an examination of L.A. Hamilton, Map of Vancouver,
    1887, and the Vancouver Assessment Roll 1888. See also G.W.S. Brooks, "Edgar
    Crowe Baker: An Entrepreneur in Early British Columbia," M.A. Thesis (University
    of British Columbia, 1976), especially pp. 202-204, and J.S. Matthews, "Early Vancou-
    ver," II, 91-110 passim.
13. Alan F.J. Artibise, Winnipeg: A Social History of Urban Growth, 1874-1914 (Montreal,
    1975), pp. 70-73. See also Pierre Berton, The Last Spike: The Great Railway, 1881-1885
    (Toronto, 1971), pp. 302-305; Max Foran, "Early Calgary, 1875-1895: The Controversy
    Surrounding the Townsite Location and the Direction of Town Expansion," in A.R.
    McCormack and Ian MacPherson, eds., Cities in the West: Papers of the Western
    Canada Urban History Conference (Ottawa, 1975), pp. 26-45; and J. William Brennan,
    "Business-Government Cooperation in Townsite Promotion in Regina and Moose Jaw,
    1882-1903," in Alan F.J. Aritibise, ed., Town and City: Aspects of Western Canadian
    Urban Development (Regina, 1981), pp. 95-120.
14. See Robert M. Fogelson, The Fragmented Metropolis: Los Angeles, 1850-1930 (Cam-
    bridge, 1937), p. 59; Frederick J. Grant, History of Seattle, p. 147. The Northern
    Pacific rejected Seattle's offer and decided to locate the terminus in Tacoma where they
    had already bought up most of the future townsite.

landowner in Vancouver, and in time their plans and decisions would shape the street layout and the general location of the city's commercial, industrial and residential areas.

The complex job of preparing a plan for the future city fell largely on L.A. Hamilton, the CPR's surveyor, and later land commissioner. He had already surveyed Regina, Moose Jaw, Swift Current and Calgary for the CPR. But Vancouver had a population of 2,000 and as the future terminus of the railway played a larger role in CPR thinking than did the prairie cities. Under his direction, along with H. Cambie and H. Abbott, the entire area was surveyed in 1885 and 1886. Existing arrangements in the old Granville townsite and the West End determined the location of many streets, but the CPR's detailed plans ultimately shaped the layout of much of the city.[15] The placement of the CPR station, office, and wharf and the Hotel Vancouver along the Granville Street axis was especially significant, for it pulled the centre of the city to the west of the existing townsite. This placement avoided the extensive building and settlement in Gastown, gave the CPR room to develop their own facilities, and also meant that real estate profits would not have to be shared with others. This western section of the city would be less heavily populated in the early years but it attracted a more prosperous clientele, commanded the highest prices and held the greatest prestige.[16]

In the largely untouched area south of False Creek, Hamilton had a freer hand. A numbered system of east-west avenues was established, with the cross streets named after trees – Arbutus, Cypress, Fir, Hemlock, Oak. Granville Street was extended across False Creek to run through the centre of the CPR land grant. While Hamilton's detailed plan carried only to Ninth Avenue, about a half mile south of False Creek, his basic street system would, in time, be extended all the way to the Fraser River, some four or five miles farther south.[17]

The long-range impact of the CPR's land grants, plans and decisions was pervasive. Vancouver's central business district with its banks, offi-

---

15. A "Plan of the City of Vancouver - 1886," by H.B. Smith, predates a better-known 1887 map by L.A. Hamilton. The earlier map is a detailed, accurate one showing the precise location of the various CPR facilities. Smith spent most of his career as an engineer and surveyor with the Vancouver Water Works and assisted Cambie in the early CPR surveys. The 1886 map by Smith is available in the Vancouver City Archives, while Hamilton's 1887 map is in the B.C. Archives.

16. Early real estate ads clearly indicate the CPR's assessment of their property as well as their long-range plans. See *Vancouver News Advertiser*, 1 June 1886, 10 February 1887, and 14 November 1888. See also Goad's *Fire Insurance Map of Vancouver, 1897*.

17. The Hamilton material is based primarily on letters which he wrote to J.S. Walker of the Vancouver Town Planning Commission in 1929 and to City Archivist J.S. Matthews in 1934 and 1936. Though written some fifty years after the events described, they are invaluable. See Matthews, "Early Vancouver," I, 328; III 207-8. See also Hamilton to Matthews, 27 April 1936 in Matthews, Uncatalogued Material, Vancouver City Archives.

ces, department stores and theatres was firmly delineated. The placement of CPR rail lines and wharves on the inlet, and the choice of the
north shore of False Creek for freight yards and repair shops, helped
shape the industrial character of both areas. Similarily the arrangements
in the street pattern of the West End set the district off as a distinct
residential area. When General Superintendent Harry Abbott built a
handsome home there the entire district was soon identified as the most
prestigious area in the city. It would sustain this pre-eminence for
twenty years until displaced by the opening of the CPR's Shaughnessy
Heights in the tract of the land grant south of False Creek. Of course,
residential development continued in other areas of the city as well. But
it is significant that the eastern edge of the CPR land grant became an
unofficial dividing line between the more affluent established west side
of the city and the less prosperous working-class east.

During the boom years from 1886 to 1892, Vancouver made a rapid
transition from primary producing centre to one offering a great variety
of goods and services. Whereas the great majority of its workers were
engaged in the lumber and logging industry in the mid-1880s, the pattern was sharply different in 1892. By that time Vancouver had a labour
force of some 5,000. Of these, about 900 persons were employed in retail
and wholesale firms; 800 worked in locally oriented bakeries, confec-
tioneries and machine shops; 750 were in the building trades; 500 were
in domestic and personal service as waitresses, cooks and janitors;
another 300 were involved in local transportation with the streetcar line
or drayage firms; finance and real estate employed about 140; and there
were 70 professionals.[18] Altogether these locally oriented activities
accounted for 65 to 70 per cent of the city's labour force.

While the sawmill worker and logger no longer dominated the city's
economy, they were still important, and helped sustain much service-
type employment. In 1892 Vancouver's nine lumber and shingle mills
employed some 900 workers, with the Hastings Saw Mill still the largest.
Benjamin Roger's B.C. Sugar Refinery had about 50 workers, while at
the peak of the salmon-canning season local canneries employed 200,
but for most of the year operated with small staffs.

Vancouver's largest single employer was the CPR. The construction
of rail lines, wharves, freight sheds, station and hotel had created considerable employment in the 1880s, and this increased with the opening
of its trans-Pacific steamship service and the development of its passenger and freight service with eastern Canada. By the early 1890s, some
500 to 600 persons were employed as CPR labourers, mechanics, train-

---

18. Williams, *B.C. Directory 1892*, pp. 788-90. *Census of Canada, 1901*, III, Manufactur-
    ing," p. 326, states that in 1891 Vancouver had 94 manufacturing plants, with a total
    of 1,084 workers.

men, conductors, freight handlers, engineers, janitors and cooks.[19] In a private capacity, CPR officials like William Van Horne, George Stephen and Donald Smith all developed residential and commercial properties.

In the exuberant atmosphere of the late 1880s, with people pouring into the city and with both newcomers and residents alike aware of the opportunities in a booming real estate market, no single organization could dominate the entire field.[20] The city had some thirty real estate firms, with Oppenheimer Bros,. Ross & Ceperley, Berwick & Wulffsohn, Innes & Richards, R.G. Tatlow and C.D. Rand being especially active. Occasionally, groups of investors found it advantageous to incorporate as land companies, since the combined assets made it possible to deal in larger units of land. The Vancouver Land and Improvement Company, predominantly a group from Victoria, was the largest organization of this type. A great variety of individual investors were also active, with Dr. I.W. Powell, C.T. Dupont, C.G. Major and J. Robson being especially prominent. But even with this diversity, the CPR was in a class by itself. The value of its land holdings was some seven or eight times as large as its nearest competitor, and no firm or individual could ignore its approach or prices.

CPR management followed a cautious, prudent, long-range policy in the sale of its extensive holdings. They sold lots only in the city proper during the 1880s and 1890s and buyers had to meet very stiff terms. The usual requirement was one-third down, one-third in six months, and the balance in twelve months. Discounts of 20 to 30 per cent were offered if buildings were erected on the property within twelve months.[21] According to C.D. Rand, the memories of the catastrophic slump in Winnipeg real estate after the great boom of 1881-82 were still very fresh, and the approach in Vancouver was to avoid the kind of "dishonest speculation" that had led to that setback.[22] Whether CPR policies actually retarded the rise in real estate prices is unknown. But it is noteworthy

19. Williams, *B.C. Directory 1889*, pp. 254-71, provides a valuable listing of all persons employed in the Pacific Division of the CPR. Of some 1,300 listed, 432 gave Vancouver as their home address. The 1892 Directory estimated that 600 persons in the city were employed by the CPR.

20. Material on real estate activity in Vancouver is voluminous. Newspaper advertisements are an ever-present source, but for more precise insights, the Vancouver City Archives has an abundance of excellent material. The Vancouver Assessment Roll 1888, the David Oppenheimer MSS, and the F.C. Innes MSS are especially valuable.

21. *New Advertiser*, 1 June 1886.

22. Douglas Sladen, "Vancouver, A Great Seaport of the Twentieth Century," *Popular Monthly*, XXIV, no. 5 (May 1890), pp. 513-22.

that while prices levelled off, and even slumped, after 1893, they did not collapse in the manner of Winnipeg or of Tacoma.[23]

Along with the CPR another important agency that helped shape the real estate market and settlement pattern was the Vancouver Electric Railway and Light Company (see Map 3). Electric street railways had been perfected in Richmond, Virginia in 1888, and like counterparts in other North American cities the Vancouver company quickly adopted the new system.[24] Tracks were first laid in 1889 and by 1891 with the completion of the Fairview Belt Line, Vancouver had a street railway system of thirteen miles. The CPR encouraged the extension by providing the street railway company with a number of free lots in the Fairview, Mount Pleasant area. It was no coincidence that the first large sale of CPR lands south of False Creek occurred in 1890, just prior to the completion of the Fairview Belt Line and the anticipated rush of population to that district.

There can be little doubt of the ramifications of the CPR on Vancouver's economic life and physical development. Whether one considers the actual migrants it carried to the city, the direct and indirect employment it generated, the sets of expectations it shaped, or the way it determined Vancouver's street layout as well as the location of commercial, residential, and industrial districts, all were important. Similarly, its real estate prices in Fairview, its wage rates for unskilled labourers, its freight charges for lumber shipments to the prairies and its room rents in the Hotel Vancouver all played a part in shaping the local economy.

II

Vancouver's politics and municipal activies in the pre-1900 era are still largely unexplored, and any attempt to assess the relative importance of the CPR in this realm must be considered tentative. During these years Vancouver was divided into five wards, with two aldermen elected from each ward and the mayor elected at large. Elections were held annually, but with few exceptions created little excitement or controversy. In fourteen mayoralty elections between 1886 and 1900, two were by acclamation, in two others the winner swept all wards, and on the average only thirty per cent of the eligible voters actually voted.[25]

With annual elections, there was considerable turnover in city coun-

23. Vancouver journalists never tired of pointing out that their real estate market was vastly different from the highly publicized, vigorously promoted patterns of Tacoma and Seattle. See for example *Western News Advertiser*, 4 June 1890, p. 4.
24. Between 1891 and 1897 Ottawa, Hamilton, Montreal, Winnipeg, St. John's, Halifax and Quebec City all established electric railway concerns. W.G. Ross, "Street Railways in Canada," *Canadian Magazine*, January 1902, pp. 276-78. For the Toronto story, see Michael Doucet's chapter in this volume.
25. See Vancouver's *Voters List 1888-1900*, and *Record of Elections* to 31 December 1924, in Vancouver City Archives.

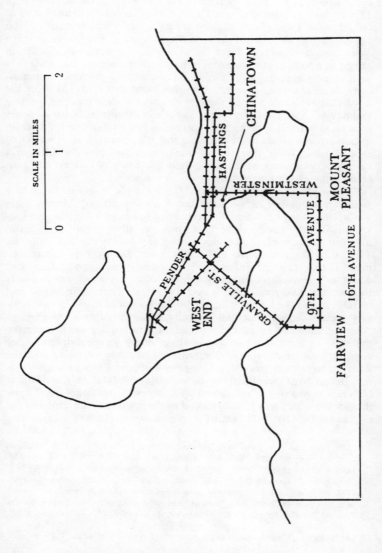

Map 3    Vancouver's Street Car System, 1891.
         SOURCE: N. MacDonald, "The Canadian Pacific Railway."

cil, yet there is no doubt that the CPR was well represented. Between
1886 and 1900, eight different CPR officials served on council, and in
only three of these years did council lack a direct representative of the
railway. As a rule most of these "CPR Aldermen" lived in or near the
West End and served Ward 1.[26] One must proceed cautiously in
attempting to assess the role and influence of this group, for, while they
undoubtedly gave the CPR a voice on council, this does not automatically show that they dominated or controlled that body. Most of the
day-by-day work of council focused on issues like the clearing, grading
and paving of streets, the construction of schools, fire halls, parks, sewer
lines and electrical systems, the supervision of police officers and health
inspectors, the issuance of licences, and the approval of a variety of
expenditures. While these issues might impinge upon the CPR in a
variety of ways, they were essentially urban issues of interest and concern to the entire community.

The problem of garbage disposal, for example, might be taken as a
fairly representative urban issue faced by city officials. In the early
1880s, when the population amounted to a few hundred, garbage disposal was of little consequence. Most persons either burned their trash,
tossed it in the inlet, or abandoned it in any nearby ditch or gully. By
1888, with the population at 8,000, the city health inspector recommended that garbage disposal be centralized and suggested the city
wharf as a convenient dumping ground. Not only was it conveniently
located but "the water here is deep."[27] A year later the problem had
increased, and the harbour master pointed out that to avoid "unnecessary stench" it was essential that city scavengers throw the garbage
"clear of the wharf stringers."[28] Apparently the request was not met, for
the right to use this wharf as a garbage dump was withdrawn. After
consideration of a variety of alternatives, the city began to rent the
wharf of the Union Steamship Company and paid $30 per month for its
use. But garbage soon filled the immediate area, and though invisible at
high tide, it was visible, messy and smelly when the tide went out.[29] Even
though the wharf was later extended to deeper water, the problem did
not disappear. By 1891, with the population at 13,000, council was

26. "CPR Aldermen" with their railway position and years on council were: L.A. Hamilton,
    land commissioner 1886, 1887; W.F. Salsbury, treasurer 1889, 1893, 1894; Dr. J.M.
    Lefevre, physician 1887, 1888, 1889; J.M. Browning, land commissioner 1890; H.E.
    Connor, freight agent 1892; J.J. Gavin, conductor 1892; H.G. Painter, accountant 1896,
    1897, 1898; H.B. Gilmour, machine shop foreman 1899. Identification was obtained
    primarily from Williams, *B.C. Directory*. The 1889 issue, pp. 254-71, was especially
    valuable. See also Patricia E. Roy, *Vancouver: An Illustrated History* (Toronto, 1980),
    p. 30.
27. Vancouver City Clerk Incoming Correspondence (cited hereafter as Van C Cl in Corr),
    1, 469.
28. *Ibid.*, 2, 1290-92.
29. *Ibid.*, 3, 3055.

warned that the garbage would soon "overwhelm us." In April of that year the Union Steamship Company offered to build a scow, capable of carrying fifty tons, which would carry city garbage once a day to deep water in Burrard Inlet. The scow cost $1,200, the services $200 per month.[30] This solved the problem for the time being, at least, and little further reference was made to it. The change from simple, inexpensive, pre-urban facilities to the complex, costly facilities required by a heavily populated area could be duplicated by many other municipal services. While the CPR might have had special interests, the issue certainly went far beyond the railway's interest alone.

While little evidence was uncovered which showed manipulation by CPR alderman on behalf of the railway, this does not mean that city council was indifferent to the CPR. Rather, during the 1880s council did virtually everything possible to cater to the company's needs and wishes, and gratefully received any favours that the CPR turned over to them. But there was nothing covert about such transactions. Council, editorial writers and the general public were all aware that Vancouver's prosperity resulted from the CPR's completion, and all agreed that any steps that would bind the railway to the city were not only sensible but eminently desirable. In 1886, 1887 and 1888 few would challenge the assertion "What's good for the CPR is good for Vancouver."

With its small population, limited tax base and lack of power or influence, the city was also in no position whatever to bargain with its powerful benefactor. The fact that council had to seek out private donations of lots for the construction of a city hall,[31] that it had to negotiate with the railroad for the purchase of school sites, firehall and locations and park lands,[32] and that it relied on the CPR for a hospital building, as well as for the free services of the company's physician to attend to the city's poor,[33] suggests a good deal about the city's weakness and limited resources. When council instructed the city clerk to ask the CPR if they would "entertain an offer of exemption from taxation on their workshops, yards, roundhouses, etc."[34] there was little doubt of the power-relationship between the two institutions.

As the years slipped by, this early relationship underwent a slow but unmistakeable change. The city became more and more assertive, independent and influential, while the relative power and influence of the railway declined. In part, this resulted from Vancouver's increased size and strength, but the sheer number of issues that had to be confronted

30. *Ibid.*, 5, 4618-22.

31. Vancouver City Council Minutes, cited hereafter as Van C.C. Min. 25 May, 7 June, 28 July, 2 August, 8 August, 15 October, 1886.

32. H. Abbott to T.F. McGuigan, 10 November 1887, in Van C C1 in Corr, 1, 35. L.A. Hamilton to D. Oppenheimer, Van C C1 in Corr. Additional vol. 1886-1891. Van C.C Min, 2 May, 7 November 1887 and 2 July 1889.

33. *Daily News Advertiser*, 10 August 1887, p. 2.

34. Van C.C. Min, 1 November 1886.

Vancouver Townscape, 1889.
Looking north from the Hotel Vancouver, there is still much vacant land, even in the rapidly growing residential districts of the West End.
SOURCE: Vancouver Public Library.

also contributed. Whether it involved the division of responsibility for the clearing and grading of roads on CPR lands, the need for protective gates at downtown railroad crossings, or the need for a draw in the CPR trestle across False Creek, city offficials pushed their case strongly.

By the 1890s, CPR officials were becoming exasperated by what they considered the assertive, self-willed, unappreciative behaviour of "their" city. General Superintendent Harry Abbott reminded council in 1891 that Vancouver property owed its value "almost entirely . . . to the existence of the CPR." He argued that the company derived only as "small income" from Vancouver business, and that the connection there was "immeasurably in favour of the city."[35]

But city officials were not impressed. In 1892 council authorized, and voters approved, the granting of a $300,000 bonus to the Burrard Inlet and Fraser Valley Railway. This local line was to go from Vancouver to the U.S. boundary at Sumas, where it would link up with the Northern Pacific Railway. The move essentially meant that the CPR would have to share Vancouver traffic with a major American rival. This was an especially painful blow to the CPR directors, who still considered Vancouver "their" city, and made doubly so by the fact that they claimed to pay 20 to 25 per cent of Vancouver's total taxes. While the contemplated line was not built at that time, the incident clearly showed that by the 1890s the needs and wishes of the city and of the railway no longer coincided. It also suggested that Vancouver had enough strength and economic clout to proceed as it saw fit. It could not ignore the CPR, but the utter dependence of the 1880s was clearly a thing of the past.[36]

### III

No single event marked the end of the great CPR boom, but by 1893 it was clearly over. During the ensuing depression the assertion of city rights and city powers continued, but the previous prosperity made the mid 1890s especially difficult, disappointing years. Immigration, employment, real estate values and tax revenues all slumped, and a whole series of glowing predictions about the future had to be revised downward. The Klondike gold rush of 1897-98 provided a brief flurry of excitement and prosperity, but it was not until the early years of the twentieth century that a vigorous new cycle of growth began.[37]

35. Abbott to City Clerk, Van C C1 in Corr, 9 February 1891.
36. See *News Advertiser*, 16 August and 21, 25, 28 September 1892. Vancouver's action had clearly shocked and pained CPR officials. Five years later in separate visits to Vancouver, both President Van Horne and Vice-President Shaughnessy referred to this incident. See *News Advertiser*, 8 September, 1897, p. 6, and 27 October 1897, p. 6. See also Roy, *Vancouver*, p. 42.
37. For an excellent overview, see Robert A.J. MacDonald, "Victoria, Vancouver and the Economic Development of British Columbia, 1886-1914," in Artibise, ed., *Town and City*, pp. 31-55.

Unlike the boom initiated by the construction of the CPR and the arrival of the first train in Vancouver, no single event announced the beginning of another expansionary phase in the early 1900s. Nor did any single organization influence development to the extent the CPR had done previously. Rather, growth during these years was a diverse, pervasive process shaped by many elements. It was sustained year after year and gradually transformed Vancouver from a community of 25,000 in 1900 to a substantial city of 115,000 in 1914, that stretched from Burrard Inlet to the Fraser River.[38]

If one factor can be said to be basic to all the changes that took place at this time it was undoubtedly rapid population growth. Vancouver's

### Table 1: Birthplace of Vancouver's Population[39]

|                      | 1891   | 1901   | 1911    | 1921    |
|----------------------|--------|--------|---------|---------|
| Total Native Born    | 6,739  | 14,452 | 43,978  | 79,920  |
| Maritimes            | 1,159  | 1,760  | 5,698   | 6,900   |
| Quebec               | 639    | 772    | 2,170   | 3,000   |
| Ontario              | 3,254  | 4,920  | 16,663  | 21,200  |
| Prairies             | 264    | 638    | 3,925   | 9,100   |
| B.C. & Others        | 1,423  | 6,362  | 15,522  | 39,720  |
| Total Foreign Born   | 6,970  | 12,558 | 56,423  | 83,300  |
| England & Wales      | 2,261  | 3,870  | 18,414  | 32,140  |
| Scotland             | 783    | 1,094  | 9,650   | 14,900  |
| Ireland              | 554    | 618    | 2,625   | 4,260   |
| United States        | 1,174  | 2,232  | 10,401  | 10,500  |
| Europe               | 505    | 1,100  | 6,141   | 5,489   |
| China & Japan        | 1,434  | 3,388  | 5,205   | 12,250  |
| Other                | 259    | 256    | 3,987   | 3,761   |
| Total Population     | 13,709 | 27,010 | 100,401 | 163,220 |

38. See Norbert MacDonald, "A Critical Growth Cycle for Vancouver, 1900-1914," in Gilbert A. Stelter and Alan F.J. Artibise, eds., *The Canadian City: Essays in Urban History* (Toronto, 1977), pp. 142-159.

39. The detailed data for 1891, 1901 are estimates only and are probably accurate to 15 per cent. They were extrapolated from New Westminster District data for 1891, and Burrard District data for 1901, with allowance made for the approximate number of British Columbian-born Indians in those districts. See *Census of Canada*, 1891, I, 332; *ibid.*, 1901, I, 418 and *Annual Report Department of Indian Affairs 1892*, 311-15; *ibid.*, 1901, 158-66.

The 1911 and 1921 data are from *Census of Canada*, 1911, II, 426-428, 440-41; *Census of Canada*, 1921, II, 315, 365-66. The 1921 data have been modified to include South Vancouver and Point Grey and are probably accurate to 5 per cent.

growth was not unique, for while it was surging ahead much the same thing was happening in Los Angeles, Seattle, Spokane, Calgary and Edmonton. All felt the consequences whether direct or indirect of an immense stream of immigrants from Europe to America, as well as a heavy westward migration of Canadians and Americans. Some idea of the size and nature of the city's population growth can be seen in Table 1.

It is clear that migration was a major factor in Vancouver's population growth, and that this was especially so between 1901 and 1911. Most of these migrants were of English-speaking background, and whether from Canada, Britain or the United States accounted for 85 per cent of the city's population. The large group of settlers from the British Isles was an especially significant feature. Such British immigrants had always played an important role in Vancouver. But after a great influx between 1900 and 1912 their importance was enhanced, and by the latter date they accounted for almost one Vancouverite in three. Whether the profound influence of these people is measured by the kinds of labour organizations and political parties they helped establish, by the types of homes and gardens they created, by the kinds of educational facilities they shaped, or by the general tone that thousands of British clerks, lawyers, merchants, and craftsmen gave Vancouver, it became well established in the opening decade of the century.

Persons of non-English speaking background were also quite numerous, with the Chinese by all odds the most distinctive. Originally introduced for construction work on the CPR, they gradually moved into a variety of menial tasks in the city's laundries, hotels, boarding houses and private homes. As in most North American cities, Vancouver's Chinatown was segregated, located not far from the CPR repair shops and the mud flats of False Creek. Whether viewed· with compassion, amused tolerance, or outright hostility, the Chinese were seen as an inferior breed well outside the main stream of society. Riots in 1887, 1907 and 1914 merely marked peaks in this antipathy.[40]

Vancouver also had a substantial sprinkling of Europeans by 1911, especially Italians, Scandinavians, Germans and Russians. Some lived with wife and family, but many were single men who lived in the city's hotels and boardinghouses only during the rainy winter months when

---

40. These riots did not go untouched. In a letter to President Theodore Roosevelt in 1907, Senator Henry Cabot Lodge observed, "I cannot help feeling a certain gentle interest in the performances now going on in Vancouver in regard to the Japanese and other Asiatics. It is a demonstration of the fact that the white people will not suffer Asiatic competition in their own country and I think it will perhaps make England a little less inclinded to preach in a patronizing way at us about San Francisco." Quoted in Thomas A. Bailey, *Theodore Roosevelt and the Japanese-American Crisis* (Stanford and London, 1934), p. 253. For more information on the Chinese in Vancouver, see Roy, *Vancouver*, *passim*.

they were not needed in the region's logging camps, shingle mills and fishing. The relatively small numbers of such immigrants, coupled with the instability of their living and working arrangements, worked against the development of distinctive ethnic neighbourhoods.

An additional symbol and yardstick of Vancouver's growth in the years before 1914, was the expansion of an urban transportation network. After the original construction of 1889-91, expansion during the depressed 1890s was limited to some three additional miles. It was not until the 1900-14 period that the major growth occurred.[41] During these hectic years numerous additional lines were laid in the downtown area, and tracks were pushed east, south and west bringing service to Hastings, Collingwood, South Vancouver, Marpole, Kerrisdale, Kitsilano and West Point Grey. By 1914 the 103-mile system of the B.C. Electric Railway Company served virtually the entire city.

A major extension of inter-urban railway lines also increased the range of urban settlement and activity. The CPR constructed the Lulu Island line in 1902, connecting the city with the fishing centre of Steveston. By 1914 three separate lines connected Vancouver with New Westminster and that city in turn was linked with the farming centres at Abbotsford and Chilliwack. The net result of all this construction was that Vancouver became the focal point of a diverse, regional transportation network. For all practical purposes anyone living within an eight-mile radius of downtown could get there in thirty-five minutes, while a rush order from Chilliwack, forty miles away, could be delivered within a few hours.

At the time the first tramway lines were built in 1889-91 the great bulk of Vancouver's population was concentrated in a small area north of False Creek, between Granville and Main Streets. Since this was precisely the area served by the original trams, one can say that in the earliest years the tramway tended to follow the population build up. But as early as 1891 as the tramways were pushed across False Creek along Broadway to complete the Fairview belt line, there is no doubt that the trackage was well in advance of settlement, and tended to influence settlement patterns. The entire district of Fairview and Mount Pleasant, for example, had only 186 households at the time it received its first rails.[42]

The districts that received tramway service in the great surge of construction carried out between 1900 and 1914 were not as thinly settled as Fairview and Mount Pleasant had been in the early 1890s. Yet these rail facilities preceded any significant population growth, and it was the expectation of a rapid rise in real estate values that influenced both the

41. H. Bartholomew and Associates, *A Plan for the City of Vancouver* (Vancouver, 1929), pp. 87-90, 131-34.
42. Patricia E. Roy, "The British Columbia Electric Railway Company 1897-1829," Ph.D. Thesis (University of British Columbia 1970), pp. 22-34.

municipal council and B.C. Electric Railway Company in their decision to go ahead. The municipality of Point Grey, for example, had only 854 dwellings and a population of 4,320 in 1911, yet in 1912 it had approximately sixteen miles of tramways.[43] Much the same pattern was evident in the development of the inter-urban lines. The region served by the Vancouver–Lulu Island line was virtually unsettled when the CPR laid its track in 1902. Yet although these lines exceeded Vancouver's immediate needs, they opened up lands in a variety of districts, and brought about a rapid dispersion of population.

Along with the development of these urban and inter-urban railway lines went an incredible amount of construction. During the CPR boom years from 1886-1892 construction averaged about $1,500,000 per year, but as the 1900s unfolded construction raced ahead at an unprecedented rate. Each year of the decade seemed to set some kind of record only to be exceeded by an even greater volume of activity the following year. An idea of the size, growth, and cyclical nature of this activity can be gathered from Figure 1.

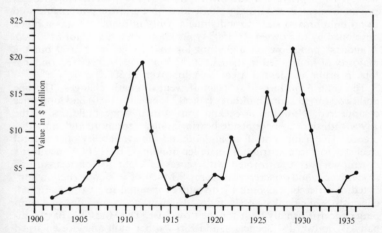

Figure 1: Value of Building Permits in Vancouver 1902-1936[44]

43. *Census of Canada* 1911, I, 249; J.A. Paton, "The Inside Story of Point Grey," *B.C. Magazine* July 1911, pp. 735-37.
44. *Vancouver Annual Report 1936*, "Report of Building Inspector," p. 91, lists the yearly value of building permits issued in the city from 1902 to 1936.

It is evident from the graph that construction activity rose steadily throughout the decade, peaked in 1912 and the dropped sharply. Some idea of its magnitude can be gained from the fact that in the entire period from 1900 to the end of the Second World War, it was only in the banner year of 1929 that the value of construction exceeded that achieved in 1912. There can be little doubt that the diverse demand created by this activity played a major role in sustaining Vancouver's growth.

Construction spread over the entire city and ranged from small bungalows on 33-foot lots to massive department stores in the central business district. The great bulk of it consisted of single-family frame houses in the West End, Kitsilano, Fairview, Mount Pleasant and Hastings. In 1906 alone, 23 houses were constructed on Sixth Avenue in Kitsilano, 35 on Seventh, 47 on Eighth, 31 on Ninth and 30 on Tenth, while Comox, Robson, Nelson, and Barclay in the West End each had about 20. In the peak year 1912, 2,224 houses, 217 factories and warehouses, 293 offices and stores, and 218 apartments were built.[45] Whether the land was bought from provincial authorities, the CPR, real estate concerns, or private owners, it was usually subdivided on a gridiron pattern and sold in an unimproved state to individuals. Only infrequently was a block developed by one owner. In 1905 typical housing lots sold for $100-200. A modest bungalow on a 33-foot lot sold for under $1,000 but the majority of houses fell in the $1,500-$2,500 range. A very few on large lots in prime residential areas sold for over $3,500.[46]

The central business district underwent dramatic changes. The one and two-storied frame buildings typical of the early 1880s had long since disappeared. Now the three and four-storied stone buildings of the 1890s with their heavy outside bearing walls, were more and more displaced. The rising cost of urban land and the availability of structural steel led to the construction of taller buildings. By 1914 Vancouver's downtown was transformed with scores of eight to fourteen-storied stone, brick, and concrete structures.[47] Much of this construction consisted of office blocks built by investors for rental to a variety of small concerns, while banks, hotels, railroads, insurance companies, and shipping concerns usually built for their own use. The Hudson's Bay Company, Woodward's, Spencer's, and Birks either built anew or enlarged their department stores while the Court House, Post Office, Hotel Vancouver, Carnegie Library, and the World Building provided prominent new landmarks.

With immigrants arriving in unprecedented numbers, tramways

45. *Vancouver Daily Province*, January 19, 1907, pp. 25-27, provides a detailed street by street breakdown of all building permits issued in the city in 1906. *News Advertiser*, January 5, 1913, p. 29 gives data on 1912.

46. *Ibid.*, January 1905, p. 6.

47. *Ibid.*, January 26, 1913, p. 21, has an excellent composite photograph of the major buildings that had recently been completed.

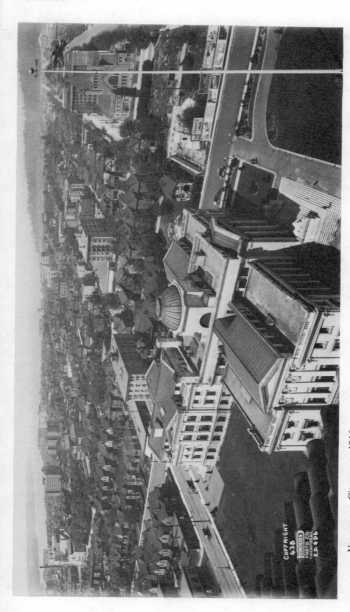

Vancouver Cityscape, 1916
Looking west from the Hotel Vancouver, after major expansion and population growth past 100,000.
SOURCE: Vancouver Public Library

opening new districts, and construction booming, real estate promotion and speculation became an integral part of the Vancouver business world. Hundreds of persons became real estate agents – the Vancouver Directory of 1910 listed 650 – and thousands of Vancouverites participated. Whether they dabbled or plunged depended on personal whim and available resources, but few were indifferent. As one exuberant promoter exclaimed, "Do you know how I would write a dissertation on 'How to be rich in Vancouver?' Take a map of the lower peninsula, shut you eyes, stick your finger anywhere and sit tight."[48]

Such activity was not a new phenomenon for the city. At the time of the original boom of the late 1880s many who had speculated in real estate reaped handsome returns. But in the period from about 1890 to the early 1900s prices rose only modestly, about six per cent per year.[49] The real surge began about 1904 and continued unabated to 1912. In an annual business survey the *New Advertiser* pointed out:

> With the real estate dealers 1909 has been a banner year in Vancouver. Three years ago it was prophesied that the real estate business had reached its apex and must decline; today prices have doubled and trebled and real estate brokers have probably made more money in 1909 than even in the early days of the great real estate movement about 3 years ago. Five years ago lots might have been purchased within the city limits for $50 each; now it would be difficult to secure a single lot anywhere in Vancouver for less than $600.[50]

With appropriate changes in prices the comment made about 1909 could be repeated for 1904, 1907, or 1912.

Among the numerous examples of profitable real estate transactions was the small 26-foot lot on Pender Street that was purchased from the CPR in 1887 for $480. The original owner held it for three months and sold it for $600. By March 1888 it sold again, for $1,500. It did not change hands again until 1904 when it sold for $4,000, and just two years later it sold once more, this time for $6,000.[51] Similarly a 52-foot lot on Hastings Street owned by ex-mayor C.S. Douglas, was sold in 1904 to the real estate firm of Martin and Robinson for $26,000. They held it for four years and sold it in 1908 to N. Morin for $90,000 and he in turn sold it just one year later for $175,000.[52]

The boom in real estate peaked in 1912, and for those who held out for still higher prices it was a painful experience to see their paper profits suddenly evaporate. While evidence abounds of the dramatic profits between 1887 and 1889 and from 1904 to 1912, it is not suprising that

48. R.J. McDougall, "Vancouver Real Estate," *B.C. Magazine*, June 1911, p. 607.
49. This was estimated from a sampling of representative house prices listed in Vancouver newspapers.
50. *News Advertiser*, January 2, 1910.
51. *News Advertiser*, January 22, 1905, p. 6.
52. McDougall, "Vancouver Real Estate," p. 603.

few people were willing to divulge the extent of their losses after 1912. Then, as now, no one liked to admit that he had failed to exploit a "sure thing" – or even worse that he had lost heavily. Yet there are some examples of the downslide. One owner of an expensive corner lot on Granville and Robson who had refused an offer of $250,000 in 1910, sold for $122,500 in 1916, while another disillusioned investor sold his Granville-Helmcken lot for $40,000 in 1917 having rejected an offer of $125,000 in 1912.[53]

Another significant process during these years was the way in which Vancouver's residential, commercial and industrial districts became firmly established. The process had begun with CPR plans and decisions in the 1880s but it was not until the mushroom growth and expansion of 1900-1912 that the entire pattern became clear. To appreciate this development one might consider the city at the height of the Klondike gold rush in 1897-98. At that time Vancouver still occupied the small peninsula of land between Burrard Inlet and False Creek. It was strongly oriented to the waterfront with about 80 per cent of its population of some 22,000 living within a mile of the CPR's depot, yards and wharf near the foot of Granville Street.[54] The tramway along Granville, Hastings and Main clearly identified the city's core and most of the city's retail businessess were concentrated in that area, especially on Water, Cordova and Hastings Streets. Two industrial districts could also be distinguished, one on the waterfront with the CPR facilities, the B.C. Sugar Refinery, and the Hastings Sawmill, another on the north side of False Creek with lumber and shingle mills, wood working plants, as well as the freight shed and roundhouse of the CPR. Homes were scattered throughout the community and some settlement had spread into East Hastings, Mount Pleasant and the West End. Legally, Vancouver covered some ten square miles, but essentially the city was concentrated in a two square mile area, and viturally everyone could walk to work or to downtown stores in fifteen to twenty minutes.

By 1912 the patterns that had been outlined in 1898 were clearly delineated, and there was no doubt about the future layout of the entire city. Population had jumped to 115,000 and was spread over much of the peninsula from Burrard Inlet to the Fraser River, or some thirty square miles in all, though the legal city amounted to sixteen square miles. Business was still concentrated in the same core, but the cramped and limited quarters of Water and Cordova had evolved into a wholesale and warehouse district. Hastings, Pender, Granville and Main were now the main business thoroughfares, and had been sharply extended and built up. The West End, early recognized as a desirable residential

---

53. J.P. Nichols, "Real Estate Values in Vancouver," *Vancouver City Archives*, (April 1954), p. 26.

54. *News Advertiser*, January 1, 1911, p. 17, provides a detailed description of the changes during the previous decade. An excellent print entitled "Panoramic View of the City of Vancouver, British Columbia, 1898," is available in Vancouver City Archives.

area, was now heavily settled. Settlement had also followed the B.C. Electric lines into Fairview, Kitsilano and Point Grey, while Fourth Avenue and Broadway had acquired a variety of small retail outlets. Although no construction had taken place, extensive lands had been set aside for the future University of British Columbia. To the south Kerrisdale was developing, while Shaughnessy Heights, in the heart of the CPR's land grant, was being hailed as the city's most elegant area. Throughout these years the prosperous, influential and socially prominent Vancouverites continued to settle in the western sections of the city, while skilled workers and labourers gravitated eastwards. In the East End lots were smaller and cheaper, with parks, playgrounds and schools more limited. Grandview, Collingwood and South Vancouver all grew sharply and the annexation of Hastings townsite in 1911 extended the eastern boundary of the city to some four miles from downtown. All of these districts were overwhelmingly residential, with single-family frame houses virtually universal.[55]

Industrial districts too were clearly delineated. Shipping facilities on the waterfront had been expanded, but most industrial activity concentrated in the False Creek Basin.[56] In the 1890s it seemed likely that industry would remain localized on the north shore of the creek, and that the higher land of the south shore would evolve into a prime residential area. A number of substantial houses were built in Fairview and Mount Pleasant, but as the decade unfolded significant changes occurred. The CPR constructed a branch line along the south shore of the creek in 1902, and this hastened the industrial development of the entire area. By 1903 the British Columbia Mill, Timber and Trading Company had shifted its major plant from the north to the south shore and this was followed by a whole series of lumber and shingle mills, machinery depots, ship yards, gravel and cement plants. The construction of additional lines and marshalling yards by the Great Northern, and the Canadian National, the filling in of the tide flats east of Main Street, and the creation of Granville Island in 1914 marked the culmination of the industrialization of the False Creek area. Marpole, on the north arm of the Fraser River, also evolved as an industrial area during these years. Its growth was aided by inter-urban rail facilities while the low land costs and the availability of river transport hastened the development of lumber and shingle mills, an abattoir, and a flour mill.[57]

The problems that this growth entailed proved a severe burden for a

55. Deryck Holdsworth provides an analysis of the distinctive characteristics of urban development in Point Grey, Shaughnessy and South Vancouver in "House and Home in Vancouver: Images of West Coast Urbanism, 1886-1929," in Stelter and Artibise, *The Canadian City*, pp. 186-211.

56. Dennis M. Churchill, "False Creek Development," M.A. Thesis (University of British Columbia,1953), pp. 48-77.

57. Jeremy Barford, "Vancouver's Inter-urban Settlements," B.A. Essay (University of British Columbia, 1906), p. 18.

host of administrative officials in city hall.[58] Whether one was responsible for the construction of roads, bridges or sewer lines, handled the intricacies of land zoning changes, designed new schools or hospitals, carried out inspections of hotels, restaurants, and laundries, or issued permits to carpenters, plumbers and electricians, the work seemed endless. Rather than being able to stay ahead of the demands placed on them, most officials worked from crisis to crisis, and it is not suprising that resignations were frequent, and turnover high.[59]

Mayor and city councils too felt the strain of the city's mushrooming growth. Meetings and committee assignments took more and more time, especially since council felt obliged to discuss and decide a host of minor issues. Yet although aldermen complained of overwork there was at the same time a sense of excitement and satisfaction with the city's growth. The inevitability of such growth had been endlessly asserted, and at long last it was actually happening. Like their counterparts in Winnipeg, Edmonton, Seattle and Los Angeles, the aldermen of Vancouver delighted in this growth, for it meant more people, more jobs, bigger payrolls, and greater opportunities for all. Since many of them were directly involved in the city's business life – of the 1912 council of sixteen, for example, at least ten were in real estate or general business – it is not surprising that they looked kindly on this growth and wanted to see it continue.

Ten different mayors and some eighty different aldermen served Vancouver between 1900 and 1914, but with the possible exception of the administration of Louis D. Taylor, the general set of ideas which guided their actions was consistent. Beyond the conviction that growth was desirable, was the equally strong conviction that property owners should be free to develop their property as they saw fit, and that private enterprise should be relied on to provide the basic municipal services and utilities. Vancouver owned and operated its own water works system, but this was the exception rather than the rule. The British Columbia Electric Railway Company provided both streetcar service and electric lighting. After 1904 when it took over companies controlled by the Mackenzie-Mann syndicate, it provided gas service as well.[60] Although complaints about inadequate service and high prices were perrennial, and although rates charged were apparently much higher than in comparable Canadian and American cities, city council's opposition to municipal ownership was shared by the public. When plebiscites were held seeking authorization to borrow funds for the purpose of these facilities they were decisively defeated.[61]

58. The following paragraphs on the activities and beliefs of various municipal officials are based primarily on an examination of Vancouver City Council Minutes, 1899-1915.
59. Vancouver City Council Minutes, September 17, 1906; April 8, April 15, May 13, 1907.
60. *British Columbia Electric Railway Co., 29 Years of Public Service* (Vancouver 1926), p. 39.
61. See *News Advertiser*, January 19, 1906, and letter to editor by C.M. Woodward in *ibid.*, January 9, 1914, p. 4.

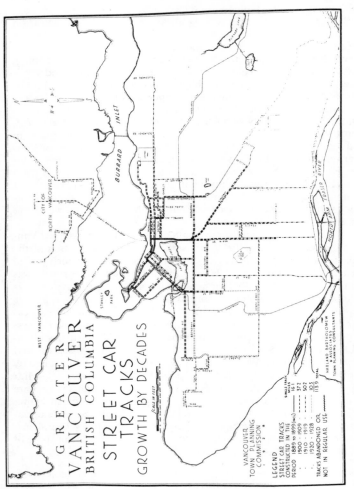

Map 4    Vancouver's street car system
         SOURCE: Harland Bartholomew, *A Plan for the City of Vancouver* (1929).

In line with the idea that private enterprise should be given free rein was the council's belief that city government should function primarily as a profitable, efficient, businesslike concern. Their task was not to direct or plan city growth, but to see that municipal services were available, that city regulations were met, and that funds were spent in an honest, impartial manner. Many considered the mayor-council system both unwieldy and inefficient, especially as new wards were added to the city and the number of aldermen increased from ten in 1902 to sixteen by 1912. Various charter amendments and new forms of city government were proposed all of which aimed at withdrawing power from the elected ward representatives and concentrating it in a small, highly centralized board.[62] In 1907-1908, for example, Mayor A. Bethune suggested that the city be managed by a board of control consisting of the mayor and three councillors. In 1911 council considered at length the possibility of establishing a commission form of government. And in 1914 the idea of a board of control was again brought forward.[63] None of these proposals were implemented, yet their recurrence suggests that municipal government was hard pressed to keep up with the demands of rapid growth and change.

While there is no doubt that city council favoured private ownership and private enterprise, there is some evidence that during these years a sense of public needs and public rights was also beginning to assert itself.[64] Throughout the decade council insisted on the need to protect the public interest, especially in its right to easy access to waterfront facilities. One significant step in this direction was the acquisition by the city of a number of private lots on English Bay for public bathing and picnic areas. The steady acquisition of a variety of parkland, the resistance to Mr. Ludgate's attempts to obtain Deadmen's Island and establish a lumber mill there, and the vigour with which both council and public responded to any indication that Stanley park might be developed

---

62. In this respect Vancouver closely followed the patterns revealed in a number of American and Canadian cities. See Samuel P. Hays, "The Politics of Reform in Municipal Government in the Progressive Era," in A.B. Callow, ed., *American Urban History* (New York, 1969), pp. 421-39; and the articles by John Weaver and James Anderson in Alan F.J. Artibise and Gilbert A. Stelter, eds., *The Usable Past: Planning and Politics in the Modern Canadian City* (Toronto, 1979).

63. Vancouver City Council Minutes, December 28, 1907; January 6, 1908; January 13, 1908; January 19, 1911; May 30, 1911; January 24, 1912; and 1914 *passim*. See also Roy, *Vancouver*, pp. 73-78.

64. *Ibid.*, November 4, 1902; December 8, 1902, May 4, 1903; February 29, 1904; April 24, 1904; and February 27, 1905.

or "improved" all indicate that the dominant philosophy did not go unchallenged.[65]

During these hectic years Vancouver gradually lost its intimate, folksy, small-town quality. The chatty, local columns that had been a staple item in the eight-page local papers of the 1890s faded into the background as the 1900s unfolded. By 1912 weekend editions ran from forty to sixty pages. National and international events dominated the news, special sections on business, sports, society, entertainment, and advertisements abounded. One might still recognize a prominent lumberman, or real estate broker as he strolled along Hastings or Granville, but this became more and more the exception as the years went by. There were just too many new companies, new offices, new faces and new millionaires. Unless one belonged to the Vancouver Club, the Terminal City Club, or was active in the Board of Trade one had little chance to know who these men were, let alone recognize them. The visiting salesman from Montreal, Chicago or London would still consider the city a comfortable, prosperous, low-keyed community, but for the local resident who had arrived at the time of the Klondike gold rush, or possibly even before the CPR, Vancouver had been transformed.

Vancouver remained a trading, transportation, and service centre with most of its work force still engaged as carpenters, clerks, school teachers, and real estate agents.[66] Yet the scope and focus of its economy had changed. No longer was it geared primarily to local and coastal needs; rather, it was more and more part of a national economy.[67] Its local entrepreneurs had to share the stage with branch managers, salesmen and officials of national concerns who implemented policies decided on in Toronto and Montreal. But whether Vancouver's businessmen promoted mines in the Kootenays, publicized investment opportunities in the Okanagan, sold timber and shingles on the prairies, or lobbied for elevators to handle Saskatchewan wheat, they operated far beyond the city's boundaries.

There is little doubt that economic and political power remained highly concentrated. Vancouver's business elite were overwhelmingly from an English speaking, Protestant background. One analysis of some three hundred business leaders active in Vancouver in 1910 showed that

---

65. Both Stanley Park of 1,000 acres, and Hastings Park of 160 acres had been granted to the city without cost in the 1880s. Between 1902 and 1912 Vancouver spent approximately a million dollars for sixteen parcels of parkland, that together amounted to about 60 acres. Bartholomew, *A Plan for the City of Vancouver*, pp. 177-78. For an example of public concern on Stanley Park, see Letters to the Editor, in *News Advertiser*, January 21, 1903, and January 22, 1903. See also W.C. McKee, "The Vancouver Park System, 1886-1929: A Product of Local Businessmen," *Urban History Review*, 3-78 (February 1979), pp. 33-49.

66. *Census of Canada* 1911, VI, "Occupations," 286-96.

67. McDonald, "Victoria, Vancouver and the Economic Development of British Columbia."

over 90 per cent of them were from Canada, Britain, and the United States. Most were Protestant and especially Presbyterian or Anglican.[68]

Business leaders of non-English-speaking background were few and far between. The one conspicuous exception during this decade was Alvo von Alvensleben, a German-born financier, promoter, and real estate broker.[69] He had the good fortune to come to Vancouver in 1904 just when the city was on the verge of dramatic expansion. As a man with some personal wealth and as an agent for European capital, Von Alvensleben prospered. By 1912 he was one of the most prominent businessmen in the entire city with a host of mining and lumbering organizations under his control. He was particularly active in promoting Kitsilano real estate, and with his Canadian-born wife and a handsome Kerrisdale residence enjoyed major prominence. With the First World War, however, his small empire disintegrated and as an enemy alien he quickly lost virtually everything. By 1915 he had disappeared into relative obscurity in the United States.

A much more representative member of Vancouver's business elite was Robert McLennan, founder of the wholesale firm of McLennan and McFeely.[70] Born in Pictou, Nova Scotia, of Scottish-Presbyterian background, McLennan left the Maritimes in his early twenties and worked briefly in Winnipeg before moving to Victoria in 1884 where he opened a small building-supply store. He did very well there and invited Jim McFeely, an old Winnipeg friend then working in Minneapolis, to join him. The young partners prospered. They were aware of the great potential for growth on the B.C. mainland, and by the mid-1890s had transformed their main centre of operations to Vancouver. Their company profited from the Klondike gold rush of 1897-98 and it was in a favourable position for the dramatic expansion of the following decade. As a member of the local elite McLennan was prominent in many phases of Vancouver's life. Not only did he head one of the biggest wholesale firms in the region, he also served on the board of governors of the University of British Columbia and McGill, ran as a Liberal candidate for a seat in the provincial Legislature, was a member of the prestigious Vancouver Club and the Terminal City Club, and served a term as president of the Vancouver Board of Trade.

An additional pattern that remained constant during these years was the sense of pride in the city and its growth. Business leaders and editors never tired of pointing out that unlike Victoria or Seattle all of the

---

68. See Robert A.J. McDonald, "Business Leaders in Early Vancouver, 1886-1914," Ph.D. Thesis (University of British Columbia, 1977), especially Chapter 8.

69. Mr. Ronald D'Altroy of the Vancouver Public Library kindly provided me with a tape recording of a speech given by Von Alvensleben to the German-Canadian Association on January 30, 1961. It is a vivid, moving statement of his meteoric rise and fall. See also *Vancouver Province*, June 19, 1909, p. 17; and *B.C. Saturday Sunset*, September 25, 1909, p. 7.

70. F.W. Howay and E.O.S. Scholefield, *British Columbia* (Vancouver, 1914), 4, pp. 1186-87.

development in Vancouver was the result of local initiative, unaided by any government subsidy or help.[71] Though this distorted the truth there is no doubt that the concept was fervently held. Similarly, although the more candid Vancouverite might admit that the city had its fair share of drunks, prostitutes, thieves, and derelicts, no one doubted that Vancouver was a decent, respectable, law-abiding city and as such differed sharply from raucous, disreputable American cities. Few challenged the view that most of the city's troublemakers came from Bellingham or Seattle.[72] Conversely, since the city police force and the park system were modelled on British institutions, they were indisputably superior to their American counterparts.[73]

IV

Although various examples of continuity and stability can be uncovered, the dominant motif in Vancouver's development from the turn of the century to the First World War was one of growth and change. The quadrupling of population, the massive influx of migrants from the British Isles with their distinctive styles and beliefs, the disappearance of the intimate small-town quality and the development of a more impersonal city with an increased national and international orientation, the construction of an elaborate street railway system and the subsequent expansion of the physical city from about two square miles to some thirty square miles, the establishment of well-defined commercial, industrial, and residential districts with distinctive shadings in price and prestige, the construction of miles of streets and sidewalks, and thousands of homes, offices, stores, and business blocks – all combined to make this one of the most significant periods in Vancouver's history.

By 1914 Vancouver's basic urban landscape had been determined. It is true that this landscape would continue to evolve and change during the First World War, the relative prosperity of the 1920s and the prolonged agony of the 1930s. Yet these changes would be modifications only. The character and appearance achieved by 1914 would still be clearly recognizable not only in 1924, but in 1934 as well as 1944. Not until the 1950s and 1960s would Vancouver again experience a transformation of comparable magnitude.[74]

71. *Weekly News Advertiser*, May 28, 1909, p. 6.
72. *Daily News Advertiser*, January 2, 1902, p. 12; January 15, 1905, p. 8; and January 28, 1908, p. 5.
73. Vancouver City Council Minutes, February 29, 1904; *B.C. Magazine*, VII (1911), pp. 558-61.
74. For a general history of Vancouver, see Roy, *Vancouver*.

# Canadian Resource Towns in Historical Perspective

GILBERT A. STELTER and ALAN F.J. ARTIBISE

In a provocative analysis of urban design, Edmund Bacon suggests that the form of cities "always has been and always will be a pitiless indicator of the state of civilization."[1] This statement is particularly applicable to the building of Canadian resource towns, for these communities clearly reflect the harsh reality of Canada's development as an industrialized and urbanized nation. This reality becomes more apparent when these communities are examined from a historical perspective. Since most resource towns remain small and static, they – more than most communities – continue to be a product of past decisions. The original site near the resource base, the early division of land, and even the substandard quality of the buildings impose a measure of permanence. So too do the political structures and value systems rooted in the initial phase of building and growth. The way in which the resource towns were planned and built usually parallelled changes in the Canadian approach to town planning. It is, therefore, crucial to know whether a town was founded before 1920, between the world wars, or after 1945, because distinct forms and institutional arrangements appeared in each period.

The study of resource towns also involves the question of whether or not they can be considered *urban* communities. One authority on the subject, sociologist Rex A. Lucas, has taken issue with the census defini-

SOURCE: This is a revised and expanded version of the article which appeared in *Plan Canada*, 18 (March 1978), pp. 7-16. Reprinted by permission of the editor and the Canadian Institute of Planners.

1. Edmund Bacon, *Design of Cities*, rev. ed. (New York: Penguin Books, 1976), p. 13.

tion which includes any community of 1,000 or more in the urban
category. He points to the fact that small, single-enteprise towns do not
possess the complex institutional life usually associated with urbanism
in large cities.[2] Most resource towns certainly would not qualify for
Louis Wirth's classic definition of a city as "a relatively large, dense,
and permanent settlement of socially heterogeneous individuals.[3] In a
similar vein, Jane Jacobs equates an urban economy only with cities, for
cities generate growth from their own local economies. Towns, on the
other hand, she says, do not generate growth from their own economies,
and the exports from a town's region (such as resources) do not produce
self-generating growth.[4]

While resource towns are not necessarily urban places, they do serve
as agents of urban metropolitan centers. Economically, the towns are
usually entrepôts, collecting staples from their region for shipment to
the metropolitan centre for final processing and, in turn, distributing the
manufactured goods received from the metropolis. Culturally, the
towns represent the metropolis in the rapid transmission of the metropolitan center's style of life to a new frontier. This involves not only the
traditional institutions such as churches and schools and the manner of
social organization, but also the physical planning and layout of the
towns.[5]

## COMMON CHARACTERISTICS OF RESOURCE TOWNS

In several respects, Canadian resource towns* resemble a host of towns
throughout the world which are based on the extraction or processing of

*The definition of resource town used in this article is sufficiently broad to include any
community whose economic base depends mainly on resource extraction or processing. The
definition includes diverse communities such as mining boom towns like Dawson City and
Cobalt and service and supply centers like Whitehorse and Sudbury. It also includes the
host of "single-enterprise" communities – mostly company towns – which are often equated
with the term "resource town" to the exclusion of the other types mentioned above.

2. Rex Lucas, *Minetown, Milltown, Railtown: Life in Canadian Communities of Single
   Industry* (Toronto: University of Toronto Press, 1971), pp. 4-11. For general bibliogra-
   phical references to Canadian resource towns, see Alan F.J. Artibise and Gilbert A
   Stelter, *Canada's Urban Past, A Bibliography to 1980 and Guide to Canadian Urban
   Studies* (Vancouver: University of British Columbia Press, 1981), pp. 12-14.
3. Louis Wirth, "Urbanism as a Way of Life," in Louis Wirth, *On Cities and Social Life*
   (Chicago: University of Chicago Press, 1964), pp. 60-83.
4. Jane Jacobs, *The Economy of Cities* (New York: Vintage Books, 1970), especially the
   summary of definitions, p. 262.
5. For an elaboration of this approach, see Gilbert A. Stelter, "The Urban Frontier in
   Canadian History," in A.R. McCormack and Ian Macpherson, eds., *Cities in the West*
   (Ottawa: National Museum of Man, 1975), pp. 270-86. See also Doug Baldwin, "Imita-
   tion vs. Innovation: Cobalt as an Urban Frontier Town," *Laurentian University Review*,
   vol. 11 (1979), pp. 23-42.

a non-agricultural resource such as minerals, forest products, or hydro-electric power. The basic common characteristic, which transcends time and place, is the resource town's status as an adjunct of an industrial enterprise. Ervin Galanty has traced towns of this type back to antiquity and concluded that "as a rule, development goals and objectives of efficient production receive priority over welfare considerations."[6] A second common characteristic is related to the first – a lack of any local control over the town's economic development. The economic base is controlled by outside corporations or governments who determine the nature and extent of the extractive or processing activity and thereby determine the size of the local work force and the degree of local prosperity or growth. Fluctuations between boom and bust depend on the vagaries of the international market in resources or corporate and government decisions, not on local initiative as is often the case with other types of communities. Recurring fluctuations generate a feeling of insecurity and impermanence, a feeling which is accentuated in mining towns by the knowledge that the resource base eventually will be exhausted.

A third common feature of resource towns is a simplified occupational structure. The middle class is relatively weak and usually confined to a small group of merchants and professionals. Several factors discourage the development of a diversified economy which would result in a more heterogeneous work force. Isolation from major markets, relatively high wages paid by resource industries, and high development costs combine to prevent the influx of secondary industry. One result is that the male-female ratio in resource towns is usually skewed heavily in favor of men since there are few employment opportunities for women.

A final common characteristic is physical appearance. Although recently built resource towns tend to resemble the new suburbs of large cities, older towns are generally ramshackle communities whose townscape is dominated by the mine or mill. Residential and commercial buildings, often of an inferior quality, are repeated monotonously, especially in the older, standardized company towns. Haste was usually necessary to house the population in the initial phases of construction. Other problems arose from the nature of the terrain near location-bound resources. Of equal importance in determining the second-rate quality of construction was the well-founded lack of confidence, on the part of corporations and workers alike, in the long-term life of the community.[7]

6. Ervin Galanty, *New Towns: Antiquity to the Present* (New York: George Braziller, 1975), p. 38.
7. For comparative examples, see R.T. Jackson, "Mining Settlments in Western Europe: The Landscape and the Community," in R.P. Beckinsale and J.M. Houston, eds. *Urbanization and Its Problems* (Oxford: Blackwell, 1968); John Coolidge, *Mill and Mansion* (New York: Columbia University Press, 1942); and James B. Allen, *The Company Town in the American West* (Norman, Oklahoma: University of Oklahoma Press, 1966).

## DISTINCTIONS AMONG RESOURCE TOWNS

### *Origin of the Population*

While Canadian resource towns have a great deal in common both with each other and with those elsewhere, it is possible to delineate several distinct characteristics. One basic distinction involves the origin of the town's population. Many of the resource-based towns of eastern Canada – the Atlantic provinces and Quebec – are located near established populations. The industrial population of the resource town is often drawn from the surrounding fishing, lumbering, and agricultural population. An example is Wabana, a coal-mining town on Bell Island, Newfoundland. The mining population retained close ties with its old maritime traditions and the old institutions continued to compete successfully for the population's loyalty against the inroads of what might be called industrial traditions.[8] A similar cultural dichotomy exists in less isolated Quebec resource towns like Asbestos and Thetford Mines. While their populations are drawn from a wider territory than is the case with textile towns such as Drummondville, the social pattern is similar: the labour force is largely native and the management alien.[9] In sharp contrast are most of the resource towns of Ontario and western Canada, where much of the work force and management is drawn from populations remote from the town or even from outside the country. These are more consciously new towns, representing metropolitan viewpoints. They are intrusions into largely uninhabited areas, isolated physically and culturally from rural connections.

### *Function*

A second major distinction among resource towns is based on the decision-making process involved in creating and maintaining the community. Some towns are the product of decisions by a single company or a government; others represent the outcome of a multiplicity of decisions by a number of companies, or by the residents of the community itself. The two types which emerge are service and supply towns, which sometimes had their origin as boom towns, and the company town, which generally remained a small, static community closely attached to one industry's operation.

### *(1) Service and Supply Towns*

Some of the service towns originated as colourful mining camps which bloomed virtually overnight. Much of the population involved with resource development in the late nineteenth and early twentieth centur-

8. Peter Neary, *Bell Island: A Newfoundland Mining Community, 1895-1966*, Canada's Visual History Series, 12 (Ottawa: National Museum of Man, 1974).

9. Everett C. Hughes, "Industry and the Rural System in Québec," in Marcel Rioux and Yves Martin, eds., *French-Canadian Society* (Toronto: McClelland and Stewart, 1964).

ies was housed in towns of this sort. Haphazardly built, these towns lacked centralized direction or control by company, government, or even an established municipal corporation. The prototype for this kind of town was Dawson City in the Klondike at the turn of the century. An "instant town" in the true sense of the word, Dawson was the product of the sudden and intense nature of gold-mining development. Harold Innis described the phenomenon as comparable to the onset of a cyclone. Stages of development which took generations in established regions of the country were telescoped into only a few months.[10] The phenomena appeared in all parts of the country, as well as in the Yukon. Good examples were the mining towns which sprang up between 1880 and 1900 in the Kootenay region of southern British Columbia. One of the mining camps was Sandon, with a population of about 2,000, including the notorious camp followers, the promoters, drifters, lawyers, gamblers, and prostitutes. Nelson and Trail were primarily smelter towns, while Kaslo was the largest town of the region (reported to be 6,000) based on its role as a transportation and distribution centre.[11] Ontario, too, had its dramatic mining camps following the discovery of silver in Cobalt in 1903 and gold in the Porcupine area in 1907.[12] Cobalt was the most spectacular, attracting a population of approximately 25,000. A contemporary observer effectively categorized it in terms of North American mining camps:

> The town is by no means typical of the true mining camp, perhaps because it is situated in Ontario, or does its comparative nearness to civilization influence it? Bret Harte would have passed it by, for "Roaring Camp' could be more to his taste. Only in the fantastic variation of its straggling structures, in the daily

10. Harold Innis, *Settlement and the Mining Frontier* (Toronto: Macmillan, 1936), pp. 173-212.
11. Paul A. Phillips, *The British Columbia Mining Frontier, 1880-1920*, Canada's Visual History Series, 21 (Ottawa: National Museum of Man, 1977). See also W.F. Brougham, "A Typical Mining Town – Nelson, B.C.," *Canadian Magazine*, 14, (1899), pp. 19-27; and I.M. Bescoby, "Some Social Aspects of the American Mining Advance into the Cariboo and Kootenay." M.A. Thesis, (University of British Columbia, 1935). For more detailed references to B.C. resource towns, see Alan F.J. Aritibise, *Western Canada Since 1870: A Select Bibliography and Guide* (Vancouver: University of British Columbia Press, 1978). For a general bibliography, see Rolf Knight, *Work Camps and Company Towns in Canada and the U.S.: An Annotated Bibliography* (Vancouver: New Star Books, 1975).
12. Gilbert A. Stelter, *The Northern Ontario Mining Frontier, 1880-1920*. Canada's Visual History Series, vol. 10 (Ottawa: National Museum of Man, 1974). For detailed references to these communities, see Gilbert A. Stelter and John Rowan, *Community Development in Northeastern Ontario: A Selected Bibliography* (Sudbury: Laurentian University Press, 1972).

blasting, and in the careless, happy-go-lucky methods of its cosmopolitan inhabitants is Cobalt typical of the new camp; in all else it is – an Ontario village.[13]

Cobalt and many of the other mining camps declined almost as quickly as they had sprung to life. But a large number of towns were able to maintain their positions and eventually grow because they found a new resource base, or because they took on the functions of a service and supply centre. Both factors are present in the development of Nanaimo, British Columbia. For about seventy years, Nanaimo's economic existence depended on coal mining. As late as 1921, almost the entire work force of the town was employed in mining. The coal mines of the area were almost depleted by the 1930s, but Nanaimo was more fortunate than most resource-based towns in that a new resource base, forest products, was developed which eventually directly employed about one-quarter of the work force. At the same time, the town's location provided the means for the development of a substantial local distributing function.[14]

Sudbury is another example of a commercial town whose existence depends directly on the resources produced in its area. For almost forty years after the townsite was laid out by the Canadian Pacific Railroad Company in 1883, Sudbury served only as a distributing centre to the numerous small towns of the nickel belt. The nickel industry expanded enormously from the 1920s, but the company towns were not enlarged. The result was an increasing proportion of the arriving work force resident in Sudbury; by 1941, one half of Sudbury's work force was employed by the mining companies. Many other miners lived in relatively unorganized suburbs on the city's periphery. The city thus became increasingly susceptible to the industry's expansion or cutbacks, for in terms of housing the work force, it had taken over some of the functions formerly performed by the company towns.[15]

13. "Cobalt As She Is," *Canada First, The Canadian Preference League Magazine* (October 1905), pp. 433-34. Douglas Baldwin, "The Development of an Unplanned Community: Cobalt, 1903-1914," *Plan Canada*, 18 (March 1978), pp. 17-29.
14. Norman Gidney, "From Coal to Forest Products: The Changing Resource Base of Nanaimo, B.C.," *Urban History Review*, 1-78 (June, 1978), pp. 18-47. See also J. Lewis Robinson, "Nanaimo, B.C.," *Canadian Geographical Journal*, 70 (1965), pp. 162-69.
15. Noel Beach, "Nickel Capital: Sudbury and the Nickel Industry, 1915-25," *Laurentian University Review*, 6 (1974), pp. 55-74.

## (2) Single-Enterprise or Company Towns

The second major type of resource town is the single-enterprise town, created by the fiat of a single authority, usually a private corporation.[16] The overriding common characteristic of these towns is their function – all were built as adjuncts of an industrial enterprise to provide a means of attracting and holding a stable work force in the face of a harsh climate, a rugged terrain, and isolation. Unlike those resource towns which eventually became service centers, the company towns' functions are narrowly defined by the founding company's needs, and they do not spontaneously or consciously develop functions beyond this circumscribed activity. The time dimension, however, is an important factor in understanding the evolution of this type of resource town. Company towns often have been associated with "paternalistic dictatorship" and "modern feudalism." But many were originally conceived as major improvements over the haphazard "sleep camps" associated with early mining, smelting, and lumbering operations. In some cases, company stores were seen as protection for the company's employees from the prices prevalent in the region. The company town of Tadanac, for example, built near Trail, British Columbia, by the Consolidated Mining and Smelting Company, opened a retail store in 1917 ostensibly to protect employees from inflationary prices in Trail.[17]

The stages of evolution of company towns can be illustrated by the example of the nickel belt towns of the Sudbury area. The original housing for the mining and smelting work force was located at haphazardly constructed camps, with company owned boarding houses and dormitories scattered between smelters and mines, as at Murray, Victoria, and Gertrude mines during the 1880s and 1890s (see Figure 1). Most of these early camps disappeared because of the impermanence of the founding company's operations in the nickel field. The towns which remained were the products of the two major companies in the industry, Canadian Copper (later Inco) and Mond. Canadian Copper's two company towns, Copper Cliff and Creighton, began life in the nineteenth century as casual camps surrounding the smelter and mining operations (see Figure 2). Not until the early twentieth century did the company initiate control, with the introduction of some planning and a trend away from private home ownership to company ownership. The eventual results were showcase towns, with comfortable homes for management and labour, and well-kept streets, parks, and recreational facilities.

16. Estimates of the number of communities which would fit this category vary, but the number is probably around 175. Ira Robinson, who excluded smelter towns, counted 161 in 1962. See Ira Robinson, *New Industrial Towns on Canada's Resource Frontier* (Chicago: University of Chicago Press, 1962); Leonard Marsh, *Communities in Canada: Selected Sources* (Toronto: McClelland and Steward, 1970), lists 160 resource towns.

17. Institute of Local Government, Queen's University, *Single-Enterprise Communities in Canada* (Kingston: Queen's University, 1953), p. 156.

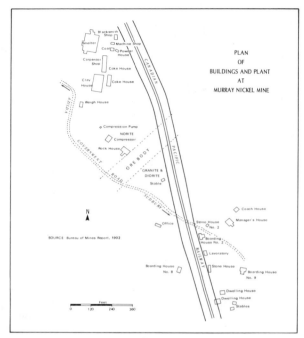

Figure 1

Figure 2    Copper Cliff in 1898
SOURCE: International Nickel Company

The Mond Company's towns of Coniston and Levack were built in 1913 to centralize their smelting and mining operations. These towns were laid out prior to settlement and, like Copper Cliff and Creighton, provided housing and amenities superior to that available in the region's commercial center, Sudbury (Figure 3). The paternalistic relationship between the companies and the towns was evident, however, in the nature of local government. Copper Cliff was incorporated as a town in 1902, but local participation in municipal government was minimal. Every mayor between 1902 and 1970 was a company official and was elected by acclamation.[18]

Changes in the nature of the company towns in the nickel belt in the post World War II period reflected changes taking place in company towns across the country. Companies began moving out of the housing field as quickly as possible, selling houses they once only rented. Municipal councils became more representative of the residents' desires. The transition was particularly evident in the nature of the more recently founded company towns, where direct company control gave way to both provincial government intervention and the organized force of local residents, often led by unions, as at Mackenzie, British Columbia, where the union successfully petitioned the provincial government to get involved in order to "establish policies that provide for the active protection of citizens from the grip of resource companies."[19]

*Form*

A third major distinction among resource towns is based on physical form. This distinction, like function, involves the question of who makes the decisions regarding planning and building. Differences in form become more apparent when examined chronologically, for the shaping of resource towns reflected the approach to town planning current in Canada in specific periods. It is possible to distinguish three generations of resource towns built since 1867: the privately built towns of the period before 1920; the holistically planned towns of the period between the wars; and the comprehensively planned third generation towns constructed since the Second World War.

*(1) The Pre-1920 Period*

The first generation resource towns built before 1920 reflected a prevailing laissez faire philosophy of planning. In the larger Canadian cities, rapid expansion was governed only by the private decisions of thousands of individuals and corporations, leading to fragmented patterns of

18. Gilbert A. Stelter, "Community Development in Toronto's Commercial Empire: The Industrial Towns of the Nickel Belt, 1883-1931," *Laurentian University Review*, 6 (1974), pp. 3-53.

19. Quoted by Norman Pressman and Kathleen Lauder, "Resource Towns as New Towns," *Urban History Review*, 1-78 (June 1978), pp. 78-95.

Figure 3     Types of company housing in Copper Cliff, 1917
      a)    official's houses under construction
      b)    club houses
      c)    workmen's houses
            SOURCE: Royal Ontario Nickel Commission, 1917

development.[20] Among resource towns, the seemingly chaotic mining camp was one characteristic form. It would be a mistake, however, to conclude that this kind of resource town was totally unplanned. Rather, planning was done on a private level, without regulation from some public body such as a municipal or provincial government.[21] Private corporations were directly involved in laying out most of the new resource towns of this period, even those which eventually became independent commercial centers. Sudbury, for example, was laid out by the Canadian Pacific which had been granted the townsite, and future development patterns were determined by this early plan (see Figure 4). The original plan of Nanaimo was drawn up by coal barons in London, England, in the nineteenth century. The spiderweb design (see Figure 5) continues to be the organizing principle of the town's central core. The mining camp of South Porcupine, with its traditional main street (see Fig. 6) exhibited a remarkable sense of order for an "instant" community. In Cobalt, on the other hand, where many small companies were involved, a sense of order was introduced only after the mining camp had already been established.

A sense of orderly planning was generally more apparent where one company's control was unquestioned. Companies showed little interest, however, in initiating some of the advanced industrial town planning in Britain where Lever's Port Sunlight and Cadbury's Bourneville were designed with the residents' health and comfort taken into consideration.[22] Planning was often restricted to little more than grid surveys, with the townsite closely associated with the mine, mill, or smelter. The town appears on some plans a something of an afterthought to the construction of the industrial plant, as illustated in the plans for Espanola, Ontario, an early pulp and paper town built in 1903, and Coniston, a smelter town. (see Figure 7).[23] In a pre-automobile age, it was necessary to make it possible for employees to walk to work, so that the distance between residential areas and the industrial plant was usually very short. A form of segregation by class and ethnicity was built into most of the early towns by relating house and lot sizes to occupation. Some immigrant groups who were less acceptable to the company were forced to congregate outside the organized limits of town.

20. See for example, Michael Doucet, "Speculation and the Physical Development of Mid-Nineteenth Century Hamilton," in this volume.
21. L.D. McCann, "The Changing Internal Structure of Canadian Resource Towns," *Plan Canada*, 18 (March 1978), pp. 46-59. John W. Reps, "Bonanza Towns: Urban Planning on the Western Mining Frontier," in R.E. Ehrenberg, ed. *Pattern and Process: Research in Historical Geography* (Washington, D.C.: Howard University Press, 1975). pp. 272-75.
22. For a brief discussion of Port Sunlight and Bourneville, see Michael Hugo-Brunt, *The History of City Planning: A Survey* (Montreal: Harvest House, 1972), pp. 158-59.
23. Eileen Goltz, "Espanola: The History of a Pulp and Paper Town," *Laurentian University Review*, 6 (1974), pp. 75-104; Stelter, "Industrial Towns of the Nickel Belt."

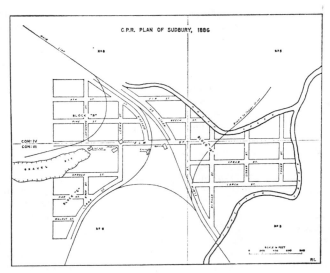

Figure 4      The Canadian Pacific's plan of Sudbury, 1886

Figure 5      Nanaimo, aerial view, 1958
SOURCE: Victoria Press Library

Figure 6        Porcupine, Ontario, 1911
                SOURCE: Peters Collection, Public Archives of Ontario

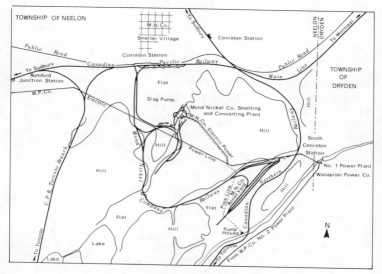

Figure 7        SOURCE: Royal Ontario Nickel Commission, 1917

## (2) The Interwar Years

Some of the second generation resource towns built between the wars reflected changes in planning thought and a wave of urban reform dating back to the late nineteenth century. At least two sources can be isolated. One was the City Beautiful Movement, exemplified by the Chicago Exposition of 1893, which visualized a civic landscape of monumental public buildings, great diagonal boulevards, squares, parks, and trees. Although most cities drew up grand plans along these lines, few of these grandiose schemes were actually put into operation in the existing cities.[24] The principles appeared, however, in some of the totally planned new resource towns. More directly influential in the planning of resource towns was the British Garden City Movement whose moving force was Ebenezer Howard. Howard advocated the founding of new self-contained settlements which would combine the virtues of town and country living by planned open space within the community, a green belt surrounding it, radial or curvilinear roads, and superblocks containing a wide variety of block shapes.[25] These ideas came to Canada via a number of sources, but particularly through the efforts of Thomas Adams, a leading exponent of Howard's ideas. Adams was brought to Canada to act as Town Planning Advisor to the federal government's Commission of Conservation in 1914. He tactfully but persistently pressed for planning statutes at the provincial level. He also acted as a private consultant to various levels of government and to a large number of municipalities, including several resource towns.[26]

The pulp and paper industry pioneered in adopting the Garden City principles to the planning of their new towns in the interwar period. Several towns were built in the two years immediately following the war which incorporated the general principles of the British Garden City, but the planning process was considerably different in each case. Temiskaming, Quebec, for example, was constructed in 1919 by a subsidiary of the Riordan Pulp and Paper Company whose officials hoped to create a model town (see Figure 8). A special town department examined "similar developments in Canada, in the United States, and in Europe," according to its director, and also requested advice directly from the town planning branch of the Commission of Conservation.[27]

---

24. Walter Van Nus, "The Fate of City Beautiful Thought in Canada, 1893-1930," in Gilbert A. Stelter and Alan F.J. Artibise, eds., *The Canadian City: Essays in Urban History* (Toronto: McClelland and Stewart, 1977), pp. 162-85.

25. Hugo-Brunt, *History of City Planning*, pp. 205-214.

26. Oiva Saarinen, "The Influence of Thomas Adams and the British New Towns Movement in the Planning of Canadian Resource Communities," in Alan F.J. Artibise and Gilbert A. Stelter, eds., *The Usable Urban Past: Planning and Politics in the Modern Canadian City* (Toronto: Macmillan, 1979), pp. 268-92.

27. A.K. Grimmer, "The Development and Operation of a Company-Owned Industrial Town," *Engineering Journal*, 17 (1934), p. 219.

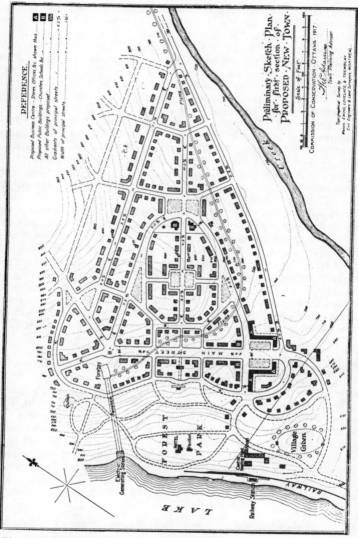

Figure 8     Thomas Adams' plan of Temiskaming, 1917
             SOURCE: Thomas Adams, *Rural Planning and Development*
             (Ottawa: Commission of Conservation, 1917).

GILBERT STELTER and ALAN ARTIBISE

Thomas Adams drew up a plan incorporating the spatial features associated with the British new town's philosophy and the company put these into practice in laying out the town. Another example was Kapuskasing, a town planned in 1921 on land owned by the Ontario provincial government, even though a single industry was to be served – the Kimberley-Clark Corporation (see Figure 9). The decision not to allow a closed company to emerge was due to Ontario Premier E.C. Drury's dislike of company towns. The government's plans, to be used "largely for guidance" by the town council, incorporated the principles Adams and others were advocating.[28]

Attempts to apply the new town concepts in transforming or adding to other pulp and paper towns were less effective, but indicate the strength of these ideas. An early example was Iroquios Falls, a town built by the Abitibi Power and Paper Company between 1915 and 1919. Company officials hired a landscape gardener from Montreal in an attempt to transform the town into what they promoted as "the wonder town of the northland." The company's plans for the town unfortunately did not provide for expansion. Population growth resulted in two fringe settlements primarily inhabited by francophones, creating an ethnic cleavage between the two portions of the community.[29] A similar pattern developed at Espanola, when the Kalamazoo Vegetable and Parchment Company took over the town in 1943 and made additions to it. The company's headquarters in Parchment, Michigan, became the model for the "improved townsite," which included the addition of curved streets to the old gridiron pattern. Nothing was done, however, about the population of an extensive fringe settlement known as "Frenchtown," an area inhabited by labourers of many ethnic groups.[30]

*(3) The Post-World War II Era*

Since the Second World War, a third generation of resource towns can be distinguished. Like those of the post-World War I period, the schemes for town building after 1945 can be regarded as part of an afterwar feeling of needing to start anew to build a better world. To a large extent, however, the models were based upon practices which had already proved successful particularly in the United State and Great Britain. These included land use segregation, green belts, separation of vehicular and pedestrian traffic, and the neighbourhood unit concept known as the Radburn Plan.[31] Another factor was a general expectation of provincial government involvement. In some cases governments and corporations jointly planned towns, such as Manitouwadge, Ontario,

28. Institute of Local Government, Queen's University, *Single Enterprise Communities*, pp. 83-99.
29. Saarinen, "Thomas Adams."
30. Goltz, Espanola."
31. Hugo-Brunt, *History of City Planning*, pp. 188, 208-14.

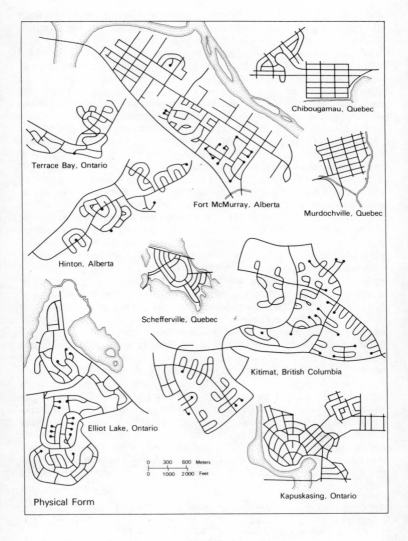

Figure 9    Physical form of selected towns
            SOURCE: L.D. McCann, "The Changing Internal Structure of
            Canadian Resource Towns," *Plan Canada*, vol. 18
            (March, 1978)

and Uranium City, Saskatchewan. Other towns were planned and built directly by a provincial crown agency, as was the case with Leaf Rapids, Manitoba, and Grande Cache, Alberta.[32] The coming together of a body of planning principles and government intervention is illustrated in the experience of Elliot Lake, Ontario. A ramshackle mining camp of trailers, tents, and shacks appeared in 1953 in the Elliot Lake area soon after the discovery of uranium set off one of the biggest prospecting rushes in Canadian history. Several mining companies were involved in developing the new deposits, and rather than constructing several competing townsites, they jointly requested the Ontario government to take the initiative for planning and developing one townsite. The result was a "combine town" serving the residential, commercial, and social needs of several mine sites, with a plan which employed the most modern planning techniques (see Figure 9).[33] The solution was certainly superior to the alternatives of allowing a boom town to develop naturally, or the more common practice in earlier resource development of constructing a small company town at the site of each company's operation.

The completely planned resource towns have offered planners, companies, and governments the opportunity to put the most advanced planning ideas into practice without the difficulties inherent in working with an existing community infrastructure. Not surprisingly, planning concepts developed for a southern climate have been transferred to northern settings and many resource towns resemble the suburbs of larger southern cities.[34] A number of recent studies reflect a growing awareness of the necessity of finding solutions which are specially adapted to the needs of these northern communities.[35]

One indication that a new approach is being taken is the planning and building of Leaf Rapids, Manitoba. The Manitoba government created a crown agency charged with the responsibility of acting as a developer for a new community in 1966 after the Sherritt Gordon Company's discovery of a copper-zinc ore body 530 miles north of Winnipeg. The town's focal point became a town centre, housing under one roof facilities normally found in a city centre – retail stores, a hotel, government offices, a school, and a library. Multiple family housing units made the community more compact, preserved as much of the natural environment as possible, and facilitated easy communication between parts of the community, an important feature in a harsh climate (see Figure

32. Norman Pressman, *Planning New Communities in Canada* (Ottawa: Urban Affairs, 1975), pp. 10-13.

33. Robinson, *New Industrial Towns*, pp. 23-25.

34. *Ibid.*, p. 2.

35. W.G.S. Shaw, "Homes, The Neglected Element in Canadian Resource Town Planning," *Albertan Geographer*, 7 (1971), pp. 43-49; L.B. Siemens, *Single Enterprise Community Studies in Northern Canada* (Winnipeg: Centre for Settlement Studies, 1973); J.D. Porteous, "Which Environment Would You Prefer," *Habitat*, 17 (1974), pp. 2-8.

10).[36] But some questions about planning such towns in the future still remain unresolved. There is no agreement, for example, on how much innovation and pre-planning should be imposed from the outside, or on the extent to which residents wish to create their own services.

In contrast to the comprehensive planning possible in the building of new, small resource towns, the complexities of handling growth in larger towns such as Nanaimo and Sudbury resemble those of the large metropolitan centres. In Nanaimo, the original design and function of the town long remained a negative legacy of "strangulation and dilapidation," according to some residents. Not until the major amalgamation of much of the surrounding territory in 1975 was Nanaimo able to begin to overcome its past and plan on a more regional scale.[37] In Sudbury, the expansion of the district's nickel industry resulted in rapid surburban development beyond the municipality's boundaries and control. An Ontario Municipal Board order to amalgamate some of the outlying unserviced suburbs in 1960 imposed a considerable financial burden on the city. Provicial grants in lieu of the right to tax local industry did not provide an adequate tax base. The initiation of regional government in 1971 allowed a measure of control over regional sprawl and introduced a more equitable taxation system.[38]

UNRESOLVED PLANNING PROBLEMS

The modernization of some of the larger service centres and the designs of some of the new towns dramatically illustrate the advances made in resource-town building since the first generation towns appeared in the nineteenth century. But regardless of the sophistication of recent planning concepts, some of the basic problems facing resource towns remain unresolved. Many have a limited lifetime and prospects for activity and growth beyond the initial function seldom materialize (Nanaimo is an exception). In some cases the resources simply run out, market conditions change, or an international corporation moves its operations for its own reasons. The results are mine or plant closures and the eventual death of a town. Hundreds, perhaps thousands, of Canadian communities have disappeared in this way – and the process continues. In other cases the industrial plants have become obsolete. In 1972, for example, Crown Zellerbach decided to abandon its operations at Ocean Falls, British Columbia, a company town built in 1908. About the same time, Canadian International Paper pulled out of Temiskaming, Quebec.

36. Kathleen Lauder, "Planning for Qualify of Life in New Resource Towns," M.A. Thesis, School of Urban and Regional Planning (University of Waterloo, 1977), pp. 131-224.

37. Gidney, *op. cit.*

38. Oiva Saarinen, "Planning and Other Developmental Influences on the Spatial Organization of Urban Settlement in the Sudbury Area," *Laurentian University Review*, 3 (1975), pp. 38-70.

Figure 10     Housing complex in Leaf Rapids, Manitoba
              SOURCE: Leaf Rapids Development Corporation

Desperate efforts were made to save both towns. At Ocean Falls, the British Columbia government purchased the plant and town, while at Temiskaming workers and executives attempted to continue operations with the help of federal and provincial government grants.[39] In both cases the future remains uncertain. Many other first and second generation towns are now approaching the age of fifty; some, such as Ocean Falls and Temiskaming, have been refurbished, but the chances for survival are still not very good.[40]

For those towns whose very existence is perhaps not in question, fluctuations between boom and bust plague attempts to plan for orderly, long-term community development. Sudbury's recent history serves as a useful example; it has witnessed several peaks and troughs in its prosperity dating back to the halcyon days of the First World War and the total collapse of the nickel market in 1921. A rapid expansion of the nickel industry's work force in the late 1960s put a tremendous strain on the provision of adequate housing and services. The boom times were quickly followed by a substantial cutback by the nickel industry in 1971, and severe cutbacks of more than 3,000 workers announced late in 1977.[41] Even the most carefully planned modern towns face the same dilemma. Kitimat, British Columbia, was hailed as the "first modern city built in the twentieth century" when it was founded in 1953. The town was planned to house a population of 50,000, but a worldwide surplus of aluminum in 1957 led to reductions in the work force and the population still has not surpassed 14,000.[42] In the case of Elliot Lake, the peak of population in 1958 may have reached 25,000, but by the early 1960s it had declined to about 6,000 as the result of an American government decision not to continue stockpiling uranium. New international demands for uranium have recently resulted in another boom town situation, but it is a prosperity coupled with increasingly severe environmental problems for workers and residents.[43]

Resource development has long been recognized as a significant fac-

---

39. William Johnson, "Reprieve – Or Town Faces Death," *Toronto Globe and Mail*, February 15, 1972, p. 31; Malcolm Gray, "When the Government's the Boss, the Workers Still Don't Like It," *Globe and Mail*, May 9, 1974, p. 47.

40. Gerald Hodge, "Age and the Resource Frontier," *Globe and Mail*, May 22, 1973, p. 7.

41. Rudy Platiel, "Mine Layoffs Have Almost Become Way of Life in Sudbury," *Globe and Mail*, October 22, 1977, p. 13.

42. Malcolm Gray, "Kitimat: The Perfect Planned Town that has Failed to Endear Itself to Newcomers." *Globe and Mail*, July 7, 1976, p. 9. See also B.J. McGuire and Roland Wild, "Kitimat – Tomorrow's City Today," *Canadian Geographical Journal*, 59 (1959), pp. 142-61.

43. Robert Pennington, "Elliot Lake: The Ghost Town is Making a Comeback," *Toronto Star*, August 20, 1976, p. E1.

tor in shaping patterns of Canadian institutions and culture.[44] Recently, James Simmons has argued that all Canadian urban growth ultimately depends on the production of staple products.[45] Resource towns have been front-line agencies in this process of staples exploitation. Government involvement at the level of design has, in recent decades, been a positive factor in improving the quality of life in these towns. However, without a more comprehensive approach to planning in the area of resource mangement, resource towns will continue to be the most unstable and precarious of Canadian communities.

44. For an excellent summary of this staples thesis, see Carl Berger, *The Writing of Canadian History* (Toronto: Oxford University Press, 1976), pp. 85-111.
45. "The Evolution of the Canadian Urban System," in Artibise and Stelter, *The Usable Urban Past*, pp. 9-34.

# Notes on Editors

Gilbert A. Stelter and Alan F.J. Artibise have co-edited and co-authored three books: *The Canadian City: Essays in Urban History* (Toronto, 1977); *The Usable Urban Past: Planning and Politics in the Modern Canadian City* (Toronto, 1979); *Canada's Urban Past: A Bibliography to 1980 and Guide to Canadian Urban Studies* (Vancouver, 1981); and a special issue of *Plan Canada* on resource towns (1978).

Stelter is Professor of History at the University of Guelph and coordinated the Guelph Urban History Conference in May 1977. His publications include numerous articles on Canadian urban historiography and frontier and resource towns. He has also co-edited *Urbanization in the Americas: The Background in Comparative Persepctive* (Ottawa, 1980). He is chairman of the Urban History Committee of the Canadian Historical Association, and a member of the editorial boards of the *Urban History Review* and the *Urban History Yearbook* (University of Leicester).

Artibise is Professor of History at the University of Victoria and General Editor of the History of Canadian Cities Series, a joint venture of the National Museum of Man and the publisher, James Lorimer. He is the author of *Winnipeg: A Social History of Urban Growth, 1874-1914* (Montreal 1975); *Winnipeg: An Illustrated History* (Toronto, 1977), and several other books and articles on Western Canadian urban history and has edited *Town and City: Aspects of Western Canadian Urban Development* (Regina, 1981). He is editor of the *Urban History Review* and a member of the editorial board of the *Journal of Urban History*.

# Notes on Contributors

**Elizabeth Bloomfield** is a post-doctoral fellow in the History Department at the University of Guelph and is studying the factors that led to the industrialization of southern Ontario cities and towns.

**Susan Buggey** is Head, Construction History Section of National Historic Parks and Sites Branch, Parks Canada, Ottawa, and the author of several articles on the buildings process in Halifax.

**Michael Doucet** is Assistant Professor of Geography at Ryerson Polytechnical Institute, Toronto. He has published widely on the history of land development, housing and transportation in Canadian cities.

**Isobel Ganton** is a Ph.D. student in Geography at the University of Toronto.

**Marc Lafrance** is Assistant Chief, Military History Section Parks Canada, Quebec City, and co-author of *Québec, ville fortifée, du XVIIe and XXe siècles* (Quebec, 1982).

**Leo Johnson**, Associate Professor of History at the University of Waterloo, is author of *A History of Ontario County* (1975) and *A History of Guelph, 1827-1927* (Guelph, 1977) and many articles dealing with the question of class in Canadian society.

**Paul-André Linteau** is Professor of History at the University of Quebec at Montreal. He is the author of *Maisonneuve ou Comment des promoteurs fabriquent une ville* (Montreal, 1981) and co-author of *Histoire du Québec contemporain* (Montreal, 1979).

**L.D. McCann** is Associate Professor of Geography at Mount Allison University, Saskville, New Brunswick. His publications include articles on urban development in Western and Maritime Canada and *Heartland and Hinterland: Canadian Regions in Evolution* (Toronto, 1982)

**Norbert MacDonald** is Associate Professor of History at the University of British Columbia. A pioneer in the comparative approach in urban history, he has published articles on both Vancouver and Seattle.

**David Thiery Ruddel** is Quebec Historian at the History Division, National Museum of Man, Ottawa. He is the co-author of *Les apprentis artisans à Québec, 1660-1815* (Montreal, 1977).

**Ronald Rudin** is Assistant Professor of History at Concordia University Montreal, and the author of several articles on urban history in Quebec.

**John C. Weaver** is Associate Professor of History at McMaster University, Hamilton. His publications include *Shaping the Canadian City: Essays on Urban Politics and Policy, 1890-1920* (Toronto, 1977).

# THE CARLETON LIBRARY SERIES